W9-CED-222

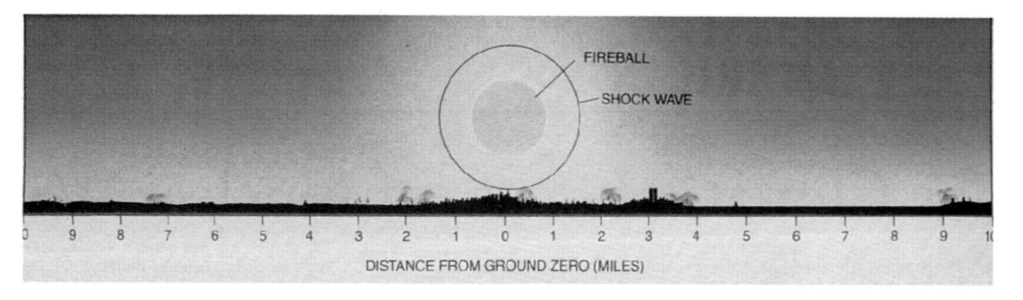


A one-megaton nuclear explosion detonates a few kilometers above New York City. The fireball radiation, traveling at the speed of light, has already ignited flammable structures ten miles and more from the city center. The shock wave, traveling at the speed of sound, has not yet reached the city. The twin towers of the World Trade Center can be seen at right.

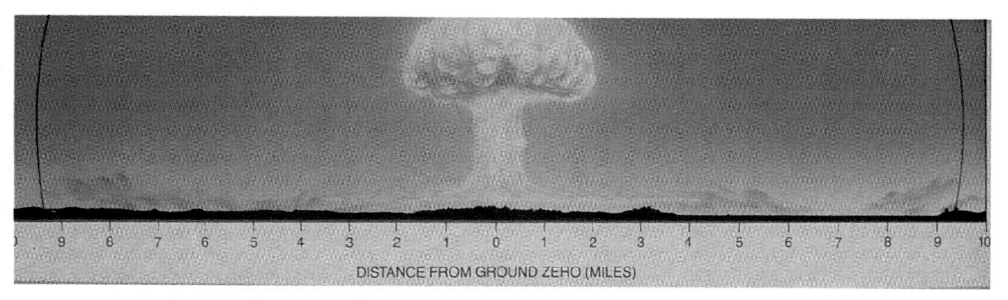


As the nuclear shock wave is leaving the city, skyscrapers and most buildings have been blown down. Fires are momentarily extinguished by the blast wave, and smoke is propelled away from the city. Looming over the scene is the mushroom cloud, which sucks debris up to high altitudes—into the lower stratosphere for a groundburst of yield greater than about 200 kilotons.

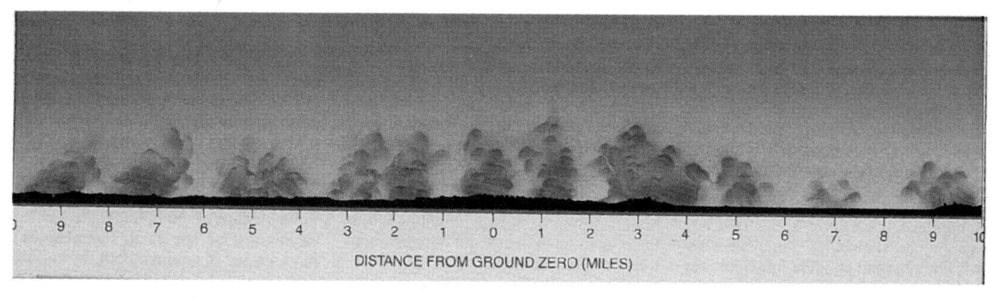


The shock wave has passed. Many fires ignited by the fireball, and others—set, for example, from broken or demolished gas mains—begin to rage.

The fires spread and merge over an area of 100 square miles or more. Great clouds of roiling black smoke rise above the fires.

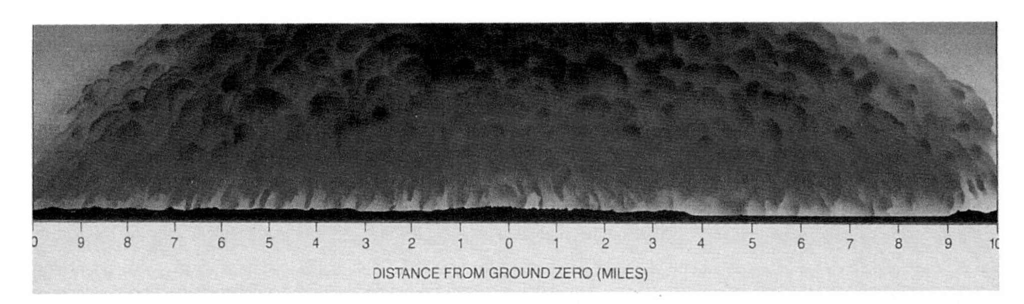


The inferno becomes a firestorm. Like a roaring fire in a fireplace with the flue open, but on a vastly larger scale, a huge column of convective air establishes itself, sucking up flames and carrying smoke to high altitudes. Winds in the firestorm can exceed hurricane force.

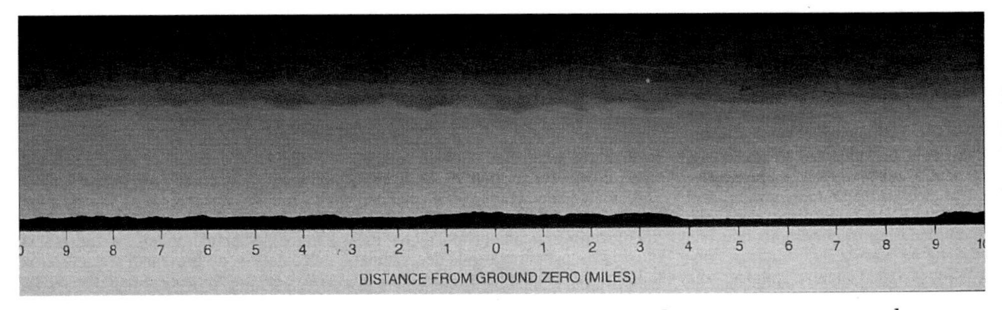


Many days later, hovering over the flattened city is a vast smoke pall extending into the stratosphere. Simultaneous development, and subsequent spreading and merging, of many such soot clouds at many altitudes can lead to nuclear winter.

[Reference: R. P. Turco, O. B. Toon, T. P. Ackerman, J. B. Pollack, and C. Sagan, "The Climatic Effects of Nuclear War," *Scientific American 251* (2), August 1984, 33–43; reprinted in Russian in *V Mire Nauki*, October 1984, 4–16. Courtesy, *Scientific American*.]

*Some Other Books by Carl Sagan:*

Intelligent Life in the Universe (with I. S. Shklovskii)

The Cosmic Connection

The Dragons of Eden

Broca's Brain

Murmurs of Earth (with others)

Cosmos

The Cold and the Dark: The World After Nuclear War (with others)

Comet (with Ann Druyan)

Contact

*Also by Richard Turco:*

Environmental Consequences of Nuclear War (with others)

# A PATH
# WHERE NO MAN THOUGHT

# A PATH WHERE NO MAN THOUGHT

CARL SAGAN RICHARD TURCO

## NUCLEAR WINTER AND THE END OF THE ARMS RACE

RANDOM HOUSE NEW YORK

Published in the United States by Random House, Inc., New York, and
simultaneously in Canada by
Random House of Canada Limited, Toronto.

Permissions acknowledgments appear on pages 469–471.

Library of Congress Cataloging-in-Publication Data

Sagan, Carl
A path where no man thought:
nuclear winter and the end of the arms race/
by Carl Sagan and Richard Turco.
p. cm.
ISBN 0-394-58307-8 — ISBN 0-679-72796-5 (pbk.)
1. Nuclear winter. 2. Nuclear warfare—Environmental aspects.
I. Turco, Richard II. Title.
QH545.N83S24 1990
304.2′8—dc20 89-43155

Design: Robert Aulicino

Manufactured in the United States of America

9 8 7 6 5 4 3 2

First Edition

To our close colleagues in the discovery of nuclear winter, Owen B. Toon, Thomas P. Ackerman, and James B. Pollack; to Paul J. Crutzen and John Birks for their inspiration and insight; and to the memory of Vladimir V. Alexandrov, who disappeared in the smoke and dust.

Great treasure halls hath Zeus in heaven,
From whence to man strange dooms be given,
Past hope or fear.
And the end men looked for cometh not,
And a path is there where no man thought . . .

—Euripides, *Medea*, Gilbert Murray, trans.
(New York: Oxford University Press, 1906)

# CONTENTS


List of Boxes xiii
List of Figures xv
Preface xvii
Prologue: "The Mere Threat of the End of the World" 3
1. Croesus and Cassandra 9
2. The Idea of Nuclear Winter 17
3. Current Scientific Knowledge of Nuclear Winter 31
4. The Witches' Brew: Poison Gas, Radioactive Fallout, Ultraviolet Light 45
5. Extinction? 61
6. Risk 79
7. Tambora and Frankenstein: What It Takes to Generate Nuclear Winter 93
8. Targeting 115
9. What Does It Take to Prevent Nuclear Winter? 129
10. Deterring Ourselves 143
11. Consequences of Execution 157
12. Nuclear Winter in Nations Minding Their Own Business 169
13. Global Policy Impact of Nuclear Winter 177
14. Darkness at Noon: Six Classes of Nuclear Winter 191
15. A Furnace for Your Foe 205
16. The Doomsday Machine 213
17. Is Infinity Enough?: Minimum Sufficient Deterrence (MSD) 221
18. What Kinds of Weapons?: Strategic Force Structures 237
19. How Do We Get to Minimum Sufficiency?: Some Milestones 257
20. Sketch of a Near-Term Strategic Path for the United States 275
21. Other Nuclear States 283
22. Abolition 295

Notes and References 303
Appendix A: Climate: The Global Energy Machine 419
Appendix B: Nuclear Winter Theory: Early Predictions Compared with Latest Findings 445
Appendix C: A Short History of the TTAPS Nuclear Winter Study 455
Index 473

# LIST OF BOXES

A City Burning 27
Wildfires, Martian Dust, and Nuclear Winter 38
Nuclear Winter: Early History and Prehistory 41
Fire and Smoke: Experiments to Simulate Nuclear Winter 50
Impact Winter 64
"The Apocalypse May Come" 68
A Modest Hope 72
SCOPE 75
Would Billions of People Really Be Killed by Nuclear Winter? 77
Improbable or Inevitable: How Likely Is Nuclear War? 83
Two Kinds of Conservative Thinking 87
Hiroshima and Nuclear Winter 106
Volcanic Winter 109
"Desired Ground Zeroes" 118
Aerial Bombing and Nuclear Testing: Are They Consistent with Nuclear Winter? 124
Vladimir Alexandrov: The First Casualty of Nuclear Winter? 135
"The Dilemma We Face" 147
Besides Deterring War, What Are Nuclear Weapons For? 151
Nuclear Winter and Famine 161
The Soviet Shelter System 164
Public Opinion on Nuclear Winter 187
Representative Human Catastrophes 198
Dr. Strangelove: Nuclear Doomsday in Popular Culture 218
What to Call It? 226
Martian Overkill 228
Thinking About the Unthinkable 234
Can Nuclear Winter Make Nuclear War More Likely? 245
The Minimum Sufficiency Debate in the U.S.S.R. 247
An Apologia for the Strategic Bomber 253
Minimum Deterrence in 1787 260
Can Citizens Make It Hard to Cheat as the Arsenals Get Smaller? 263

"Mutual Preclusion": The Joint Chiefs of Staff on Conventional War in Europe 272
"Back to Being Civilized" 278
Taking Yes for an Answer 280
Mao Zedong and Blowing Up the Earth 287

# LIST OF FIGURES

Figure 1. Nuclear Winter and Climatic Regimes 25
Figure 2. How High the Smoke Gets 36
Figure 3. Thresholds 97
Figure 4. Land Temperatures and Darkness Following Nuclear War 105
Figure 5. World Petroleum Refining Capacity 120
Figure 6. Severity of Nuclear Winter for Various Targeting Strategies 200
Figure 7. Possible Evolution Toward Minimum Sufficient Deterrence (MSD) 293
Figure 8. Energy Budget of the Atmosphere Before and After Nuclear War 430
Figure 9. Surface Darkness from Smoke and Dust 438
Figure 10. Summary of Nuclear Winter Calculations by Many Researchers 451

# PREFACE

This is a book about a disquieting scientific discovery. It is also about the prospects of life and death for everyone on Earth. In the end, and mainly, though, it is about the unexpected opening of a path that can, we hold, lead to a far safer world. This is a hopeful and optimistic book, but one grounded in a human and technical reality that seems fearful and heartless. Its subject engages powerful—sometimes unexamined—beliefs, doctrines, and prejudices. The prospect of what we have called "nuclear winter" challenges political, economic, social, and religious ideologies. It has been taken as a rebuke of what for many years passed as the conventional wisdom. Lewis Thomas has called it "extraordinary good news." Nuclear winter seems to leave some people despairing, some rejecting the prospect out of hand, and others fired up to make political change. Few who consider the matter are left indifferent.

We do not pretend to be dispassionate observers ourselves. We have been deeply involved in the discovery and development of the science of nuclear winter, and in the debate on its policy implications. We have been forced to contemplate what nuclear war would be like, and we find the experience profoundly disturbing. We have a point of view. But we believe this point of view is not prejudice, but what might be called postjudice—a judgment made not before but after examining the evidence. At a time of swiftly moving U.S./Soviet relations and of an emerging consciousness of the need to protect the global environment, we propose that the prospect of nuclear winter has much to teach us.

As this book was being completed, arms control and the elimination of at least some nuclear weapons systems were not only being seriously discussed, but actually implemented. The present comparative warmth in the relations between the United States and the Soviet Union stands in sharp contrast to the chill of the Cold War. Understandably, there is now a tendency to think of the problem of nuclear war as solved, or at least as in the process of being solved, so we can at last ignore it and turn our attention to the vast array of other pressing problems. This opinion is surprisingly widespread. It blossoms especially when superpower summit meetings are cordial. It is, we believe, a dangerous illusion.

For all the genuine goodwill in the present attitudes of the superpowers and the profound changes in their relationship, the simple fact is that, at this moment, over 10,000 nuclear weapons on each side are, with fine premeditation, aimed at specific targets on the other. Some of those "targets" have millions of people in them. In the nosecones of missiles and in the bombracks of aircraft the weapons wait—faithful, obedient servants, awaiting orders. If they are activated, they will fly away, halfway around the planet it may be, sent on their one-way missions by the merest word. These are the strategic weapons, designed to travel from one homeland to another. Then there are nearly 35,000 tactical nuclear weapons, with more modest objectives. The bombs that destroyed Hiroshima and Nagasaki were on such a scale. Altogether there are nearly 60,000 nuclear weapons in the world. Behind the welcome improvements in rhetoric and relations, the machinery of mass murder still waits, purring and attentive. It is no exaggeration, no hyperbole to say that billions of people are at risk. It is a little early for complacency.

Of all the perils facing the human species, nuclear war and nuclear winter—overwhelmingly, as we will show—pose the greatest dangers. As long as such a multiply redundant, hair-trigger capability for mutual annihilation exists, all assurances of safety or security will ring hollow. *Challenger* and Chernobyl remind us that high-technology systems into which enormous amounts of national prestige are invested can go disastrously wrong. The politics of the United States and the Soviet Union are unpredictable—as recent events have richly demonstrated. We do not know who will accede to power in the coming years and decades. And nuclear weapons, like diseases, proliferate. The longer the major nuclear powers dally about substantial mutual arms reductions, the less moral authority and political credibility they bring to preventing proliferation of these weapons to other nations, and the broader becomes the set of issues and national interests that could ignite a nuclear war. Safely reversing the nuclear arms race should have, in Andrei Sakharov's words, "absolute priority over all other problems of our times."

An era of improved relations between the United States and the Soviet Union is the optimum time to work to reassess military doctrine and policy, to reconsider weapons systems on order, to reverse the arms race. No significant reversal is possible, however, without far-reaching changes in the attitudes that each nation bears toward the other. But such changes are, by the beginning of the last decade of the twentieth century, clearly under way. Improved U.S./Soviet relations make possible, and derive from, fair and verified arms reductions. There is a positive feedback here; the political and arms control processes drive one another.

We believe that nuclear winter provides a compelling incentive for reversing the arms race—an incentive embracing not only the nuclear-armed nations, but the entire human community. It also offers important clues on how to go about arms reductions and what lower levels to aim for.

A reawakened passion for democracy is sweeping across our planet. It comes at a time when issues of science and technology, some of unprecedented difficulty, have moved to center stage. Our well-being requires informed citizens and informed policymakers. On issues of this importance it is not enough for citizens and policymakers to rely on experts (much less "authorities," of which, in science, there are, and should be, none); they need to inform *themselves*. There is no other way to make responsible decisions.

As with the many other urgent matters on the national and global political agendas, nuclear winter has a basic scientific and quantitative aspect. It is fully possible to understand the fundamental ideas and to debate the policy implications of nuclear winter without having any background at all in science or mathematics. But our understanding is improved if we take a little trouble to consider the science. In this book, we describe some of the principal scientific issues in different ways—in words, in graphs, and, to a limited extent, in numbers. Even readers with no knowledge of physics and a phobia for mathematics will, we hope, have little trouble following the argument: We have tried to describe the key ideas more than once, and from several standpoints. We have also interdigitated the science and the policy, especially early in the book, to lay stress on their interaction.

In a book that crosscuts so many fields and that engages so many strong emotions, we are well aware that errors of fact or judgment can be made. Through vigorous criticism of each other's arguments during the writing of this book, and through the criticism and advice of many others, we have sought to minimize such errors. We trust we will hear from readers of any that remain. If we may judge by past experience, we will also hear objections of a more philosophical and ideological bent. We hope this book may make a contribution toward discussion, debate, and action on what is still the most urgent and fateful issue of our times.

We are indebted to all those who read and commented on earlier drafts, including Desmond Ball, The Australian National University, Canberra; McGeorge Bundy, New York University; Ingvar Carlsson, Prime Minister of Sweden; Christopher Chyba, Cornell University; Paul Crutzen, Max Planck Institute for Chemistry, Mainz, West Germany; Ted Doty, University of California at Los Angeles; Freeman Dyson, Institute for Advanced Study, Princeton; Jerome D. Frank, Johns Hopkins University School of Medicine, Baltimore; Richard L. Garwin, Thomas J. Watson Research Center, IBM Corporation; Admiral Noel Gayler, U.S.N. Ret.; Gyorgi S. Golitsyn, Institute of Atmospheric Physics, Soviet Academy of Sciences, Moscow; Lester Grinspoon, Harvard University Medical School; Ray Kidder, Lawrence Livermore National Laboratory; Rear Admiral Gene La Rocque, U.S.N. Ret., Center for Defense Information, Washington; Herbert Lin, House Armed Services Committee; Jon Lomberg; Robert C. Malone, Los Alamos National Laboratory; Michael C. MacCracken, Lawrence Livermore National Laboratory; Robert S. McNamara; Roland Paulsson, Alva and Gunnar Myrdal Foundation, Stockholm; Gov. Russell W. Peterson; A. Barrie Pittock, Commonwealth Scientific and Industrial Research Organization, Australia; Frank Press, President, National Academy of Sciences, and members of the National Research Council Staff; Alan Robock, University of Maryland; Stephen H. Schneider, National Center for Atmospheric Research, Boulder; Richard Small, Pacific Sierra Research Corporation, Los Angeles; Jeremy Stone, Federation of American Scientists, Washington; Frank von Hippel, Princeton University; and several present or former military officers and government officials who wish to remain anonymous. We have made an effort to reflect all their opinions in this book, although not all the reviewers will subscribe to all our opinions. A much earlier incarnation of this book, intended mainly for policymakers, was circulated with a request for suggestions about the most appropriate medium for publication. Freeman Dyson's strong recommendation that we make it available for general audiences played an important role in the subsequent evolution of the manuscript, and in the publication of the present volume.

Others who have helped us in various ways—especially by broadening and deepening our understanding of nuclear war, nuclear winter, and their implications, or by stimulating our thinking on these issues—are too many to cite fully, but include: Vladimir Alexandrov; Robert L. Allen; Luis Alvarez; Jean Audouze; Gyorgi Arbatov; David Auton; Hans Bethe; John Birks; Harold Brode; Helen Caldicott; Cary Caton; Robert Cess; Yevgenii Chazov; Tom Cochran; Samuel Cohen; Stephen Cohen; Curt Covey; Philip Dolan; Paul Ehrlich; Tom Eisner; Daniel Ellsberg; Alton Frye; Jack Geiger; Forrest Gilmore; Iain Gilmour; Newt Gingrich; Alexander Ginzburg; Barry Goldwater; Al Gore; Kurt Gottfried; Stephen J. Gould; Leon Gouré; Kennedy Graham; Irving Gruber; James Hansen; Mark Harwell; Alan Hecht; John Holdren; Franklyn D. Holzman; Eric Jones; Sergei Kapitsa; Amron Katz; George F. Kennan; Glenn A. Kent; George Kistiakowsky; Andrei Kokoshin; Valmore LaMarche, Jr.; David Lange; Ned Lebow; Robert Lelevier; Bernard Lown; John Maddox; Jesse Marcum; Carson Mark; Ali Mazrui; Philip Morrison; Paul Nitze; Olof Palme; Kevin Pang; Richard Perle; William Perry; Vladimir Petrovsky; John L. Pickitt; David Pimentel; George Porter; Theodore Postol; William Proxmire; Michael Rampino; George Rathjens; Peter Raven; Glenn Rawson; Irwin Redlener; John Rhinelander; Walter Orr Roberts; Estelle Rogers; Igor Rogov; Joseph Rotblat; Roald Sagdeev; Abdus Salaam; James Sanborn; Jacob Scherr; Pete Scoville; Brent Scowcroft; Charles Shapiro; Dingli Shen; Stephen Shenfield; Steven Soter; John Steinbruner; Arthur Steiner; Gyorgi Stenchikov; Theodore B. Taylor; Edward Teller; Starley Thompson; Charles Townes; Pierre Trudeau; Yevgenii Velikhov; Frederick Warner; Paul Warnke; Mark Washburn; Thompson Webb III; Viktor Weisskopf; Carl Friedrich von Weizsäcker; Robert R. Wilson; Tim Wirth; Albert Wohlstetter; Lowell Wood; Roy Woodruff; George Woodwell; Andrew T. Young; Ya. B. Zeldovich; Solly Zuckerman; and of course our TTAPS colleagues Brian Toon, Tom Ackerman, and

Jim Pollack. They are not to be held responsible for the opinions expressed herein. There are many others to whom we are indebted for firm encouragement and valuable advice.

One of us (C.S.) was involved in the early 1980s in a never-completed television project called *Nucleus*. Some of the ideas that were first formulated during that project have come to fruition in this book. He wishes to thank the distinguished advisory board of *Nucleus* and his co-writers, Ann Druyan and Steven Soter, on that project.

We thank Scott Meredith and Jack Scovil of the Scott Meredith Literary Agency, Bob Aulicino, Joni Evans, Derek Johns, Hugh O'Neill, and Becky Saletan of Random House, and our families, for their understanding and support. Ann Druyan made major and substantive contributions over many years to the content of this book. We are deeply grateful to her. Chapter 1 was taken from Carl Sagan's Oersted Medal Acceptance Address, American Association of Physics Teachers, Atlanta, January 23, 1990.

Robert Nevin has ushered this manuscript through its various incarnations with great skill. We are especially grateful to him. We also thank Shirley Arden, Nancy Palmer, and Eleanor York for their help.

# A PATH
# WHERE NO MAN THOUGHT

# PROLOGUE

# "THE MERE THREAT OF THE END OF THE WORLD"

Although physicians frequently know their patients will die of a given disease, they never tell them so. To warn of an evil is justified only if, along with the warning, there is a way of escape.

—Cicero, *Divination*, II, 25

BY 1982 IT HAD BECOME CLEAR TO THE AUTHORS OF THIS book—and to a few others—that the consequences of nuclear war might be far worse than had been acknowledged or understood by the civilian and military establishments of the contending nation-states. We had entered into the research on what we later were to call nuclear winter with few preconceptions—if there were any, they were that nuclear war might, at most, produce a small ripple in the global climate. This had surely been the prevailing wisdom since the invention of nuclear weapons. But now our calculations had revealed the possibility of a global climatic catastrophe, even from a "small" nuclear war. We felt an obligation to call these findings to the attention of those in charge of nuclear strategy and policy.

Late in 1983, we were able to organize a small meeting of senior government advisers and officials from past and, as it turned out, future Administrations. This was at a time in the Reagan years when "fighting" and "winning" a nuclear war was considered feasible, and merely describing the dangers of nuclear war—nuclear winter aside—was judged, if not unpatriotic, then at least eroding the will of the American people to oppose Soviet tyranny and therefore naive and foolish. No Administration officials would attend our meeting. Present, though, were retired senior military officers, a past and future Presidential National Security Adviser, past members of the National Security Council, a former Director of the National Security Agency, analysts from leading think tanks, and Ambassador Averell Harriman, who had negotiated the 1963 aboveground nuclear test ban treaty. This closed meeting was held in an ornate, windowless conference room on the Senate side of the Capitol. Using a slide projector, we presented our scientific findings at what we hoped was a comprehensible level.

Nuclear winter was so serious, it seemed to us, as to carry major implications for nuclear strategy, policy, and doctrine, and indeed for the underlying, often unexamined, attitudes shared by almost all American and Soviet officials toward the Cold War. Briefly, we outlined some of these implications as well.

As you might imagine—because our findings were so unexpected and because their implications ran so much at cross-purposes to what then passed for prevailing wisdom—there was a fairly spirited discussion. The remark that we found most memorable, as well as most useful, was uttered by one senior practitioner of dark arts: "Look," he said, "if you believe that the mere threat of the end of the world is enough to change thinking in Washington and Moscow, you haven't spent much time in those cities."

Since then, we've spent considerable time in Washington and Moscow and many other places where nuclear war is planned and weighed. The remark was particularly helpful because it reminded us of how abstracted many officials and strategists are from the horrors they plan for, and how resistant to fundamental change the principal political and military establishments and the weapons laboratories had become. If nuclear winter were to inform, much less change, national policies, it would take time.

Today, the peoples and leaders of the world seem far more aware of the dangers of nuclear war than was the case in the early 1980s when nuclear winter was discovered. We think it possible, for reasons we will describe, that nuclear winter had something to do with this change in attitudes and awareness, that "the mere threat of the end of the world"—or something approaching it—*has* finally begun to help change things. We also believe that the significance of nuclear winter is still incompletely grasped and that its greatest influence is yet to come.

If we were today to give a briefing to officials of the nuclear-armed nations on nuclear winter, the present book is the basis of what we would say. It describes what determines the global climate of the Earth, and how nuclear war could change that climate; what the long-term consequences of nuclear war would be like, for individuals and societies; and how nuclear winter can help chart a path to take us from the present obscenely bloated nuclear arsenals into a world which, if not wholly freed from the scourge of nuclear war, at least is far safer than our present world—which is a world made in almost total ignorance of the most serious consequences of actually using the weapons that, at great cost, we have painstakingly accumulated in order to keep us "secure."

We are now in a time when the nuclear superpowers, with great trepidation, are actually contemplating, perhaps as early as the next few years, the fearsome step of reducing the world nuclear arsenals from nearly 60,000 weapons to only a little less than 50,000 weapons. But if we are to escape the threat that nuclear winter portends—of a global climatic catastrophe and the deaths of billions of people—we will have to do much better.

CHAPTER 1

# CROESUS AND CASSANDRA

I prophesied to my countrymen all their disasters.

—Cassandra, in Aeschylus, *Agamemnon*

APOLLO, AN OLYMPIAN, WAS GOD OF THE SUN. HE WAS also in charge of matters other than sunlight, one of which was prophecy—that was one of his specialties. Now the Olympian gods could all see into the future a little, but Apollo was the only one who systematically offered this gift to humans. He established oracles, the most famous of which was at Delphi, where he sanctified the priestess. She was called the Pythia. Kings and aristocrats—and occasionally ordinary people—would come to Delphi and beg to know what was to come.

Among the supplicants was Croesus, King of Lydia. We remember him in the saying "rich as Croesus." Part of the reason he was so rich is that he was one of the people who invented money—the first coins were Lydian, and were minted during Croesus' reign. (Lydia was in Anatolia, contemporary Turkey.) His ambition could not be contained within the boundaries of his small nation. And so, according to Herodotus' *History*, he got it into his head that it would be a good idea to invade and subdue Persia, the superpower of the 7th century B.C. Cyrus had united the Persians and the Medes and forged a mighty Persian Empire. Naturally, Croesus had some degree of trepidation.

In order to judge the wisdom of invasion, he dispatched emissaries to consult the Delphic Oracle. You can imagine them laden with opulent gifts—which, incidentally, were still on display in Delphi a century later, in Herodotus' time. The question the emissaries put on Croesus' behalf was "What will happen if Croesus makes war on Persia?"

Without hesitation, the Pythia answered, "He will destroy a mighty empire."

"The gods are with us," thought Croesus, or words to that effect. "Time to invade!"

Licking his chops and counting the satrapies shortly to be his, Croesus gathered his mercenary armies, invaded Persia—and was humiliatingly defeated. Not only was Lydian power destroyed, but he himself became, for the rest of his life, a pathetic functionary in the Persian court, offering little pieces of advice to often indifferent officials—a hanger-on ex-king. It's a little bit as if the Emperor Hirohito had lived out his days as a consultant on the beltway in Washington, D.C.

Well, the injustice of it really got to him. After all, he had played by the rules. He had asked for advice from the Pythia, he had paid handsomely, and she had done him wrong. So he sent another emissary to the Oracle (with much more modest gifts this time, appropriate to his diminished circumstances) and asked, "How could you do this to me?" Here, from Herodotus' *History*, is the answer:

> The prophecy given by Apollo ran that if Croesus made war upon Persia, he would destroy a mighty empire. Now in the face of that, if he had been well advised, he should have sent and inquired again, whether it was his own empire or that of Cyrus that was spoken of. But Croesus did not understand what was said, nor did he make question again. And so he has no one to blame but himself.

If the Delphic Oracle were only a scam to fleece credulous kings, then of course it would have needed excuses to explain away the inevitable mistakes. Disguised ambiguities were a stock in trade. Nevertheless, the lesson of the Pythia is germane: Even of oracles we must ask questions, intelligent questions—even when they seem to tell us exactly what we wish to hear. The policymakers must not blindly accept; they must understand. And they must not let their own ambitions stand in the way of understanding. The conversion of prophecy into policy must be done with care.

This advice is fully applicable to the modern oracles, the scientists and think tanks and universities. The policymakers send, sometimes reluctantly, to ask of the oracle, and the answer comes back. These days the oracles often volunteer their prophecies even when no one has asked. In either case, the policymakers must then decide what, if anything, to do in response. The first thing to do is to understand. And because of the nature of the modern oracles and their prophecies, policy-makers need—more than ever before—to understand science and technology.

There's another story about Apollo and oracles, at least equally relevant. This is the story of Cassandra, Princess of Troy. (It begins just before the Greeks invade Troy to start the Trojan War and therefore is also set in what is now Turkey.) She was the smartest and the most beautiful of the daughters of King Priam. Apollo, constantly on the prowl for attractive humans (as were many of the Greek gods and goddesses), fell in love with her. Oddly—this almost never happens in Greek myth—she resisted his advances. She refused the overtures of a god. So he tried to bribe her. But what could he give her? She was already a princess. She was rich and beautiful. She was happy. Still, Apollo had a thing or two to run by her. He promised her the gift of prophecy. The offer was irresistible. She agreed. *Quid pro quo.* Apollo did whatever it is that gods do to create seers, oracles, and prophets out of mere mortals. But then, scandalously, Cassandra reneged.

Apollo was not amused. But he couldn't withdraw the gift of prophecy, because, after all, he was a god. (Whatever else you might say about them, gods keep their promises.) Instead, he condemned her to a cruel and ingenious fate: that no one would believe her prophecies. (What we're recounting here is largely in Aeschylus' play *Agamemnon.*) So Cassandra predicted to her own people the fall of Troy. Nobody paid attention. She predicted the death of the leading Greek invader, Agamemnon. Nobody paid attention. She even predicted her own early death, and still no one paid attention. They didn't want to hear. They made fun of her. They called her—Greeks and Trojans both—"the lady of many sorrows." Today perhaps they would dismiss her as a "prophet of doom and gloom."

There's a nice moment in the play when she can't understand how it is that these urgent predictions of catastrophe—some of which, if believed, could be prevented—are being ignored. She says to the Greeks, "How is it you don't understand me? Your tongue I know only too well." But the problem isn't her pronunciation. The answer (we're paraphrasing) is, "You see, it's like this. Even the Delphic Oracle sometimes makes mistakes. Sometimes its prophecies are ambiguous. We can't be sure. And if we can't be sure, we're going to ignore it." That's the closest she gets to a substantive response.

The story was the same with the Trojans: "I prophesied to my countrymen," she says, "all their disasters." But they ignored her prophecies and were destroyed. Soon, so was she.

The resistance to dire prophecy that Cassandra experienced is equally stubborn today. Faced with an ominous prediction involving powerful forces that may not be readily influenced, we have a natural tendency to reject or ignore the prophecy. Mitigating or circumventing the danger might take time, effort, money, courage. It might require us to alter the priorities of our lives. And not every prediction of disaster, even among those made by scientists, is fulfilled: Most animal life in the oceans did not perish from insecticides; despite Ethiopia and the Sahel, worldwide famine was not a hallmark of the 1980s; supersonic aircraft do not threaten the ozone layer—although all these predictions had been made by serious scientists (ref. 1.1).* So when faced with a new and uncomfortable prediction, we might be tempted to say: "Improbable." "Doom and Gloom." "We've never experienced anything remotely like it." "Trying to frighten everyone." "Bad for public morale." What's more, if the factors precipitating the predicted catastrophe are longstanding, then the prediction itself is to us an indirect or unspoken rebuke. Why have we permitted this peril to develop? Shouldn't we have informed ourselves about it earlier? Don't we bear complicity, since we didn't work to ensure that government leaders took appropriate action? And since these are uncomfortable ruminations—that our own inattention and inaction may have put us and our loved ones in danger—there is a natural, if sometimes maladaptive, tendency to reject the whole business. It will need much better evidence, we say, before we can take it seriously. There is a temptation to minimize, dismiss, forget. Psychiatrists are fully aware of this temptation. They call it "denial." The rock group Dire Straits has a line in one of their songs: "Denial ain't just a river in Egypt."

* Notes and references, ordered in sequence by chapters, will be found at the back of the book. Ref. 1.1 is reference 1 of Chapter 1. Ref. 14.6 would be reference 6 in Chapter 14.

The stories of Croesus and Cassandra represent the two extremes of policy response to predictions of deadly danger—Croesus himself representing the pole of credulous, uncritical acceptance, propelled by greed or other character flaws; and the Greek and Trojan response to Cassandra representing the pole of stolid, immobile rejection of the possibility of danger. The job of the policymaker is to steer a prudent course between these two shoals.

Suppose a group of scientists claims that a major environmental catastrophe is looming. Suppose further that what is required to prevent or mitigate the catastrophe is expensive: expensive in fiscal and intellectual resources, but also in challenging our way of thinking—that is, politically expensive. At what point do the policymakers have to take the scientific prophets seriously? There are ways to assess the validity of the modern prophecies—because in the methods of science, there is an error-correcting mechanism, a set of rules that have repeatedly worked well, sometimes called the scientific method. There are a number of tenets: Arguments from authority carry little weight ("Because I said so" won't work); quantitative prediction is an extremely good way to sift useful ideas from nonsense; the methods of analysis must yield no results inconsistent with what else we know about the universe; vigorous debate is a healthy sign; the same conclusions have to be drawn independently by competing scientific groups for an idea to be taken seriously; and so on. There are ways for policymakers to decide, to find a safe middle path between precipitous action and impassivity.

We sometimes hear about the "ocean" of air surrounding the Earth. But the thickness of most of the atmosphere—including all of it involved in the greenhouse effect—is only 0.1% of the diameter of the Earth. Even if we include the high stratosphere, the atmosphere isn't even 1% of the Earth's diameter. "Ocean" sounds massive, imperturbable. But the thickness of the air, compared to the size of the Earth, is something like the thickness of a coat of shellac on a schoolroom globe. Many astronauts have reported seeing that delicate, thin, blue aura at the horizon of the daylit hemisphere and immediately, unbidden, thinking about its fragility and vulnerability. They have reason to be worried.

Today we face an absolutely new circumstance, unprecedented in all of human history. When we started out, hundreds of thousands of years ago, say, with a population density averaged over the Earth of less than a hundredth of a person per square kilometer, the triumphs of our technology were stone axes and fire. We were unable to make major changes in the global environment. The idea would never have occurred to us. We were too few and our powers too feeble. But as time went on, as technology improved, our numbers increased exponentially, and now here we are with an average of some ten people per square kilometer, our numbers concentrated in cities, and an awesome technological armory at hand—the powers of which we only incompletely understand and control. The inhibitions placed on the irresponsible use of this technology are weak, often half-hearted, and almost always, worldwide, subordinated to short-term national or corporate interest. We are now able, intentionally or inadvertently, to alter the global environment. Just how far along we are in working the various prophesied planetary catastrophes is still a matter of scholarly debate. But that we are able to do so is now beyond question.

There are three key indicators of technology-driven global atmospheric change: nuclear winter, ozone layer depletion, and greenhouse warming. This book is about the first, although, as we will see, all three are intimately connected. There may be—indeed we think it is inevitable that there are—other global environmental catastrophes driven by our technology that we are not yet wise enough to recognize. Perhaps the science and policy debates on nuclear winter will be useful for addressing these still undiscovered perils as well.

CHAPTER 2

# THE IDEA OF NUCLEAR WINTER

Smoke rises, the mist
is spreading.
Weep, my friends,
and know that by these deeds
we have forever lost our heritage.

—One of the last Aztec poems, written in 1521 on the eve of the destruction of Aztec civilization. From *Poems of the Aztec Peoples*, Edward Kissam and Michael Schmidt, trans. (Ypsilanti, Mich.: Bilingual Press, 1983)

WORLD WAR II HAD ENDED WITH THE EXPLOSION OF THE fission or atomic bomb, the most devastating weapon until then invented by the human species. Seven years later a weapon a thousand times more powerful was devised—the fusion or hydrogen bomb, so potent that it employed the fission bomb only as a trigger, as a match to set it off. In the decades since Hiroshima and Nagasaki the number, variety, and power of nuclear weapons increased. Many nations felt it essential to acquire them. They became entrées to international respectability; the admission fee to big power status; means of intimidating other nations, of unleashing patriotic pride, of manufacturing domestic political success. They worked wonders. Means were contrived to carry them in long-range aircraft; to launch them atop rockets from hardened concrete holes in the ground or from submarines sitting at the ocean depths; or to convey them in pilotless air-breathing vehicles that fly close to the ground under the radar, following every geographic contour. Brilliant, dedicated scientists and engineers labored to squeeze as many as a dozen of them—each directed to a different target—into the nosecone of a single missile; and to learn how to pack so many such missiles into a single submarine that one boat could destroy 200 cities of some faraway nation. The accuracy of these "delivery systems" improved. Some nuclear weapons could hit a football field halfway across the planet. Each could obliterate an area far larger than a football field and burn hundreds or thousands of square kilometers.

The world accumulated tens of thousands of nuclear weapons—always in the name of peace. Our side—whichever side we happened to be on—was always stable, cautious, peace-loving. The other side was always unpredictable, dangerous, warlike. Each side needed its vast arsenal, or so those in power told their citizens, only to deter the other side from using *its* vast arsenal. Their hands were tied. It was all the fault of the adversary. Trillions of dollars were spent.

The military establishments of the various nuclear-armed nations had, of course, an obligation to assure that at least their national leaders—if not their citizens, in whose name all this was being done—understood the consequences of nuclear war. Hundreds of nuclear weapons were exploded above and below ground and their effects monitored: blast, fire, radiation. There were some surprises, some ways in which nuclear explosions were *unexpectedly* dangerous. In many cases the new facts were discovered accidentally; and they were often classified as state secrets, so as not to erode public support for the nuclear arms race (ref. 2.1). Radioactive fallout was worse than had been guessed. High-altitude nuclear explosions were discovered to attack the protective ozone layer. The electromagnetic pulse from an explosion in space caused surprising malfunctions in electronic equipment in distant satellites and on the ground below. These unanticipated side effects should have been a warning that there might be other, still more serious, undiscovered consequences of nuclear war. But for nearly four decades no military scientist, no defense intellectual, no policy analyst ever seriously thought of anything like nuclear winter.

With our colleagues Brian Toon, Tom Ackerman, and Jim Pollack, it was our fate to be the first to calculate what the climatic consequences of nuclear war might be. From our last names (ref. 2.2), others gave our little research team the acronym "TTAPS"—appropriate, perhaps, given the nature of our findings.* All of us had studied the atmospheres and environments both of the Earth and of other worlds. We were used to thinking globally, trying to understand the big planetary picture. The story of how the discovery was made is told in Appendix C.

Our "baseline" case was a nuclear war in which less than half the strategic nuclear weapons (and none of the tactical weap-

* In U.S. military parlance, "Taps" is a bugle call, sounded at night, as an order to put out the lights. It is also played at military funerals. The melody was composed in July 1862 by Gen. Daniel Butterfield. But it is sung as well. In one version, perhaps still taught in summer camps, it begins, "Day is done, gone the Sun,/From the lake, from the hills, from the sky . . ."

ons) were detonated, many—but by no means most—over cities. In most surveys of the subject, only this baseline case is mentioned. But we also calculated some fifty other cases, covering the range of uncertainty both in physics and in targeting. Our scenarios ranged from a small war in which not one city burned, to a 25,000-megaton "future war" that would require more weapons than in the entire present world arsenals. Naturally, the severity of the results varied with the case chosen—from negligible to apocalyptic. But in too many of the cases, including those we considered most plausible, the predicted climates were much more severe than we had guessed. We were surprised and upset. We tried to be cautious. We were well aware of the preliminary nature of our early findings:

> Our estimates of the physical and chemical impacts of nuclear war are necessarily uncertain, because we have used one-dimensional models, because the data base is incomplete, and because the problem is not amenable to experimental investigation. . . . Nevertheless, the magnitudes of the first-order effects are so large, and the implications so serious, that we hope the scientific issues raised here will be vigorously and critically examined (ref. 2.2).

We tried to find errors in our calculations. (There were those who volunteered to assist us in this task.) Many of our estimates of input parameters turned out to be correct. In a few other cases our choices were inaccurate, but as it turned out the errors tended to cancel each other. In no case, we believe, did we get the fundamental physics wrong. A comparison of our original conclusions with modern results is given in Appendix B. Much progress has been made since our 1982/1983 work, much more accurate estimates of nuclear winter are now available, and much deeper insights into this fascinating and doleful subject are now at hand.

Checking for potential errors was an exercise in self-knowledge. We discovered in ourselves a wrenching ambivalence. When a potential source of error did not materialize we were elated; we had done the calculations right. But that feeling was soon replaced by another: The consequences for humanity that kept emerging were so dire that repeatedly we found ourselves hoping we *had* made a mistake. Unfortunately, or perhaps fortunately (an ambivalence persists), the central thesis of nuclear winter seems more valid today than ever before—unfortunately, because if we are so foolish as to permit a nuclear war, we now know it might constitute a worldwide disaster unparalleled in the history of our species; but fortunately, because the consequences are so serious, and so widespread, that a general understanding of nuclear winter may help in bringing our species to its senses.

Life on Earth is exquisitely dependent on the climate (see Appendix A). The average surface temperature of the Earth—averaged, that is, over day and night, over the seasons, over latitude, over land and ocean, over coastline and continental interior, over mountain range and desert—is about 13°C, 13 Centigrade degrees above the temperature at which fresh water freezes. (The corresponding temperature on the Fahrenheit scale is 55°F.) It's harder to change the temperature of the oceans than of the continents, which is why ocean temperatures are much more steadfast over the diurnal and seasonal cycles than are the temperatures in the middle of large continents. Any global temperature change implies much larger local temperature changes, if you don't live near the ocean.

A prolonged global temperature drop of a few degrees C would be a disaster for agriculture; by 10°C, whole ecosystems would be imperiled; and by 20°C, almost all life on Earth would be at risk.* The margin of safety is thin.

It is a central fact of our existence that the Earth would be some 35°C colder than it is today if the global temperature were to depend only on how much sunlight is absorbed by the Earth. This is a calculation routinely performed in introductory astronomy and climatology courses: You consider the intensity of sunlight reaching the top of the atmosphere, subtract the fraction of sunlight that's reflected back to space, and let the remainder—which is mainly absorbed by the Earth's surface—account for our planet's temperature. You balance the amount of radiation heating the Earth with the amount that is radiated (not reflected) by the Earth back to space. The temperature you derive is, disturbingly, some 35°C colder than the actual surface

* These temperature drops correspond respectively to 5 to 10°F, 18°F, and 36°F. Remember, these aren't the temperatures themselves, but the amounts by which the temperature falls.

temperature of the Earth. If this were all there were to the physics, the average temperature of the Earth would be below the freezing point of water; the oceans, still kilometers thick, would be made of ice; and almost all familiar forms of life—ourselves included—would never have come to be.

The missing factor, what we have ignored in this simple calculation, is the increasingly well-known "greenhouse" effect. Gases in the Earth's atmosphere, mainly water vapor and carbon dioxide, are transparent to ordinary visible sunlight but opaque to the infrared radiation that the Earth radiates to space as it attempts to cool itself off. These greenhouse gases act as a kind of blanket, warming the Earth just enough to make the clement and agreeable world we are privileged to inhabit today. Were the greenhouse effect to be significantly meddled with—turned up or down, much less turned off—it would constitute a planetwide disaster. This is in part what nuclear winter is about.

In a nuclear war, powerful nuclear explosions at the ground would propel fine particles high into the stratosphere. Much of the dust would be carried up by the fireball itself. Some would be sucked up the stem of the mushroom cloud. Even much more modest explosions on or above cities would produce massive fires, as occurred in Hiroshima and Nagasaki. These fires consume wood, petroleum, plastics, roofing tar, natural gas, and a wide variety of other combustibles. The resulting smoke is far more dangerous to the climate than is the dust. Two kinds of smoke are generated. Smoldering combustion is a low-temperature, flameless burning in which fine, oily, bluish-white organic particles are produced. Cigarette smoke is an example. By contrast, in flaming combustion—when there's an adequate supply of oxygen—the burning organic material is converted in significant part to elemental carbon, and the sooty smoke is very dark. Soot is one of the blackest materials nature is able to manufacture. As in an oil refinery fire, or a burning pile of auto tires, or a conflagration in a modern skyscraper—more generally, in any big city fire—great clouds of roiling, ugly, dark, sooty smoke would rise high above the cities in a nuclear war, and spread first in longitude, then in latitude.

The high-altitude dust particles reflect additional sunlight back to space and cool the Earth a little. More important are the dense palls of black smoke high in the atmosphere; they block the sunlight from reaching the lower atmosphere, where the greenhouse gases mainly reside. These gases are thereby deprived of their leverage on the global climate. The greenhouse effect is turned down and the Earth's surface is cooled much more.

Because cities and petroleum repositories are so rich in combustible materials, it doesn't require very many nuclear explosions over them to make so much smoke as to obscure the entire Northern Hemisphere and more. If the dark, sooty clouds are nearly opaque and cover an extensive area, then the greenhouse effect can be almost entirely turned off. In the more likely case that some sunlight trickles through, the temperatures nevertheless may drop 10 or 20°C or more, depending on season and geographical locale. In many places, it may at midday get as dark as it used to be on a moonlit night before the nuclear war began. The resulting environmental changes may last for months or years.

If the greenhouse effect is a blanket in which we wrap ourselves to keep warm, nuclear winter kicks the blanket off. This darkening and cooling of the Earth following nuclear war—along with other ancillary consequences—is what we mean by nuclear winter. (A more detailed discussion of the global climate and how nuclear winter works is given in Appendix A.)

A typical temperature for a point on the *land* surface of the Earth, averaged over latitude, season, and time of day, is roughly 15°C (59°F). If there were no greenhouse effect whatever, the corresponding temperature would be about −20°C (−4°F). The difference between the planetary environment with the greenhouse effect and without it is the difference between clement conditions and deep freeze. Tampering with the greenhouse effect—especially in ways that reduce it—can be very risky.

These two temperatures, with and without the greenhouse effect, are shown near the top in Figure 1. If we were to double the present concentration of the greenhouse gas carbon dioxide in the Earth's atmosphere—as will happen in a few decades if present trends continue—the surface temperature will likely increase by a few degrees, as the diagram shows. Following a major volcanic explosion the temperature can *de*crease by as much as a few degrees. During an Ice Age, the global temperatures are a few degrees colder yet, approaching the freezing

FIGURE 1

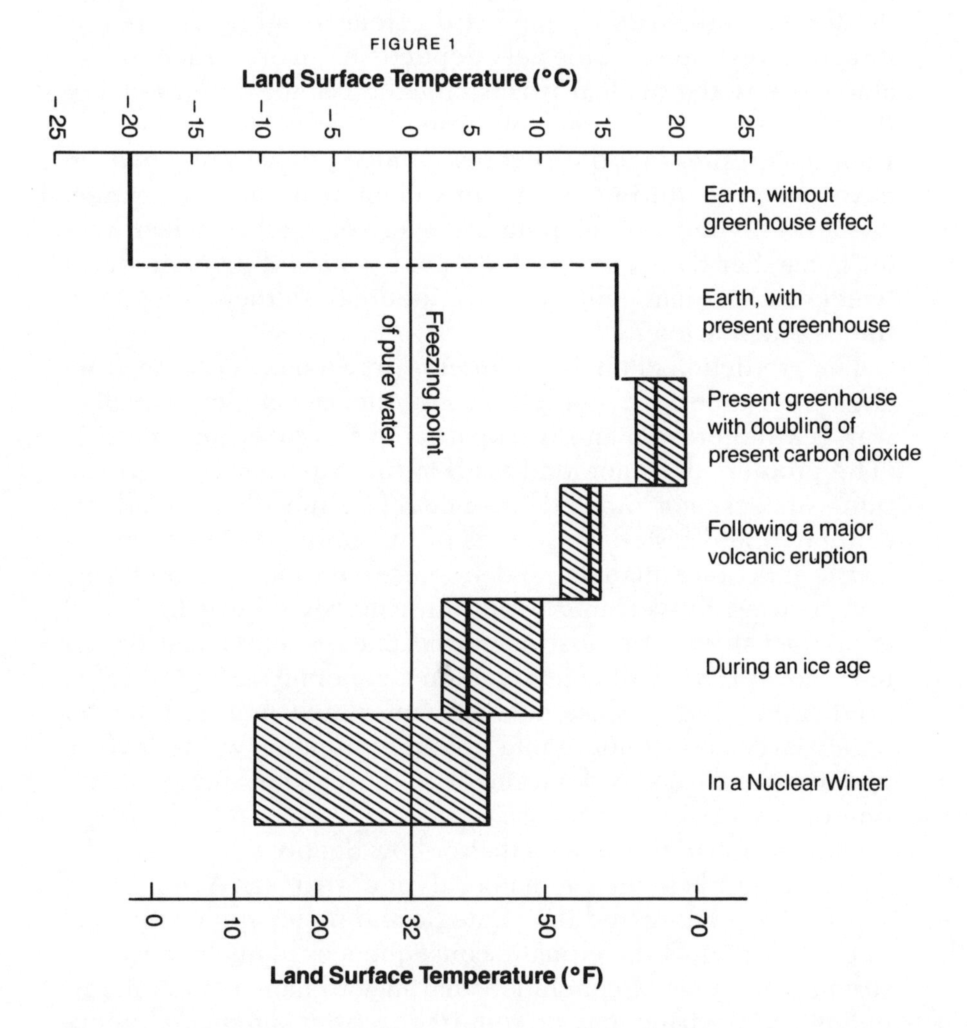


Possible climatic regimes of the planet Earth. Land temperatures are shown in degrees Centigrade (or Celsius), and in degrees Fahrenheit, averaged over latitude, season, and time of day. The plausible range of temperatures for each case is shown by cross-hatching. The heavy lines indicate the most likely value. In the case of nuclear winter, minimum temperatures beneath the smoke about a week after a July war are illustrated. The range of about equally likely values is indicated. All but the mildest nuclear winters represent a larger alteration in the Northern Hemispheric climate than any that has been experienced since the origin of the human species.

point of water. And in a nuclear winter, depending on severity, the temperatures can become still colder, ranging well below freezing. Just how cold it gets depends on many variables, including how the nuclear war is "fought," as we describe later. But even the middle range of these nuclear winter effects (see Figure 1) represents the severest climatic catastrophe ever to have occurred during the tenure of humans on this planet. Even in the range of temperature overlap, a mild nuclear winter is harsher than a severe Ice Age, because of its rapid onset (weeks rather than centuries or millennia)—although its duration is much briefer.

The prediction of nuclear winter is drawn not, of course, from any direct experience with the consequences of global nuclear war, but rather from an investigation of the governing physics. (The problem does not lend itself to full experimental verification—at least not more than once.) The models derived are calibrated and tested by studies of the ambient climate of the Earth and other planets, and by observed climatic perturbations caused by volcanic explosions, massive forest fires, and great dust storms. Because scientific analyses of nuclear winter have now pretty well converged on a generally accepted set of predictions, and because nuclear winter holds implications for policy issues now undergoing urgent rethinking, we believe an updating and reconsideration of both science and policy is timely.

Conventional wisdom, no matter how deeply felt, may not be a reliable guide in an age of apocalyptic weapons. A number of studies have addressed the strategic and policy implications of nuclear winter. If the climatic consequences of nuclear war are serious, many have concluded that major changes in strategy, policy and doctrine may be required. A brief summary of these early studies is given in references 2.3 through 2.6; some related commentary is given in references 2.7 and 2.8. In this book we reappraise both science and policy, and conclude that nuclear winter has strong implications—in some cases primary, in many others at least reinforcing—for nearly every area of nuclear strategy, doctrine, policy, systems, deployment, and ethics. This broad impact stems from two basic and connected facts about nuclear winter: (a) its occurrence would present an unacceptable peril for the global civilization and for at least most of the human species; and (b) it puts at risk in the devastating aftermath of nuclear war not only survivors in the combatant nations, but also enormous numbers in noncombatant and far-distant nations—people, most of them, wholly uninvolved with whatever quarrel or fear precipitated the war.

Since we have not yet had a global nuclear war, our conclusions must remain inferential and therefore necessarily incomplete. Some counsel that policy should not be decided on the basis of incomplete information. But policy is *always* decided on incomplete information. Nuclear winter has now attained standards of completeness and accuracy at least comparable to those on which many vital real world policy decisions are made. In the following pages, we summarize the scientific basis of nuclear winter, emphasizing key issues regarding the interpretation and uncertainty of the theory; then analyze the implications of this new understanding of nuclear war for a range of weapons systems and policies; and, finally, offer a set of objectives in policy and force structure for governments and alliances in the emerging new nuclear age.

## A CITY BURNING

Apart from the blast, radiation, and prompt radioactive fallout of a nuclear war, cities burn. The consequences can be devastating (see frontispiece). A classic description of a massive urban fire appears in Jack London's articles on the events following the Richter 8.2 San Francisco earthquake of 1906 (over 20 times more severe than the 1989 San Francisco quake). His first article was published in the April 18, 1906, number of *Collier's Magazine*. Through the overturning of lamps and lanterns and the rupturing of gas mains, earthquakes can set fire to cities. (Nuclear weapons exploding on the ground, or—a current trend in weapons development—underground, will also be able to burn a city, even if they have no other effect besides blast.) The flammable parts of San Francisco in Jack London's time were made mainly of wood. In modern cities there are enormous concentrations of flammable plastics and other synthetics—which produce darker smoke.

London's account gives us some distant glimpse of what

nuclear war would be like. (Note his description of the firestorm that carries fine debris to great altitudes.)

> The earthquake shook down in San Francisco hundreds of thousands of dollars' worth of walls and chimneys. But the conflagration that followed burned up hundreds of millions of dollars' worth of property. . . .
>
> Within an hour after the earthquake shock the smoke of San Francisco's burning was a lurid tower visible a hundred miles away. And for three days and nights this lurid tower swayed in the sky, reddening the sun, darkening the day, and filling the land with smoke. . . .
>
> On Wednesday morning at a quarter past five came the earthquake. A minute later the flames were leaping upward. . . .
>
> Before the flames, throughout the night, fled tens of thousands of homeless ones. Some were wrapped in blankets. Others carried bundles of bedding and dear household treasures. Sometimes a whole family was harnessed to a carriage or delivery wagon that was weighted down with their possessions. Baby buggies, toy wagons, and go-carts were used as trucks, while every other person was dragging a trunk. . . .
>
> . . . A rain of ashes was falling. The watchmen at the doors were gone. The police had been withdrawn. There were no firemen, no fire engines, no men fighting with dynamite. The district had been absolutely abandoned. I stood at the corner of Kearney and Market, in the very innermost heart of San Francisco. Kearney Street was deserted. Half a dozen blocks away it was burning on both sides. The street was a wall of flame, and against this wall of flame, silhouetted sharply, were two United States cavalrymen sitting [on] their horses, calmly watching. That was all. Not another person was in sight. In the intact heart of the city two troopers sat [on] their horses and watched.
>
> . . . Here and there through the smoke, creeping warily under the shadows of tottering walls, emerged occasional men and women. It was like the meeting of the handful of survivors after the day of the end of the world.
>
> . . . I watched the vast conflagration from out on the bay. It was dead calm. Not a flicker of wind stirred. Yet from every side wind was pouring in upon the city. East, west, north, and south, strong winds were

> blowing upon the doomed city. The heated air rising made an enormous suck. Thus did the fire of itself build its own colossal chimney through the atmosphere. Day and night this dead calm continued, and yet, near to the flames, the wind was often half a gale, so mighty was the suck.
>
> —Jack London, "The San Francisco Earthquake," in Stuart Hirschberg, ed., *Patterns Across the Disciplines* (New York: Macmillan, 1988), 86–90.

The following year saw an outbreak of plague in San Francisco—sufficiently serious for the mayor to wire the President for assistance. Unsanitary conditions in the burned districts of the city were blamed.

Here, in an eyewitness account of the great Chicago fire of October 1871—the one allegedly set by Mrs. O'Leary's cow—is another depiction of an urban firestorm:

> Everybody was mad, and everything was hell. The earth and sky were fire and flames: the atmosphere was smoke. A perfect hurricane was blowing, and drew the fiery billows with a screech through roads and alleys, between the tall buildings, as if it were sucking them through a tube: great sheets of flames flapped in the air. The sidewalks were all ablaze and the fire ran along them as fast as a man could walk. Roofing became detached in great sheets and drove down the sky like huge blazing arrows. There was fire everywhere, underfoot, overhead, around. It ran along tindery roofs, it sent out curling wisps of blue smoke from under eaves, it smashed glass with an angry crackle and gushed out in a torrent of red and black: it climbed in delicate tracery up the fronts of buildings, licking up with a serpent tongue little bits of woodwork: it broke through roofs with a rattling rush and hung out blood-red signals of victory. The flames were of all colors, pale pink, golden, scarlet, crimson, blood-hued amber. The flames advanced like a great army.
>
> —Hugh Clevely, *Famous Fires* (New York: The John Day Company, 1957), 157.

# CHAPTER 3

# CURRENT SCIENTIFIC KNOWLEDGE OF NUCLEAR WINTER

For almost half my lifespan
I have lived on this planet,
but I still do not understand
the wicked beauty of the atom,
or the Easter Island of the heart.

But I understand winter.

—"Nuclear Winter," in Diane Ackerman, *Jaguar of Sweet Laughter: New and Selected Poems* (New York: Random House, 1991)

THE THEORY OF NUCLEAR WINTER, FIRST INTRODUCED IN 1982, has been a subject of controversy (ref. 3.1). Debate is common when new scientific ideas are introduced, and healthy. However, much of the controversy over nuclear winter has been artificially generated at the borderline where science and policy intersect. Some has been fueled by confusion among nonspecialists over certain technical findings, and by comparisons of various computer models without sufficient care having been taken to resolve, or even to note, differences in initial assumptions. Among the troubling issues, laden with ideological connotations, raised by the nuclear winter theory are the possibilities that a major consequence of nuclear war eluded the American and Soviet nuclear arms establishments for thirty-seven years (ref. 3.2); that a "small" nuclear war might have widespread, perhaps even global, catastrophic climatic consequences; that distant nations would be in jeopardy, even if not a single nuclear weapon were detonated on their soil; that massive retaliation, and equally, attempts at a disarming first strike, in a variety of policy frameworks, would be disastrous for the nation employing such policies (and for its allies)—independent of its adversary's response; and that the size and nature of the present nuclear arsenals as well as the central role of nuclear weapons in the strategic relations of the United States and the Soviet Union may be not merely imprudent, but a policy mistake unprecedented in human history.

"Pentagon officials are plainly worried about the nuclear-winter problem," wrote Thomas Powers in late 1984 (ref. 2.6),

> and plainly at a loss over what to do about it. In conversation with officials at the nuts-and-bolts level one picks up interesting nuances of reaction: a wistful hope that "more

> study" will make the nuclear-winter problem go away, embarrassment at having overlooked it for nearly forty years, resentment that the peacenik doom-mongers might have been right all these years, even if they didn't know why they were. Above all, one finds a frank dismay at what the nuclear-winter problem does to a defense policy based on nuclear weapons. Being only human, officials are probably hoping to turn up uncertainties enough to justify more study forever, or at least until the next Administration.

Nuclear winter seems to challenge a wide range of well-established interests and beliefs. As Powers predicted, some critics have sought to minimize the significance of nuclear winter or the urgency of its policy implications by pleading unresolvable uncertainties, or emphasizing less severe effects (ref. 3.3). We shall argue here that neither approach is any longer tenable.

In the years since the original TTAPS (ref. 2.2) study, the scientific basis of the nuclear winter theory has been extended, refined, and strengthened. Findings have been published by scientists in the U.S., U.S.S.R., U.K., both Germanys, Japan, China, Brazil, Australia, New Zealand, Canada, and Sweden, among other nations. Data from many fields have been analyzed and applied to the problem. Important insights into related problems in the atmospheric sciences have emerged because of nuclear winter (ref. 3.4). It has provided an impetus for the rapid evolution of computer models of the three-dimensional general circulation of the Earth's atmosphere; these models have later proved important for studies of greenhouse warming (ref. 3.5). Yet, the central points remain unchanged: The key climatic predictions of the original nuclear winter theory have been generally confirmed, and the potential societal and human impacts remain extremely serious on a global scale.

The foregoing statements are at variance with some commentaries on nuclear winter that have appeared in print (ref. 3.6). However, we are talking here about a scientific theory that has evolved in an orderly manner (ref. 3.7). In science, valid criticisms are—on the weight of the evidence, and through common consent—rapidly integrated into a theory, or cause it to be superseded. Invalid criticisms are eventually rejected, as has been the fate of many early critiques of nuclear winter. So, for example, the criticisms that there is less to burn in cities and that a given amount of burning releases more soot than TTAPS estimated are valid, and the newer numbers are used in modern calculations. (Note that these two changes have offsetting consequences.) But the criticism that the great preponderance of the smoke would be promptly washed out by rainfall is invalid, and in modern calculations much of the smoke persists for months or years.

An example of the distortion of legitimate scientific findings has been the misinterpretation, both in some parts of the scientific community and in the press, of the "nuclear autumn" simulations of our colleagues Starley Thompson and Stephen Schneider (ref. 3.8) at the National Center for Atmospheric Research. It has been claimed that these climate model calculations show a fundamentally new environmental outcome of nuclear war; namely, much less severe temperature declines (around 10°C) rather than the 20 to 25°C of the TTAPS results (ref. 2.2)—and in some commentary, these conclusions are said to make the climatic effects of nuclear war nearly trivial. After all, who's afraid of autumn?

In the first place, the difference between 10°C and 20 to 25°C is not a central issue. What matters is any temperature drop of more than a few degrees. These values are sufficiently close to one another as to be mutually reinforcing. They are derived from the same physics. Both values represent extreme climatic changes (see Figure 1). There is a very real sense in which the "autumn" calculations confirm nuclear winter theory.

Moreover, careful comparison (ref. 3.9) between the TTAPS and the "autumn" models reveals general agreement in the most important predictions, including the severity of average land temperature decreases beneath extended smoke clouds. The purported differences are due in part to different *starting conditions*. For example, as compared to TTAPS, the autumn model assumed that nuclear war–generated smoke is inserted at much lower altitudes and is removed much more rapidly and efficiently. Figure 2 compares some of the smoke injection profiles used or recommended for nuclear winter assessments. The "autumn" smoke of Thompson and Schneider has been injected closer to the surface than is consistent with the known physics of massive smoke plumes (refs. 3.10, 3.11). Their assumed low-altitude smoke injection and highly efficient removal make the predicted surface cooling noticeably milder

FIGURE 2

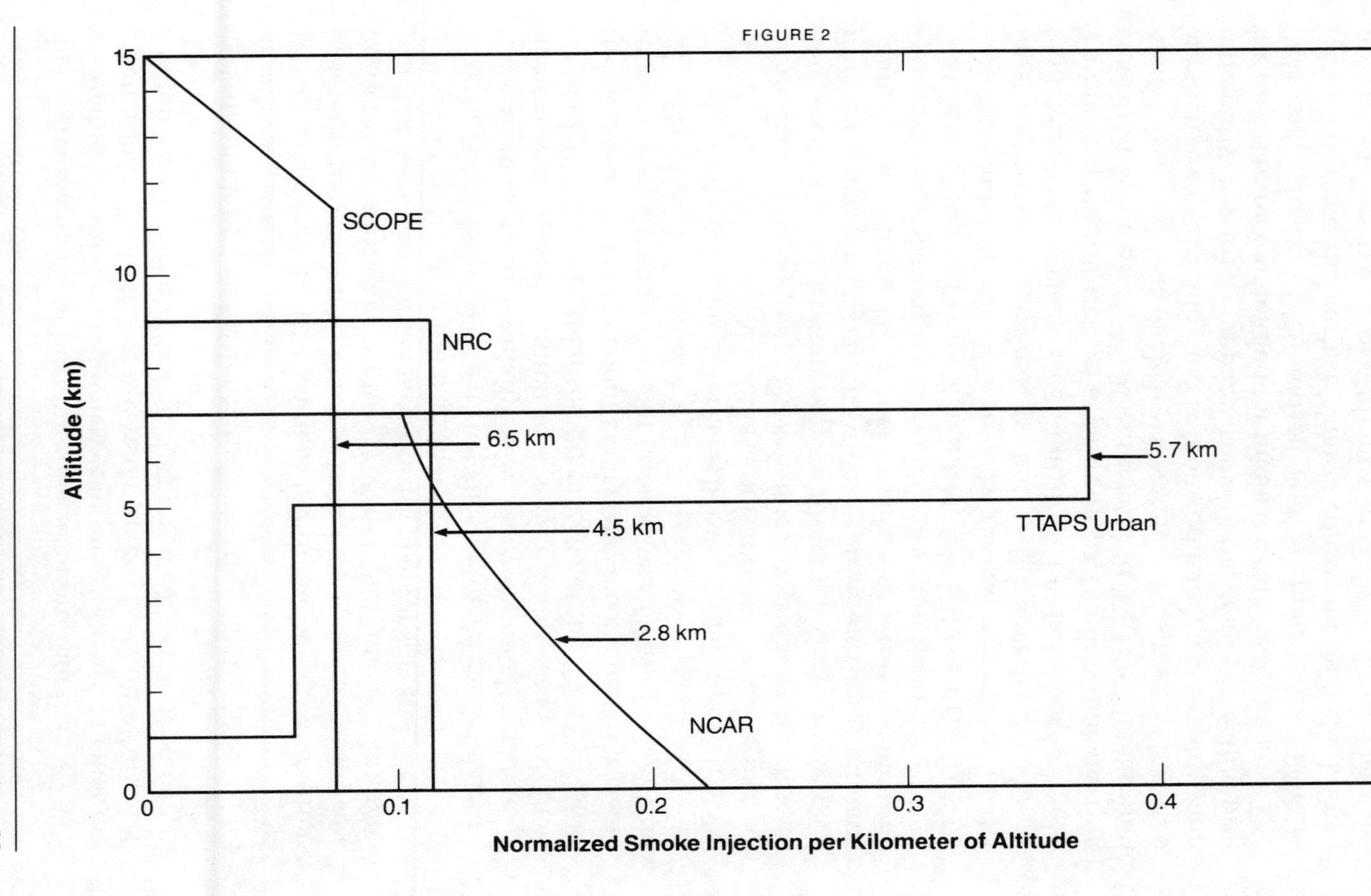

Altitude (km)
15
10
5
0
SCOPE
NRC
6.5 km
5.7 km
TTAPS Urban
4.5 km
2.8 km
NCAR
0
0.1
0.2
0.3
0.4
0.5
Normalized Smoke Injection per Kilometer of Altitude

Smoke injection profiles (how much smoke is put up at what altitude) used in a number of nuclear winter studies. Profiles are measured relative to a "unit" smoke injection, expressed as the fraction of the total amount of smoke injected per kilometer (km) of altitude. So for the "TTAPS Urban" profile, for example, about 38% of the smoke is between 5 and 6 km, another 38% between 6 and 7 km, and the remainder between 1 and 5 km. For each profile, the altitude above and below which half the smoke is injected (centroid) is indicated. The "NCAR" profile corresponds to injection between 0 and 7 km with uniform mixing through this region of the atmosphere, as assumed by Thompson and Schneider (ref. 3.8). Since the air is denser in the lower atmosphere, they take the amount of smoke to be greater there as well. The "TTAPS" (ref. 2.2), "NRC" (ref. 3.10), and "SCOPE" (ref. 3.11) profiles all correspond to injections at considerably higher altitudes, as detailed fire-plume studies indicate is likely to occur in a nuclear war. A different study [R. D. Small, *Ambio 18* (1989), 337–383] shows a range of smoke injection profiles for different fire areas and intensities. Those centroids vary from 2.3 to 6.7 km, consistent with what is shown in this figure. In many cases significant amounts of smoke reach the stratosphere, sometimes to 15–30 km altitude even without self-lofting (heating of the smoke by sunlight making it rise through the air). The stratosphere begins at about 12 km (actually between 8 and 15 km, depending on latitude, with the highest stratospheric altitudes at the lowest latitudes).

# WILDFIRES, MARTIAN DUST, AND NUCLEAR WINTER

A major forest fire, terrifying in itself, can also generate vast quantities of dark smoke and, if the smoke persists at altitude, an effect very like nuclear winter. Here is an eyewitness account of a large forest fire:

> Whole townships had to be evacuated: roads were crowded with blinded and hysterical refugees. Hospitals were filled with men and women suffering from fire blindness and burns. Manning the fire lines, the fire fighters struggled against hopeless odds. Burning bark, carried long distances, continually ignited fresh fires which, in turn, blazed into conflagrations. A pall of black smoke covered hundreds of miles of countryside, reducing visibility to a few yards.

Harry Wexler, then Director of the U.S. Weather Bureau, noticed that the smoke pall from the Alberta, Canada, forest fires of 1950 had briefly lowered surface temperatures in Washington, D.C., by 4 to 6°C. In Ontario, more than 1000 kilometers from the fires, it was dark as midnight near noon. The smoke was plainly visible to the naked eye in Western Europe, from Scandinavia to Portugal. N. N. Veltishchev, A. S. Ginsburg, and G. S. Golitsyn of the Institute of Atmospheric Physics of the U.S.S.R. Academy of Sciences discovered weather records that reveal similar temperature drops following the great Siberian forest fires of August 1915. All of central Siberia was enveloped by smoke, in some locales day turned into night, and unusual behavior was noted among wild animals—e.g., "Bears and wolves appeared near Krasnoyarsk." Smoke from large forest fires has been tracked by satellite imaging 5,000 kilometers from its source. Alan Robock of the University of Maryland has analyzed satellite-based meteorological records of forest fires in Canada, China, California, and Wyoming and finds cooling of up to 20°C beneath the smoke. Chinese wildfires in May 1987 reduced daytime temperatures in Alaska by 2 to 6°C. Given the experimentally determined properties of forest fire smoke, and the observed height and optical thickness of the smoke layers, all of these observed temperature responses are consistent with existing nuclear winter models—although the absorption by smoke in all such cases is less than expected in nuclear winter, and so the temperature declines are generally less as well. Robock describes these findings as providing "observational confirmation of a portion of the nuclear winter theory."

Also, scientists in the Department of Physics of Ahmadu Bello University in Zaria, Nigeria, found that a dense Saharan aerosol carried by the low-altitude Harmattan wind caused marked temperature drops—as much as 6°C on Christmas Day, 1977. G. S. Golitsyn and A. K. Shukurov, in an analysis of some fifty dust storms in Tadzhikistan, U.S.S.R., found that daytime surface temperatures under the dust clouds were depressed by as much as 10 to 12°C.

The climate models employed in nuclear winter studies have been successfully tested in predicting the cold and dark produced by the atmospheric debris from large volcanic explosions, as well as the atmospheric structure and surface climate of Mars and Venus, worlds very different from our own. There are also data more directly relevant to nuclear winter from observations of other planets: *Viking* lander observations indicate average temperature drops of several degrees C during global dust storms on Mars; because the Martian greenhouse effect provides only a 5 to 10°C warming, even thick dust clouds cannot provide a large anti-greenhouse cooling there. The fact that Martian surface temperatures drop while atmospheric temperatures increase during a global Martian dust storm was observed in 1971 by the American *Mariner 9*, the first spacecraft to orbit another planet. This helped lead, by a set of slow and indirect steps, to the discovery of nuclear winter, as outlined in Appendix C.

Three-dimensional general circulation models of the Earth's atmosphere, used in nuclear winter studies, have also been successful in reproducing unfamiliar regional climates—such as the existence of lakes in the Sahara—deduced from the geological record over the last 18,000 years (ref. 3.19).

(refs. 2.2, 3.8). When smoke is inserted at more realistic altitudes and with more plausible removal lifetimes, "autumn" begins to chill into "winter." Schneider himself now describes nuclear autumn as "edging back toward winter" (ref. 3.12).

Calculations by many scientists are compared in Appendix B. The basic climate changes predicted by the original nuclear winter theory have been upheld by later research, including research with significantly more sophisticated models (refs. 3.13, 3.14). Recent analyses of chills, freezes, obscuration of the Sun, and crop failures caused by wildfire smoke (see box) and volcanic aerosols (see below), support the theory. A modern reappraisal of nuclear winter theory can be found in reference 3.14. Expert summaries of the field have now been published by the Scientific Committee on Problems of the Environment (SCOPE) of the International Council of Scientific Unions (ref. 3.11), by the World Meteorological Organization (ref. 3.15), and by the United Nations (ref. 3.16).

One way of calibrating the seriousness of the average global coolings predicted for nuclear winter is to compare them with the slow global warming attributed to the increasing greenhouse effect. The decade of the 1980s has witnessed, on global average, the five hottest years of the preceding 130 (ref. 3.17). A few investigators have proposed that these years, and especially the sweltering summer of 1988, provide the first clear climatic signature of the increasing greenhouse effect. By the beginning of the 1990s, the entire global temperature increase since the industrial revolution is estimated at about 0.5°C. This is a planetary average, over latitudes, seasons, and time of day. It seems small, but it can have profound local consequences. It constitutes the highest average global temperatures in the last 120,000 years (ref. 3.17 and Figure 1). This is one way of calibrating the meaning of the 10 to 25°C temperature declines predicted for the baseline nuclear winter: Nuclear winter constitutes 20 to 50 times the maximum temperature changes attributed so far to the increasing greenhouse effect, about which there is—and properly so—grave concern (ref. 3.18). And the nuclear winter climatic changes would occur thousands of times faster.

In this book, we do not claim that a given sort of nuclear war will inevitably produce a given severity of nuclear winter; the irreducible uncertainties are too large for that. What we do claim is that the most likely consequences of many kinds of nuclear war constitute climatic and environmental catastrophes much worse than the worst our species has ever encountered—and that prudent national policy should treat nuclear winter as a probable outcome of nuclear war.

## NUCLEAR WINTER: EARLY HISTORY AND PREHISTORY

Paul Crutzen and John Birks made the first estimates of the huge quantities of smoke generated by the fires of nuclear war and were the first to note that the smoke could, over large regions, obscure the Sun and perturb the atmosphere ("Twilight at Noon: The Atmosphere After a Nuclear War," published in *Ambio*, a journal of the Swedish Academy of Sciences, Volume 11, 1982, 114–125):

> We point especially to the effects of the many fires that would be ignited by the thousands of nuclear explosions in cities, forests, agricultural fields, and oil and gas fields. As a result of these fires, the loading of the atmosphere with strongly light-absorbing particles . . . would increase so much that at noon solar radiation at the ground would be reduced by at least a factor of two and possibly a factor of greater than one hundred.

While Crutzen and Birks mentioned the possibility of city fires, their calculations were restricted to forest fires and fires in gas and oil facilities, and they did not estimate (or even mention) temperature drops resulting from the smoke. The TTAPS team introduced the term and concept of "nuclear winter"—involving darkening, cooling, enhanced radioactivity, toxic pollution, and ozone depletion (ref. 2.2, published in *Science*, a journal of the American Association for the Advancement of Science), and were the first to calculate the magnitude and duration of surface cooling (done for almost fifty different scenarios, covering many aspects of the uncertainties in the character of the nuclear war as well as in incompletely known physical parameters). A brief first announcement of the nuclear winter findings was published, also in 1982, in the American Geophysical Union periodical, *EOS*, Volume 63, 1018, as "Global Consequences of Nuclear 'Warfare,' " by R. P. Turco, O. B. Toon, J. B. Pollack, and C. Sagan. It reads in part:

> We have performed a variety of sensitivity studies to define a range of possible outcomes of a full-scale nuclear exchange. In some cases we predict long-term effects which are small in comparison to the primary nuclear destruction due to blast, thermal pulse, and local radioactive fallout. However, a significant number of cases show potentially devastating global effects. In these instances, a combination of stresses caused by severe climate perturbations (surface coolings of 10° C or more), radiation doses in the tens of rem, and tenfold increases in uv-B solar radiation exposures, together with widespread shortages of food and potable water, epidemics, serious injuries, and lack of medical facilities and supplies, cumulatively imply widespread death in man and possible extinction of numerous land and marine species.

The various streams of research whose confluence led to the TTAPS findings are summarized in Appendix C.

By analogy with volcanic explosions, it has long been suspected that the dust from many simultaneous nuclear explosions might affect the climate (e.g., L. Machta and D. L. Harris, "Effects of Atomic Explosions on Weather," *Science 121,* 75–80, 1955). In 1955 Congressional testimony, the well-known scientist John von Neumann estimated that the dust from around 100 multimegaton groundbursts could "bring back the conditions of the last ice age," after a lag of ten or twenty years. ("Health and Safety Problems and Weather Effects Associated with Atomic Explosions," Hearings, Joint Committee on Atomic Energy, U.S. Congress, April 15, 1955 [Washington: U.S. Government Printing Office, 1955].) The magnitude, the duration, and the lag of the effects of dust were all greatly overestimated by von Neumann, and the central importance of smoke was overlooked. But this was an analogy, not a calculation. Still, considering the eminence of the source, it is surprising that his opinion was soon almost wholly forgotten or ignored.

Since the discovery of the nuclear winter phenomenon, we have tried to trace any earlier forerunners. The earliest premonition of nuclear winter we can find in nonfiction is a paper by Ben Hur Wilson, a Joliet, Illinois, high school teacher ("Behavior of the Atmosphere Under Atomic Disruption," *Popular Astronomy* 57 [7], 1949, especially pp. 320–322). Wilson envisioned, as "the most profound [climatic] effect" of a large nuclear war, "the cutting off of the sun's radiant energy coming down through the atmosphere by huge smoke and dust clouds which would be carried upward by the ascending currents. As these clouds drift around the world their interference would be considerable." Both smoke and dust from a nuclear war were anticipated by Paul R. Ehrlich to lower temperatures and destroy agriculture at least on the scale (perhaps 1 to 2° C) of the sequelae of a major volcanic explosion (Ehrlich, "Population Control or Hobson's Choice," in L. R. Taylor, ed., *The Optimum Population for Britain* [London: Academic Press, 1970], 154; see also T. Stonier, *Nuclear Disaster* [Cleveland: World Publishing, 1964], who mentions dust only).

The earliest suggestion of nuclear winter in science fiction seems to be Carl W. Spohr's "The Final War," serialized in *Wonder Stories* in 1932—long before Hiroshima and Nagasaki. Strategic nuclear weapons overwhelm an extensive but porous SDI defensive system (just as they would were SDI deployed in our time), and after the resulting nuclear explosions, "the world was swallowed in black, raging darkness. . . . Dust clouds blotted out the world."

Neither Wilson nor Spohr seems to have noted the severe temperature declines that such clouds would imply. An excellent survey of nuclear war as portrayed in science fiction prior to 1945 is H. Bruce Franklin's *War Stars: The*

*Superweapon and the American Imagination* (New York: Oxford University Press, 1989).

"The Curse," a 1947 short story by Arthur C. Clarke, described the sky as "wholly darkened" following a nuclear war [reprinted in Clarke, "Reach for Tomorrow" (New York: Ballantine, 1956)].

The first science fiction account that combines nuclear weapons, soot in the air, and catastrophic temperature declines is "Torch," by Christopher Anvil (*Analog Science Fact/Science Fiction*, April 1957, 41–50). Underground oil seams are ignited in the test of an Earth-penetrator nuclear warhead; the fires spread subsurface and release huge quantities of fine soot into the atmosphere. Then the clouds spread:

> Temperatures of a hundred degrees below zero are being reported. . . . We don't know when these fine particles will settle. The heavier particles of relatively large diameter settle out unless the air currents sweep them back up again, and then we have these soot showers. But the smaller particles remain aloft and screen out part of the sun's radiation. Presumably they'll settle eventually, but in the meantime it's a good deal as if we'd moved the Arctic Circle down to about the fifty-fifth degree of latitude.

Note also the following speculation, in a book on nuclear war by Bertrand Russell; he is ruminating on the savants of the aerial island of Laputa in Jonathan Swift's *Gulliver's Travels*:

> The philosophers of Laputa reduced rebellious provinces to obedience by causing the shadow of their island to plunge the rebels into perpetual night. It should become possible, before very long, to secure that some large enemy region should have either too much or too little rain, or that its temperature should be lowered to a point where it would no longer produce useful crops.
>
> —Russell, *Common Sense and Nuclear Warfare* (London: George Allen & Unwin, 1959), 17.

# CHAPTER 4

# THE WITCHES' BREW

## POISON GAS, RADIOACTIVE FALLOUT, ULTRAVIOLET LIGHT

This earthly air . . . is terribly infected with the nameless miseries of the numberless mortals who have died exhaling it.

—Herman Melville, *Moby Dick* (1851), Chapter 27

THE PRINCIPAL AND MOST WIDELY DISCUSSED ASPECTS OF nuclear winter are the cold and the dark. But when we introduced the term, we intended it to encompass other serious long-term consequences of nuclear war, of which we identified three: the production of heavy, ground-hugging clouds of toxic gases released during the destruction of modern cities; the worldwide distribution of radioactive fallout, attached to some of the same fine particles that block the sunlight; and the assault on the protective ozone layer that ordinarily blocks deadly ultraviolet sunlight from reaching the surface of the Earth. The physics of each of these ancillary catastrophes is related to the machinery of nuclear winter. For example, recent studies show that the heating by sunlight of high-altitude clouds of soot and dust work to deplete the ozone layer; the main consequences transpire after the obscuring particles have fallen out of the atmosphere, but before the ozone layer has had time to heal itself. It is worth noting that—like the cold and the dark of nuclear winter—all three of these effects were overlooked or minimized by the world's military establishments.

## Pyrotoxins: Poisons from Fires

When we read of people dying in the burning of a skyscraper, we are told they have been "overcome by smoke." In fact, in most cases, they have died—like soldiers at Ypres, Belgium, in 1915—of poison gas (ref. 4.1).

A man's wool suit, when thoroughly burned, gives off enough cyanide to kill seven people (ref. 4.2). The most dangerous aspect of fire in modern buildings is the production of dense, deadly pyrotoxins (from the Greek for "fire poisons"). Fire generates a variety of toxic compounds ranging from simple gases such as carbon monoxide (CO), hydrogen cyanide (HCN), and hydrogen chloride (HCl) to slightly more exotic poisons such as acrolein ($C_3H_4O$) and vinyl chloride ($C_2H_3Cl$). These compounds are produced largely from the synthetic materials that are increasingly employed in construction and interior furnishings—in particular, through expanding usage of plastics and synthetic fibers. Zyklon B, the active agent in the gas chambers of the Nazi death camps, was the brand name for a crystalline powder that generated hydrogen cyanide (ref. 4.3). (HCN is still considered a convenient and quick-acting means of executing those condemned to death.) Other common structural materials—insulation, for instance—can be rich in organic compounds such as formaldehyde. Such gases, when released into the atmosphere from storage tanks, or during low-temperature smoldering combustion, react to form a heavy, ground-hugging, and potentially lethal smog.

Many chemicals in widespread use—for example, the polychlorinated biphenyls (PCBs) and chlorinated benzenes used as insulators in electrical transformers—are not only toxic themselves, but can generate even more dangerous compounds when burned. Transformer fluids, for example, create polychlorinated dibenzofurans and dioxins when burned. Although the precise toxicological effects of these compounds are not yet certain, the dioxins and furans are considered to be among the most dangerous organic compounds known. "Agent Orange"—used by the United States to "defoliate" Vietnam, and the subject of numerous suits by disabled American and Australian Vietnam War veterans (and their heirs) against their governments—is a dioxin. The production of such gases from modern urban waste, nuclear winter aside, has become a practical matter since the discovery that trash-burning incinerators manufacture dioxins (ref. 4.4).

In a nuclear war, the sources of such toxic materials would be widespread. Mass fires in urban areas would introduce unprecedented quantities of pyrotoxins into the atmosphere (ref. 4.5). Industrial zones subjected to nuclear blast and fires would release much of their exotic chemical stores into the surrounding air, land, and water. The accident at the Union Carbide pesticide plant in Bhopal, India, on December 3, 1984, released a relatively small quantity of methyl isocyanate into the atmosphere; thousands of people were killed and hundreds of thousands were injured (see ref. 6.2). Such events illustrate the potential horrors of inadvertent chemical mass murder as an inadequately studied by-product of nuclear war.

Some hazardous chemicals—including chlorine, ammonia, ethylene, sulfuric acid, nitric acid, phosphoric acid, benzene, and xylenes—are produced and stored in huge quantities measured in the millions of tons. If these vats and tanks were ruptured or destroyed by nuclear explosions nearby, the toxic clouds raised would cut poisonous swaths across the landscape. At many oil refineries, enormous stocks of sulfur recovered from processed fuels would be exposed to direct ignition by nearby nuclear bursts; combustion of the sulfur would create a plume of sulfuric acid that would poison the air and acidify clouds and rainfall far downwind of the refineries.

Nuclear explosions over urban areas could also generate an extensive pall of deadly asbestos fibers from the pulverization of buildings long ago insulated with this substance. The fine asbestos fibers would drift over large areas, exposing multitudes to the long-term prospect of the deadly cancer mesothelioma. Pyrotoxins would be a major hazard for populations that fled cities, and for all those downwind of cities, petroleum facilities, and chemical storage depots, for at least as long—and it might last a week or more—as the fires smoldered (ref. 4.6). The local environment—soil and water, including vulnerable estuary systems—could be poisoned for much longer periods by concurrent spills, runoff, and deposition of industrial chemicals.

## Radioactive Fallout

The detonation of a nuclear weapon near the Earth's surface raises enormous quantities of dust into the atmosphere and causes deadly radioactive fallout. Nuclear fission of plutonium (and uranium), the process that triggers all nuclear explosions, creates dozens of unstable atomic nuclei that decay over periods of hours to years into more stable forms. In the act of decaying, the unstable nuclei release alpha, beta, and gamma radiation. Of these, the gamma rays—a very energetic but invisible form of light—are the most dangerous. Typically, gamma rays can penetrate a foot of concrete, one or two feet of dirt, or two or three feet of water. They come from two principal

# FIRE AND SMOKE: EXPERIMENTS TO SIMULATE NUCLEAR WINTER

Full-blown nuclear winter cannot be experimentally reproduced on any scale short of nuclear war itself. However, many of the fundamental physical principles on which the nuclear winter theory is based *can* be tested experimentally. To this end, a number of fires on very different scales have been set in the name of nuclear winter. Some were more successful than others. A chemist and his graduate student at the University of Colorado, for example, devised a clever technique to study the chemical reaction between ozone (as in the stratospheric ozone layer) and sooty smoke particles (as from burning cities). Could the ozone react with the soot and dissipate the dark nuclear winter smoke pall? Their apparatus looked as if it had been assembled from spare parts and scrap hardware, and occupied a small space in one corner of a cluttered laboratory. By contrast, a team of researchers at a prestigious Eastern research laboratory constructed an elaborate, gleaming, high-technology device to measure the same effect—and, after years of work, dismally failed to obtain any useful results. The University of Colorado experiment (without direct financial support from any federal agency) demonstrated clearly that the ozone/soot reaction is much too slow to mitigate the effects of nuclear winter, as some had hoped.

In other laboratories, all sorts of materials—wood, plastics, liquid fuels—were burned to analyze the smoke they generated. As might be expected, a number of odd and elaborate experiments were proposed, and a few actually were carried out. One particularly energetic and persuasive young technician employed by a large aerospace firm managed to spend nearly a quarter of a million dollars (of government and company money) fabricating what could be billed as the most expensive beer cooler in the world—a large metal tank attached to a refrigeration unit in which cold smoke experiments were to be conducted. No experiments were ever carried out; the basic design concept turned out to be flawed.

The most spectacular nuclear winter experiments by far involved large burns of dead forest carried out by the Canadian and U.S. Forest Services. One of us (R.T.) witnessed the first of these experiments in the Canadian province of Ontario near the sleepy little town of Chapleau in August 1985. Although this exploratory burn was meant to be low-key, the press got wind of it and invaded the town. Amid helicopter shuttles and press conferences, about 1600 acres of dead trees were ignited in a mass conflagration that pushed smoke some 6 kilometers into the atmosphere, created a pall stretching more than 100 kilometers downwind, and blanketed Chapleau in a depressing gray shroud that spoiled an otherwise pleasant Canadian summer weekend. The fire plume dominated the local landscape and left many observers gaping at the huge mushroom cloud.

Over the next few years, the Chapleau experiment was repeated several times, with the addition of numerous instruments to obtain hard data. Lasers were used to probe the smoke. Intrepid scientists braved lurching flights through the fire column and through dark, turbulent clouds to find out the truth about nuclear winter. The understanding obtained is reflected in this book.

On the way to learning more about nuclear winter, the scientific community also uncovered new knowledge about the environment. In one serendipitous discovery, a research team from the University of Washington found that brush fires in the hills surrounding the Los Angeles basin released into the atmosphere unusually large quantities of oxides of nitrogen and chlorofluorocarbons—apparently through the resuspension of air pollutants deposited over months or years on vegetation and soil. This must be true for many urban areas that would be in flames in a nuclear war. These gases, especially the CFCs, could further thin the ozone layer.

sources: the initial "prompt" gamma rays produced during the nuclear explosion itself, and the "delayed" gamma rays emitted during the radioactive decay of residual unstable chemical elements synthesized in the explosion. The prompt gammas irradiate the region already subject to intense thermal (heat) radiation and crushing blast effects. For this reason, their lethal effects are comparatively unimportant. Dead is dead; it doesn't matter if those killed by falling buildings or burned to death are also fried by gamma rays.

The delayed gammas, however, are emitted by debris that can be carried by winds hundreds or thousands of miles from the explosion site before falling out or raining out of the air. The radioactive elements involved tend to condense onto dust particles. In the rising fireball of a surface nuclear detonation ("groundburst"), the intimate mixing of surface particles swept into the fireball with the newly generated radioactive elements scrubs most of the radioactivity out of the air and onto the dust. Hence, the radioactivity is distributed over a large area as the dust settles downwind of the detonation, creating an extensive field of fallout. The intensity of the radioactivity gradually fades as the fallout ages; the intensity will decline tenfold for every sevenfold increase in time. (So there's one-tenth as much radioactivity after a week as after a day; only one-tenth of *that* after 7 weeks; another 90% gone by $7 \times 7 = 49$ weeks; etc.).

Early calculations of casualties from fallout in a nuclear war were based on weapons explosions before the 1963 Limited Test Ban Treaty. The arsenals in those days were skewed to powerful, high-yield bursts that carried the fallout well up into the stratosphere. From there, it took months to years to fall out, and by the time the fallout arrived on the ground, much (but not all) of the dangerous radioactivity had decayed. Since then the superpowers have reduced the average yield of their strategic weapons. Ironically, this means that less of the radioactivity goes up into the stratosphere and more of it is carried only to the upper troposphere (just below the stratosphere), where it is distributed by winds and falls out in weeks—before as much of the radioactivity has had time to decay to safer levels. For many years calculations of radioactive fallout after a nuclear war were based on anachronistic arsenals. The result was that the predicted (and widely advertised) world fallout burden was something like a tenth of the actual value. By tracking, in theoretical models, the distribution and timescale for delayed fallout, the study of nuclear winter has made a major contribution toward putting this error right.

Fashionable current estimates are that the prompt fallout from a Soviet/American strategic nuclear exchange would kill up to 50 million people. Although such estimates are highly uncertain, there are a number of reasons to believe they are still overly conservative. Many victims who are not killed outright by radioactivity tend to succumb to secondary illnesses that take hold because radiation compromises the human immune system. Also, it is well documented that individuals suffering burn or trauma injuries are much more susceptible to death from radiation exposure. Moreover, intermediate and longer-term exposure to lower levels of radioactivity—from external gamma radiation, as well as from radioactive materials inhaled with air and ingested with food and water—can induce fatal disease long after the war. All in all, we estimate that the total number of casualties from radioactivity effects of all sorts following a major nuclear war could approach 300 million (ref. 5.9).

There would be other sources of radioactivity in a nuclear-devastated world besides the weapons fallout. Explosions near many likely strategic and tactical targets would release additional radioactivity into the environment. Such targets include plutonium and uranium refineries, nuclear weapons assembly plants and storage facilities, military nuclear power reactors (especially in ships), and civilian nuclear power reactors. An enormous amount of radioactive material—more radioactivity than is held in all the nuclear weapons in the world—is stored in extremely vulnerable pools and shallow burial sites near potential nuclear targets. The creation of large stretches of radioactive wasteland—with soil and water poisoned for hundreds or thousands of years—is a real prospect, given the present co-deployments of nuclear weapons and nuclear power facilities.

A glimpse of the potential radioactive disaster that nuclear war holds is provided by the Chernobyl accident of April 26, 1986—by far the worst nuclear accident ever to occur. The total release of radioactivity was roughly equivalent to that produced by a 0.01 kiloton nuclear fission explosion. This is about one-thousandth that of the Hiroshima bomb and one-billionth the total explosive yield in the world's nuclear arsenals—perhaps ten million kilotons of energy. Thermonuclear weapons have a lower fission yield than fission weapons, so the full world nuclear arsenals represent perhaps 10% of the radioactivity of a billion Chernobyls. What does the radioactivity from one hundred-millionth of a full nuclear war do? At the power plant itself and in the subsequent cleanup, some 250 people exposed to high levels of fallout died. The town of Pripyat near the Chernobyl reactor was evacuated almost immediately, and other villages later. A radioactive cloud drifted from the Chernobyl site in the Soviet Ukraine across much of Europe. In some areas the radioactivity from the cloud was intense, and precipitation brought dangerous levels of fallout to the ground. In Finland and Sweden, reindeer herds were severely contaminated. Elsewhere in Europe, some food was tainted, and panic and confusion resulted in the destruction of much more foodstock. The Chernobyl cloud was even detected over the western United States, by which time, however, it had become so diluted as to pose (probably) only a minor health hazard. Cleanup of the site and protection of surrounding communities strained the civil defense and emergency facilities of the entire Soviet Union and cost some tens of billions of rubles. Substantial areas are uninhabitable into the indefinite future. We can barely imagine what a full nuclear war, the equivalent of a hundred million Chernobyls, would bring in radioactivity alone.

Nevertheless, the radioactive fallout from a nuclear war is unlikely even to come close to killing everyone on Earth—as was depicted in Nevil Shute's novel *On the Beach* and in the haunting motion picture of the same name. But it could kill a few percent of the global population, and render many others vulnerable to disease, famine and other unfolding consequences of nuclear winter. The projections of low casualties from fallout, and official instructions to the citizenry to dig holes, cover them with doors and dirt and hide underneath in the event of a nuclear attack (ref. 4.7; also cf. 2.1), had the effect of pacifying the citizens of a democracy who might otherwise have objected to government policy.

### Depletion of Stratospheric Ozone by Nuclear War

The atmosphere contains a fragile layer of ozone, essential to almost all life on our planet. It seems to have been first formed one or two billion years ago—when oxygen from the newly evolved process of green-plant photosynthesis began to accumulate in the ancient atmosphere. An ozone molecule is a form of oxygen ($O_3$) distinct from the ordinary molecular oxygen ($O_2$) which we breathe. This variation—three oxygen atoms instead of two—makes all the difference in the world to us. Ozone efficiently absorbs the deadly ultraviolet rays of the Sun falling at wavelengths between 250 and 330 nanometers (one nanometer is one billionth of a meter). Ordinary oxygen is transparent at these wavelengths. Most naturally occurring ozone is found in the stratosphere, which extends roughly from altitudes of 15 kilometers to 50 kilometers. The stratospheric ozone layer —or ozonosphere—envelops our globe in a protective shield. Yet the shield is extraordinarily vulnerable. If all of the ozone in the stratosphere were compressed into a gas at the sea-level pressure of the atmosphere, the layer would be no thicker than a pencil point. Providentially, the ozone layer is dispersed high in the atmosphere, safely away from meddling humans. Or at least it used to be.

Our technology has contrived to manufacture new substances, heretofore unknown in nature, that can track down ozone anywhere in the atmosphere and destroy it. The invention and widespread use of chlorofluorocarbons (CFCs) has led us to the threshold of significant global ozone depletions, and has triggered the formation of a deep ozone "hole" over the Antarctic continent. Of course, CFCs are useful and practical compounds for refrigeration, air conditioning, cleaning solvents, as the propellant in aerosol spray cans, and for many other purposes. Nature, however, has never manufactured these compounds, and cannot readily dispose of them—down here near the ground. But there is a way. The CFCs are eventually carried upward in currents of air, reaching the middle stratosphere where the ozone layer begins to thin. Here, intense solar ultraviolet radiation is available to break up the CFCs into their elemental constituents—the most important of which is chlorine, now identified as the leading culprit in attacking ozone.

In 1974, Sherwood Rowland and Mario Molina—then both at the University of California, Irvine—hypothesized that chlorine generated during the breakdown of CFCs would attack the ozone layer. The theory is complex in its details, but has been essentially confirmed by fifteen years of intensive scientific research, and is fully accepted in the scientific community. In recognition of this scientific consensus, the landmark Montreal Protocol, which limits the production and use of some CFCs, was signed by 33 nations in 1988. While the original Protocol did not include all culprit molecules, and mandated phasing out their production at too leisurely a pace, it nevertheless represents an important precedent. It is the first international agreement restricting a chemical compound—and its associated industry—because it poses a danger to the global environment. Stricter protocols have since been adopted. Conceivably, similarly enlightened political/scientific processes will one day deal with other pressing global environmental issues, including greenhouse warming and nuclear winter.

Ozone is formed when ultraviolet sunlight breaks down the oxygen molecule ($O_2$) into its constituent oxygen atoms, each symbolized by the letter O. An O plus an $O_2$ then join to form $O_3$. Ozone is naturally destroyed by a series of reactions, some of the most important of which involve trace amounts of hydrogen, nitrogen, and chlorine compounds. Ozone is continuously produced and destroyed. At any given time, the global ozone layer is in a kind of steady state, or balance, between ozone synthesis and ozone loss. If additional ozone-destroying compounds are added to the atmosphere—for example, CFCs—the ozone equilibrium will shift toward lower total ozone amounts. The thinned ozone shield now allows more of the harmful ultraviolet sunlight (wavelengths between about 290 and 330 nanometers, the so-called UV-B radiation) to penetrate to the ground.

If ozone-corrosive compounds are removed from the atmosphere over time—for example, by phasing out the use of all CFCs and related materials—the ozone equilibrium gradually shifts back toward its natural level. In other words, ozone depletion persists as long as the depleting compound is present, but recovers if the pollutant is removed. Unfortunately, some of the most common chlorofluorocarbons have a lifetime of about 100 years once they are emitted into the atmosphere. The

CFCs released during the closing decades of the twentieth century therefore represent an environmental legacy for the next four or more human generations, not to mention the other residents of our planet. (See also ref. 6.3, below.)

The traditional understanding of how a nuclear weapon affects the ozone layer is this: If the yield (explosive energy) of the weapon exceeds about 200 kilotons, the rising fireball reaches the stratosphere. The fireball is hot enough to burn air, producing out of the nitrogen ($N_2$) in the air oxides of nitrogen —symbolized as $NO_x$, where x can be a variety of numbers. $NO_x$ is deposited at high altitude, where it attacks and depletes the stratospheric ozone. Accordingly, the recent U.S. trend in reactivating high-yield weapons (ref. 4.8) is, as regards ozone depletion, movement in the wrong direction.

But in a nuclear war, the atmosphere would be so perturbed that our normal way of thinking about the ozone layer needs to be modified. To help refocus our understanding, several research groups have constructed models that describe the ozone layer following nuclear war. The principal work has been carried out by research teams at the National Center for Atmospheric Research and at the Los Alamos National Laboratory (ref. 4.9). Both find that there is an additional mechanism by which nuclear war threatens the ozone layer. With massive quantities of smoke injected into the lower atmosphere by the fires of nuclear war, nuclear winter would grip not only the Earth's surface, but the high ozone layer as well. The severely disturbed wind currents caused by solar heating of smoke would, in a matter of weeks, sweep most of the ozone layer from the northern midlatitudes deep into the Southern Hemisphere. The reduction in the ozone layer content in the North could reach a devastating 50% or more during this phase. As time progressed, the ozone depletion would be made still worse by several effects: injection of large quantities of nitrogen oxides and chlorine-bearing molecules along with the smoke clouds; heating of the ozone layer caused by intermingling of hot smoky air (as air is heated, the amount of ozone declines); and decomposition of ozone directly on smoke particles (carbon particles are sometimes used down here near the ground to cleanse air of ozone).

Not all of these factors are taken into account in the new calculations. The eventual depletion of the ozone layer in the

Northern Hemisphere following a nuclear war could reach 70%: that is, only 30% of the present ozone would be left. In the Southern Hemisphere, where less than 15% of the human population lives, the ozone content would initially *increase* by 30% or more, due to the arrival of Northern Hemisphere ozone. Later, some reduction would occur—although whether to less than prewar levels is currently unknown. The resulting ultraviolet hazards are very serious, including a greatly enhanced incidence of skin cancer, especially in light-skinned people; cataracts; and a further assault on the human immune system. These effects, of course, would be restricted to those who venture out-of-doors—but in the aftermath of a nuclear war, a large number of survivors would have to be out-of-doors. By far the most serious consequence of such severe depletion of the ozone layer, however, applies to people indoors and outdoors, because everyone has to eat:

Ozone depletion threatens the food chains on which almost all life on Earth depends. In the oceans, there are tiny microscopic plants, called phytoplankton, which are highly vulnerable to increases in ultraviolet light; and which, directly or indirectly, other animals in the marine food chain—including humans—eat. Land plants, including crops, are also vulnerable to increased ultraviolet light, as are most microbes, including those essential for the food chain. (Ultraviolet lamps were once used in hospital operating rooms to kill potential disease microorganisms.) We are far too ignorant of the global ecological interactions to understand fully what propagating biological consequences an assault on the ozone layer would entail (refs. 4.10, 6.3). But it doesn't take a great depth of understanding to recognize that if you rip up the base of the food chain, you may generate a disaster among the beings that totter precariously near the pinnacle. Recovery of the ozone shield would probably take several years. By then enormous damage would have been wrought.

The emergence of our ancestors onto the land may have had to await the formation of the ozone layer—life there being, in earlier epochs, too dangerous without the protection offered by seawater and floating organic debris from the searing solar ultraviolet light. It now appears that nuclear war could, at least in part, bring back those primordial environmental conditions.

In its total effect on the ozone layer, nuclear war could gouge out a hemisphere-sized ozone hole. At the bottom of this hole the intensity of ultraviolet radiation would attain levels that are deadly for many organisms, levels that pose extremely grave ecological problems for humans and other living things.

Beyond the cold and the dark, nuclear winter entails pyrotoxins, radioactive fallout, and intense solar ultraviolet light—a witches' brew of deadly assaults on life on Earth. An almost wholly unexplored subject is the concatenation, or "synergism," of effects. What happens to crops and to natural ecosystems when they are subjected simultaneously to a dimming of ordinary sunlight, to substantial drops in temperature, to doses of pyrotoxins, to radioactive fallout, and then, later, to intense ultraviolet radiation? The answer is that nobody knows. Worse, nobody is trying to find out (ref. 4.11).

Much public debate has centered on a few very specific and very striking issues raised by nuclear winter: Is the extinction of the human species possible in a nuclear war? Is there a threshold for nuclear winter? Does nuclear winter imply that a "disabling" first strike (on the adversary's land-based retaliatory forces) leads to climatic suicide for the aggressor nation? We have ourselves helped to raise these issues. But, we've found, by focusing the discussion on such dramatic questions the broad significance of nuclear winter has in some discussions been overshadowed. Yet, these specific questions are essential. We will address them before proceeding to other matters.

# CHAPTER 5

# EXTINCTION?

Into the eternal darkness, into fire, into ice.

—Dante Alighieri, *The Divine Comedy*, Inferno, Canto I, line 87

Until the patient Earth, made dry and barren,
Sheds all her herbage in a final winter,
And the gods turn their eyes to some far distant
Bright constellation.

—George Santayana, "Odes," in *The Complete Poems of George Santayana*, William C. Holzberger, ed. (Lewisburg, Pa.: Bucknell University Press, 1979), Part III, lines 65–68

WHAT IS THE WORST THAT NUCLEAR WAR COULD DO? OUR technology—while capable of enormous devastation—is wholly unable to alter the orbit of the Earth, change the tilt of the rotation axis, boil the oceans, or blow up the planet. Even exploding 60,000 nuclear weapons simultaneously could not do any of that. It seems highly unlikely that, even intentionally, we could destroy all life on Earth. There are hardy insects and grasses that are resistant to nuclear radiation and know how to close up shop, even for a very long winter, in order to resume business later. There are submarine worms that reside in hot vents at the ocean floor, living out their lives by altering the oxidation state of sulfur, impervious to whatever cold and dark, pyrotoxins and radioactivity and ultraviolet light might be stalking their distant cousins far above on the Earth's surface.

There is so much life on Earth, with so many diverse adaptations, that we cannot destroy it all. Cold comfort for us—because it is well within our powers to destroy the global civilization, other species, and perhaps ourselves. We are already, every day, rendering species of life on Earth extinct without nuclear war. Extinction of many more species may be possible in the wake of a nuclear war. But for us, understandably, an important question is whether we can make *humans* extinct.

By far the majority of species that ever existed on Earth are now extinct. There is no guaranteed tenure for any species on this planet—even one that judges itself particularly clever.* The history of the Earth has been punctuated by massive, episodic, and seemingly indiscriminate extinction events. For ex-

* Our scientific name for ourselves is *Homo sapiens*—we've defined ourselves to belong to the genus *Homo* and the species *sapiens*. Wise man, it means. It's something to aspire to.

# IMPACT WINTER

Dust in the air suspended
marks the place where a story ended.

—T. S. Eliot, "Little Gidding," II, *Four Quartets*

Excess amounts of the rare metal iridium (rare on Earth, but much more common on small nearby worlds) were found in certain thin layers in the geological record by the father-and-son team of Luis and Walter Alvarez, and their colleagues. This was the first real evidence that the Cretaceous mass extinctions of most of the species then alive were caused by the impact of an asteroid or a cometary nucleus. (A similar conclusion was drawn for the Eocene extinctions.) The favored mechanism, suggested in that seminal work, is the impact generation of a vast cloud of fine particles that cooled and darkened the Earth, analogous to nuclear winter. From the amount of iridium in the layer, it follows that the impacting body was about 10 kilometers across—a mountain falling out of the sky. Geologists describe the abrupt change in the record in the rocks at this layer as marking the end of the Cretaceous Period and the beginning of the Tertiary Period, 65 million years ago. The first detailed calculations of the Cretaceous cooling and darkening were performed by O. B. Toon, J. B. Pollack, T. P. Ackerman, C. P. McKay, and R. P. Turco. This cataclysmic event—which seems to have ushered the dinosaurs off the world stage and made possible the rapid evolution of the mammals and the origin of the human species—is naturally of interest to us.

The study by Toon, Pollack et al. estimated how much fine dust would be ejected by the impact, how it would be distributed in altitude and over the Earth, how much sunlight would be blocked, and how much the Earth's temperatures would fall. It proved to be another step toward the discovery of nuclear winter (see Appendix C). The mass of aerosols lofted by the Cretaceous impact was a thousand or ten thousand times more than in a nuclear war, but its climatic effect may not have been significantly worse. Because of how much darker soot is than dust, nuclear winter has a disproportionate climatic impact. Also, the nuclear winter effects tend to last longer: When the sky is dense with dust particles from such an impact, they collide, coagulate, and more rapidly fall out.

Later studies of the geological sediments marking the end of the Cretaceous pointed to an enormous quantity of soot particles generated at about the time of the giant impact, perhaps due to delayed worldwide burning of vegetation dried from the heat of the collision. Subsequent work suggests that "a single global fire," triggered by the impact, began "before the ejecta had settled"—i.e., within a few years. One calculation suggests that the radiation from the impact debris, burning up as it fell back to Earth, was about 100 times brighter than the Sun, "equivalent to a domestic oven set at 'broil' " for hours—more than enough to ignite wildfires worldwide. The resulting soot is also more than enough to plunge the Earth into utter darkness, well below freezing—effects much more severe than the most severe possible nuclear winter. As in nuclear winter, there would have been a massive depletion of the ozone layer persisting after the dust and soot all fell out, a few years later. There would have been no radioactive fallout, but—unlike nuclear winter—rains of concentrated acids and a major subsequent greenhouse warming (because, the evidence suggests, most of the vegetation on Earth burned up all at once and carbonate sediments would have released carbon dioxide in the impact).

It's hard to see how such profound environmental changes could have avoided rendering many species extinct. But not all paleontologists are convinced that the impact—which they do not dispute occurred—was the *main* cause of the late Cretaceous extinctions, in which the dinosaurs and some 75% of all the species then alive were destroyed. We stress that if the dinosaurs became extinct through some cause other than obscuring aerosols and increased ultraviolet sunlight, it would by no means call into question the argument for nuclear winter—although if aerosols were the cause, then the Cretaceous-Tertiary event may well shed light on nuclear winter.

The idea that extraterrestrial bodies striking (or making close passes by) the Earth might have brought on the terminal Cretaceous catastrophe was discussed well before the discoveries by the Alvarezes and their colleagues. In 1973 the Nobel Laureate Harold Urey wrote: "It does seem possible, and even probable, that a comet collision with the Earth destroyed the dinosaurs and initiated the Tertiary division of geologic time." In 1978 the astronomers Fred Hoyle and Chandra Wickramasinghe imagined a near-collision with the Earth by a comet, with tiny bright particles from the cloud that surrounds the nucleus of the comet entering the Earth's atmosphere. If the total mass of such particles were as large as 100 megatons (100 million tons), and if the particles were more efficient at absorbing and scattering light in the visible than in the infrared, then, they correctly concluded, there would have been significant attenuation of sunlight—although the surface of the Earth would still be able to cool itself by radiating into space. We were unaware of this insightful paper until recently. It is the first mention in the scientific literature, so far as we know, of the anti-greenhouse effect (called by Hoyle and Wickramasinghe the "inverse" greenhouse effect). They proposed that such a near collision could—through falling temperatures and diminished sunlight for photosynthesis—have produced mass extinctions, even if the particles took no more than a year to fall out of the Earth's atmosphere. Although their research report is not about nuclear war and nowhere mentions it, it clearly anticipates much of the argument of Alvarez et al. and of Toon, Pollack et al. on the "impact winter" explanation of the late Cretaceous extinctions—all before any direct evidence of such an impact was in hand.

If the worst possible consequences followed a global thermonuclear war, then in the subsequent geological record there would be a thin, highly radioactive, sooty boundary layer of worldwide distribution marking widespread extinctions: Fossil and other remains of species found below the layer would be absent above it. It would, except for the radioactivity, be like the Cretaceous/Tertiary boundary. But instead of the white calcareous remains of the foraminifera that teemed in the warm Cretaceous seas, there would be license plates and wedding rings. A visitor from another world would have little difficulty piecing together what had happened (ref. 5.18).

ample, in the Permian catastrophe, 245 million years ago, more than 75% of all the genera (plural of genus) and more than 90% of all the species on Earth disappeared. In the late Cretaceous catastrophe, some 65 million years ago, 50% of the genera and 75% of the species disappeared. In the late Eocene event, some 35 million years ago, 16% of the marine genera became extinct (ref. 5.1). A number, but by no means all, of these catastrophic extinction events seem to have been caused by high-speed collisions with the Earth by hurtling mountain-sized worlds from space. It is not completely determined how such collisions caused extinctions, but the prevailing view is that it was something like nuclear winter (see box, "Impact Winter").

Can a nuclear war result in the extinction of the human species? From the very outset of the nuclear age, there were those who feared that we had put our species at risk (see box), and those who decried such fears as "loose talk" (ref. 5.2), or worse. Perhaps it is because most specialists on nuclear war live in the Northern midlatitude target zone that the mere mention of the possible extinction of the human species sometimes arouses their irritation: They know that they and huge numbers of their co-nationals would, very likely, be killed in the first exchange. Extinction does not much increase their own risk or that of their loved ones. But talking about it can be interpreted as at least guarded criticism of existing policy. If you take the extinction possibility seriously, it's harder to go on with your daytime job —especially when your job has something to do, if you look at it from that point of view, with the implementation of extinction:

> I believe that both the United States and NATO would reluctantly be willing to envisage the *possibility* of one or two hundred million people . . . dying from the immediate effects, even if one does not include deferred long-term effects. . . . [But] if [nuclear war] happened to involve explicitly the annihilation of all humanity it would also be totally immoral; one doubts if it could long remain an important part of United States policy (ref. 5.3).

# "THE APOCALYPSE MAY COME"

Among those who have taken seriously the possibility that nuclear war might mean human extinction—even without a detailed and compelling scientific argument—are many knowledgeable scientists and statesmen, including a representative cross-section of the inventors of nuclear weapons:

> If we focus our attention on the next twenty-five years we may say that development is likely to reach some point intermediate between the first bomb detonated over Hiroshima and processes which once initiated might put an end to all life on earth. Just what intermediate point will be reached within twenty-five years no one can tell.
>
> —Leo Szilard, the first scientist to realize that a nuclear weapon was really possible, in an address given September 21, 1945, at the "Atomic Energy Conference," University of Chicago. From Spencer R. Weart and Gertrud Weiss Szilard, eds., *Leo Szilard: His Version of the Facts (Special Recollections and Correspondence)*, (Cambridge, Mass.: The MIT Press, 1978), 234.

> It is not even impossible to imagine that the effects of an atomic war fought with greatly perfected weapons and pushed by the utmost determination will endanger the survival of man.
>
> —Edward Teller, "How Dangerous Are Atomic Weapons?" *Bulletin of the Atomic Scientists*, February 1947.

> The extreme danger to mankind inherent in the proposal [by Edward Teller and others to develop thermonuclear weapons] wholly outweighs any military advantage.
>
> —J. Robert Oppenheimer et al., *Report of the General Advisory Committee*, U.S. Atomic Energy Commission, October 1949.

> The fact that no limits exist to the destructiveness of this weapon makes its very existence and the knowl-

> edge of its construction a danger to humanity . . . It is . . . an evil thing.
>
> —Enrico Fermi and I. I. Rabi, Addendum, *ibid.*

> A very large nuclear war would be a calamity of indescribable proportions and absolutely unpredictable consequences, with the uncertainties tending toward the worse . . . All-out nuclear war would mean the destruction of contemporary civilization, throw man back centuries, cause the deaths of hundreds of millions or billions of people, and, with a certain degree of probability, would cause man to be destroyed as a biological species.
>
> —Andrei Sakharov, "The Dangers of Thermonuclear War," *Foreign Affairs*, Summer 1983.

> Much has been said of the prospect that man, along with many other forms of life . . . , would disappear as a species. In time, not a long time, that may come to be possible. What is more certain and more immediate is that we would lose much . . . that has made our civilization and our humanity . . . The threat of the apocalypse will be with us for a long time; the apocalypse may come.
>
> —J. Robert Oppenheimer, "Science and Our Times," *Bulletin of the Atomic Scientists 12* (7), September 1956, 236.

> We witness today, in the power of nuclear weapons, a new and deadly dimension to the ancient horror of war. Humanity has now achieved, for the first time in its history, the power to end its history.
>
> —President Dwight D. Eisenhower, speech, September 19, 1956.

In his famous September 26, 1961, speech, President John F. Kennedy proposed that nuclear war and its effects,

> spread by winds and waters and fear, could well engulf the great and the small, the rich and the poor, the committed and uncommitted alike. Mankind must put an end to war or war will put an end to mankind.

In private conversation during the Cuban missile crisis he is depicted as brooding that nuclear war might "engulf and destroy all mankind." (Robert F. Kennedy, *Thirteen Days* [New York: Harper & Row, 1965], 84.) "The possibility of the destruction of mankind was always in his mind." This apocalyptic prospect may have helped bolster his inclination toward restraint in responding to the Soviet emplacement of strategic weapons in Cuba—and to advice from members of the Joint Chiefs of Staff to attack Soviet installations in Cuba and to use nuclear weapons in so doing (*ibid.*, 36, 48). General Secretary Leonid Brezhnev also declared that "mankind might be wholly destroyed" in a nuclear war (speech to the Polish Sejm, July 21, 1974; cited in Foreign Broadcast Information Service *Daily Report: Soviet Union,* July 22, 1974, No. 141, D17), and Premier Nikita Khrushchev earlier made similar remarks.

Just after Hiroshima and Nagasaki, Albert Einstein was thinking about apocalyptic prospects should a large nuclear war break out. But he concluded,

> I do not believe civilization will be wiped out in a war fought with the atomic bomb. Perhaps two-thirds of the people on Earth might be killed. . . .
>
> —"Einstein on the Atomic Bomb," *Atlantic Monthly*, November 1945.

Two-thirds of the present population of the Earth is between 3 and 4 billion people.

Ten years later, after the U.S. nuclear stockpile had grown enormously, and an arms race with the Soviet Union had begun, Einstein became more pessimistic. The Einstein-Russell manifesto (*New York Times*, July 10, 1955, 25) referred to "the species man, whose continued existence is in doubt." It was signed by Bertrand Russell, Albert Einstein, Percy W. Bridgman, H. J. Muller, Cecil F. Powell, Joseph Rotblat, Frédéric Joliot-Curie, Leopold Infeld, Hideki Yukawa, Max Born, and Linus Pauling. In the correspondence leading to the manifesto, Russell wrote to Einstein, on February 11, 1955:

> War may [now] well mean the extinction of life on this planet. The Russian and American govern-

ments do not think so. They should have no excuses for continued ignorance on the point.

—Otto Nathan and Heinz Norden, eds., *Einstein on Peace* (New York: Simon and Schuster, 1980), 625–637.

The following is extracted from Einstein's last written words:

[The] conflict that exists today is no more than an old-style struggle for power, once again presented to mankind in semireligious trappings. The difference is that, this time, the development of atomic power has imbued the struggle with a ghostly character; for both parties know and admit that, should the quarrel deteriorate into actual war, mankind is doomed.

—Einstein, in an unfinished address drafted after a meeting on April 11, 1955, with Israeli Ambassador Abba Eban and Consul Reuven Dafni in Princeton, New Jersey; in Otto Nathan and Heinz Norden, eds., *Einstein on Peace* (New York: Simon and Schuster, 1960), 641.

After the 1984 showing in the United States of "Threads" (ref. 13.18), the BBC television dramatization of nuclear war and nuclear winter, a number of newspaper reviews argued that—while the program was powerful, moving, or brilliant—Americans had become saturated with portrayals of this sort. In fact, in the history of American television, there had been to that time only a handful of such dramas. It didn't take much to saturate the reviewers. How much closer to their saturation point for horror must be the strategists and contingent practitioners of nuclear war. For many of them, at least emotionally, nuclear winter changes nothing (ref. 5.4).

Still, it is not too much to ask that our leaders do not pose a threat to the human species. People of many different political persuasions believe they see a vast difference between killing, say, nine-tenths of the human species and killing everybody. Of course there is. If there are survivors, there is some chance of the regeneration of the human population. Extinction means there will be no more humans forever. We confess to having difficulty understanding why the prospect of killing everybody would bring about more protest against government policies

## A MODEST HOPE

If we do not soon destroy ourselves, but instead survive for a typical lifetime of a successful species, there will be humans for another 10 million years or so. Assuming that our lifespan and numbers do not much grow over that period, the cumulative human population—all of us who have ever lived—would then reach the startling total of about a quadrillion (a 1 followed by 15 zeros). So, if nuclear winter could work our extinction, it is something like a million times worse (a quadrillion divided by a billion) than the direct effects of nuclear war—in terms of the number of people who would thereby never live.

Jonathan Schell described extinction in these terms:

> Only by a process of gradual debasement of our self-esteem can we have lowered our expectations to this point. For, of all the "modest hopes of human beings," the hope that mankind will survive is the most modest, since it only brings us to the threshold of all the other hopes. In entertaining it, we do not yet ask for justice, or for freedom, or for happiness, or for any of the other things that we may want in life. We do not even necessarily ask for our personal survival; we ask only that we *be survived.* We ask for assurance that when we die as individuals, as we know we must, mankind will live on. [Schell, *The Fate of the Earth* (New York: Knopf, 1982).]

than the prospect of killing almost everybody; nevertheless, that increased protest (and public scrutiny) is what some analysts feared from nuclear winter and has been, we believe, behind some of the media and political fixation on nuclear extinction (ref. 5.5).

People concentrate themselves in large cities, so killing them there has become easy in the nuclear age (ref. 5.6). But people also live in towns and in the countryside. This is why killing a quarter of the population of a nation through the direct (or "prompt") effects of nuclear weapons is much easier than killing, say, half or three-quarters. That's where nuclear winter comes in. Nuclear winter is a way for nuclear weapons to find and kill those who live far from cities.

Certainly, the casualty estimates from prompt effects in a nuclear war are appalling: The U.S. nuclear war protocol (Single Integrated Operational Plan, SIOP) of 1960 vintage would have destroyed every city in the Soviet Union and China, with estimated direct fatalities around 400 million (ref. 5.7). Presidential Review Memorandum 10 (February 18, 1977) estimated some 250 million fatalities in a U.S./U.S.S.R. central exchange (ref. 5.8). Since then, estimates of the dangers of radioactive fallout have had to be revised—to take account of the tenfold underestimate of intermediate timescale fallout radiation doses in official publications, and the consequences of attacks on military and commercial nuclear fuel facilities; global casualties from radioactivity alone are now estimated at 80 to 290 million (refs. 5.9, 5.10), with the higher numbers, in our opinion, more likely. Thus, several hundred million prompt fatalities may occur in a full-scale nuclear exchange, with up to a billion more fatalities if urban centers and nuclear fuel facilities worldwide are heavily targeted (ref. 5.11); separate, longer-term fatalities —especially from nuclear winter-related crop failures and resulting malnutrition and starvation—might amount to several billion (ref. 3.11). Many others would die from the collapse of the society (the unavailability of physicians, hospitals, and medicines, for example), the spread of disease, and (later) the increased ultraviolet radiation. Under these, perhaps pessimistic, estimates, the sum of prompt and long-term fatalities approaches the total human population of over 5 billion. A key issue, addressed below, is survival in the midlatitudes of the Southern Hemisphere.

With the technological base in ruins, and accessible key resources depleted, recovery of the global civilization after nuclear war is in doubt. There would also be, in the words of Andrei Sakharov, "the rise of a savage and uncontrollable hatred of scientists and 'intellectuals' . . . , rampant superstition, ferocious nationalism, and the destruction of the material and informational basis of civilization"; it would introduce a new "age of barbarism" (ref. 5.12).

Destruction of the global civilization is very different, though, from extinction of the human species. However, the multiple stresses on biological systems, and likely interactions (synergisms) among these stresses, could fundamentally alter ecological relationships on which humans now depend. Considering a nuclear winter scenario at the severe end of the spectrum of possibilities, a distinguished group of ecologists and biologists argue that massive species extinctions—especially but not exclusively at tropical and subtropical latitudes where there are few adaptations to cold—would ensue (ref. 5.13). They conclude:

> It seems unlikely, however, that even in these circumstances *Homo sapiens* would be forced to extinction immediately. Whether any people would be able to persist for long in the face of highly modified biological communities; novel climates; high levels of radiation; shattered agricultural, social, and economic systems; extraordinary psychological stresses; and a host of other difficulties, is open to question.

The SCOPE report (ref. 3.11), the most comprehensive analysis of the biological implications of nuclear winter, does not explicitly address human extinction, but it does indicate that the death of several billion people, mainly from starvation, is possible in the climatic aftermath of a large-scale nuclear war. That would be added to the estimated prompt casualties of many hundreds of millions, severe post-traumatic stress on the survivors (ref. 5.14), and a range of as yet undiscovered synergisms among the individually adverse environmental consequences. Small groups of survivors would be particularly vulnerable to accidental unfavorable fluctuations in the physical or biological environment (ref. 5.15). The conclusion remains: Human extinction is by no means excluded (ref. 5.16).

But the issue is of such complexity and is so alien to our experience that it is beyond our present ability to predict reliably. We simply do not know.

Nuclear war was certainly considered serious business even before the discovery of nuclear winter, even in the absence of any credible demonstration that extinction was possible (ref. 5.17). At the very least, nuclear winter underscores the extreme danger of nuclear war. But it is surely incorrect to infer, as some

## SCOPE

If there is anything like a global parliament of scientists, it is the Paris-based International Council of Scientific Unions. In some sense ICSU can be said to speak—although usually very quietly—for the scientists of the planet Earth. Perhaps its best-known activity was the organization of the International Geophysical Year (1957–1958) that ushered in the Space Age. Here is the way it works: In a given discipline—astronomy, say—there are national organizations of professional scientists. In the United States the principal such group is the American Astronomical Society. With its fellow astronomical organizations from many dozens of other countries, the American Astronomical Society adheres to the International Astronomical Union. Once every three years the world's astronomers gather together under IAU auspices to exchange research findings and to discuss policy matters important for astronomy. Similarly, the American Geophysical Union belongs to the International Union of Geodesy and Geophysics, and so on for physics, chemistry, mathematics, biochemistry, and many other sciences. These international unions, in turn, are adherents to and compose the International Council of Scientific Unions. It is a society of societies of societies. By its nature and tradition, ICSU is very conservative.

ICSU's Scientific Committee on Problems of the Environment (SCOPE) organized a massive interdisciplinary and international study, chaired by Sir Frederick Warner, on the environmental consequences of nuclear war. The study involved hundreds of scientists from more than a dozen countries working over three years. Meetings were held in Australia, Canada, China, England, France, India, Japan, the Netherlands, New Zealand, Sweden, Switzerland, Thailand, the U.S.S.R., the U.S.A., and Venezuela. Some of the conclusions are very direct for so retiring an organization. The two-volume report (ref. 3.11) warns of

> . . . a potential loss of about one to a few billion humans from long-term consequences; this wide range incorporates a wide range of different potential environmental and societal disturbances. This calculation does not count the losses from direct effects. . . .
>
> The total loss of human agricultural and societal support systems would result in the loss of almost all humans on Earth, essentially equally among combatant and non-combatant countries alike. . . . This vulnerability is an aspect not currently a part of the understanding of nuclear war; not only are the major combatant countries in danger, but virtually the entire human population is being held hostage to the large scale use of nuclear weapons. . . .
>
> As representatives of the world scientific community, drawn together in this study, we conclude that many of the serious global environmental effects are sufficiently probable to require widespread concern. Because of the possibility of a tragedy of an unprecedented dimension, any disposition to minimize or ignore the widespread environmental effects of nuclear war would be a fundamental disservice to the future of global civilization. . . .
>
> A fundamentally different picture of global suffering among peoples in non-combatant and combatant countries alike must become the new standard perception for decision-makers throughout the world if the visions portrayed in this study are to remain just intellectual exercises and not the irreversible future of humanity.

have, that if nuclear winter does not guarantee the extinction of the human species, it has no policy implications and no influence on deterring nuclear war.

In what follows we will consider the wide range of nuclear winter severities consistent with the governing physics. We will neither presume nor exclude the extinction of the human species.

# WOULD BILLIONS OF PEOPLE REALLY BE KILLED BY NUCLEAR WINTER?

The methods of the SCOPE biological assessment have been assessed and confirmed by a special expert scientific and agricultural panel convened by the White House Office of Science and Technology Policy (William H. Tallent et al., CIRRPC Science Panel #5, Executive Office of the President, "Review of SCOPE 28, Vol. II," March 1988). Indeed, the panel noted that the SCOPE analysis had been too conservative—ignoring several factors that would make the outcome of nuclear war still worse. Among the conclusions:

> A spring- or summer-onset heavy frost induced by a nuclear exchange could kill all native and crop species in Northern Hemisphere temperate regions. . . . Serious losses in agricultural productivity could occur in the tropics. . . . These consequences could be expected whether [nuclear winter] . . . or the less extreme "nuclear autumn" climatic scenarios . . . are used in the analysis.

The chairman's cover letter to the Executive Office of the President summarizes:

> Crops growing in the mid-latitudes of the Northern Hemisphere could be totally destroyed or production severely reduced for at least the first growing season after a nuclear exchange, if the resulting atmospheric perturbations were to cause temperature decreases on the order of 5 to 15°C for even short periods of time. However, the panel believes that several important factors were not adequately treated

in the SCOPE study, and that inclusion of these factors would make matters worse:

> Especially noteworthy are the loss of large areas of irrigated agricultural land due to destruction of dams,

> [and] severe disruption of production, processing and distribution caused by destruction of the complex infrastructure so necessary for the U.S. food and agricultural system.

But when he transmitted the Tallent Report to the Secretary of Agriculture on March 16, 1988, the President's Science Adviser, William R. Graham, gave no hint in his cover letter of the principal conclusion: that the biological effects of nuclear war were likely to be *worse* than the SCOPE study had estimated. This is a small example of the widespread tendency, still evident in the U.S. Government, to minimize the consequences of nuclear war. (Cf. ref. 8.22.)

# CHAPTER 6

# RISK

Everything happens to everybody sooner or later if there is time enough.

—George Bernard Shaw, *Back to Methuselah* (1921), Part V, 192

Even an unloaded rifle can fire once a decade. And once a century, even a rake can produce a shot.

—Old Russian saying, quoted by Marshal Nikolai V. Ogarkov, Chief of the Soviet General Staff, March 16, 1983

COMPARED WITH OTHER POTENTIAL CATASTROPHES—and there are getting to be a fair number of them—how big a risk of nuclear war are we running today? How much effort should we put into trying to prevent it—given all the other crises clamoring for our attention?

There is a range of possible outcomes of nuclear war, each with an associated risk. The risk, in the simplest terms, can be estimated as the probability that the event occurs multiplied by its cost—cost in lives, in misery, in lost knowledge, in the destroyed artifacts of our cultures and the sensibility of our civilization, cost measured by any standard we like. Even remote contingencies must be taken seriously if their consequences are sufficiently apocalyptic—a view traditionally embraced by military planners and nuclear strategists (ref. 6.1). War is too serious a matter, they have been telling us for generations, to base our plans merely on the most probable actions of a potential enemy. We must plan on capability, not intention. We must prepare, we are told, for the worst case. The stakes are too high to do anything else. It would be tellingly inconsistent if this ancient military doctrine were to be abandoned at the very moment we confront the ultimate worst case.

Insurance companies understand this very well, and it is the basis of the idea of actuarial risk. Every year the annual probabilities of accidental deaths in the United States are routinely tabulated. For example, there is roughly one chance in ten thousand of being killed in an automobile accident; one chance in a million of being electrocuted; one chance in ten million of being fatally struck by lightning. Likewise for property damage. In determining the premium you must pay for home insurance against some dread event—flood, say, or fire or earthquake—the insurance companies multiply the low probability of the event by the high replacement cost of your house. The likelihood that the event will transpire is known very poorly. The replacement cost of your home is known to much higher accuracy. Neither of them alone determines the premium. Both are essential. In the case of global nuclear war, the probability of the event is also known extremely poorly—unlike floods, fires, and earthquakes, we have never experienced even one. The "replacement cost" of our civilization is perhaps known better. In this case as well, multiplying probability by cost might be useful; it could, for example, give us some measure of the level of effort appropriate to prevent nuclear war and nuclear winter from happening.

The prompt destruction, loss of life, agony, and return to barbarism would be so awful, it is sometimes argued, that *this* anticipated "cost" alone should effectively preclude a nuclear war. Thus, we arrive—there are many routes to this door—at the fundamental paradox of the nuclear age: Nations must be ready to fight a nuclear war in order to prevent one, while such readiness itself may lead to nuclear war despite the best intentions of those who imagine themselves in control. The larger the arsenals, the more secure "deterrence" is assumed to be; but the larger the arsenals, the more devastating the war will be should deterrence fail.

The probability of a nuclear war within the next decades is unknown. But because of the enormous numbers of nuclear weapons and their delivery systems, and the intrinsic imperfections of machines and people, nuclear war is not only possible, but if we wait long enough, it may be inevitable (see box). One purpose of this book is to assess whether the new knowledge of nuclear winter (and related effects) can lead to changes in weapons systems, policies, and doctrine that substantially reduce the probability *and* the severity of nuclear war.

The naive homeowner, presented with the possibility that his house would be washed away by flood, might dismiss the notion as "only a theory" and not even purchase insurance. He might argue, "My house has been here fifty years and there's never been a flood in all that time." But if the river is in danger of overflowing its banks, attentive homeowners not only try to buy insurance; they may help build levees or dikes, or even try to reroute the river. When the water is rising, prudent people take action.

For policymakers who have a full plate of potential disasters

# IMPROBABLE OR INEVITABLE: HOW LIKELY IS NUCLEAR WAR?

If nuclear weapons have kept the peace, it would seem to follow that any meddling with this proven deterrent to war is foolish and dangerous: "If it ain't broke, don't fix it." And yet, even its ardent supporters argue that, by its very nature, deterrence is not foolproof. Here, for example, are the words of Bernard Brodie, who in 1946 (ref. 10.1) first formulated the idea of nuclear deterrence:

> We have ample reason to feel now that nuclear weapons do act critically to deter wars between the major powers, and not nuclear wars alone but any wars. That is really a very great gain. We should no doubt be hesitant about relinquishing it even if we could. We should not complain too much because the guarantee is not ironclad. It is the curious paradox of our time that one of the foremost factors making deterrence really work and work well is the lurking fear that in some massive confrontation crisis it might fail. [Brodie, *War and Politics* (New York: Macmillan, 1973), 430–431.]

But how do we balance the benefits of deterrence against what those deterrent arsenals will bring should deterrence fail? How much weight do we give to Brodie's "lurking fear"? Do we give it more weight if we discover that the consequences of nuclear war are much worse than we thought?

Except for the demolition of two cities in the waning days of World War II, when there was no prospect of nuclear retaliation, we humans have never witnessed a nuclear war. We have little experience on which to base our estimates of its likelihood. Some look at the fact that there has been no nuclear war since 1945 and conclude, with Brodie, that deterrence works, that nuclear weapons prevent nuclear war (as well as conventional war between nuclear-armed nations). This is not some well-intentioned but naive hope, they say; this is a conclusion based on real historical evidence—the absence of world wars since 1945 and such instructive incidents of backing off from the brink as the Cuban Missile Crisis of 1962. They prefer what they consider to be the proven reliability of the present nuclear deterrent to the unknown dangers of any alternative arrangement.

Others note the apocalyptic capabilities of the nuclear arsenals, such technological disasters as *Challenger* and Chernobyl, the long sequence of U.S. and Soviet naval accidents and blunders in the late 1980s, and the occasional incompetence or madness of national leaders, and are amazed that nuclear war has not yet happened; it is only a matter of time, they believe. The decades without world war mean little, they argue; there have been similar periods long before the invention of nuclear weapons, including nineteenth-century Europe between the Congress of Vienna and the Franco-Prussian War.

The optimists sometimes look on the pessimists as doommongers, needlessly frightening a weak-minded and fickle public. The pessimists sometimes look on the optimists as they would on a man falling from the top of a skyscraper who cries out through an open window to a startled office worker: "So far, so good. . . ." The downward trajectory, they hold, is clear.

Consider a deadly peril so improbable that it can happen only once in a thousand trials, or time periods. We don't specify what the time period is; maybe it's a week or a month or a year. Many time periods go by without the peril materializing. The probability that it doesn't happen in a single trial is $999/1000 = 0.999$, very nearly 1. (A probability of 1 is an ironclad guarantee.) The chance that it won't happen in two independent trials is $(999/1000)^2$, or 0.998 —still very close to a guarantee. But as the number of trials increases, the probability goes down. If you make a thousand independent trials, then the laws of chance tell us that the probability of the event not happening becomes $(999/1000)^{1000}$, or 0.37. The chances are now better than even that the disaster strikes. By the time you make a few thousand trials, the chance of avoiding disaster becomes very small. Equivalently, the probability that the disaster occurs becomes very close to 1. With enough trials, an improbability becomes an inevitability.

It is because of arguments like this, tracing back to inexorable laws of probability, that some analysts believe the nuclear arsenals are a catastrophe waiting to happen (cf., e.g., ref. 13.4). Others argue that an accidental or unauthorized explosion of even a few weapons would not necessarily lead to global nuclear war; or point out that if the probability of the event declines fast enough in each trial —through continued improvements in reliability and safety—it is possible in principle to beat the odds. The discussion then turns to the distinction between having possession of nuclear weapons and being able to make them blow up—the fail-safe encryption keys, called "permissive action links" (PALs). They are designed to prevent inadvertent or unauthorized use of nuclear weapons. Evolving from four-digit locking devices, they are said now to incorporate "a limited-try capability that renders the weapon unusable if incorrect codes are repeatedly inserted" (ref. 8.10, p. 138). But the safety of PALs is not subject to public scrutiny. We have no way of knowing that assurances of their safety are any more reliable than those confidently issued before the *Challenger* and Chernobyl disasters. The U.S. Navy does not even *have* permissive action links on its nuclear weapons—because, it says, in a crisis PALs might prevent the timely use of those weapons (ref. 6.7). Nor is it likely that U.S. decision makers are able to make independent, knowledgeable judgments about the safety precautions taken by the Soviet Union and some other nations.

As for us, we are deeply impressed by the human penchants for weaving comfortable and reassuring illusions and for overlooking the dangers of high technology. For us, only the most compelling arguments would be sufficient when the stakes are so high. The burden of proof is on those who argue that there's nothing to worry about. But because of the secrecy into which all questions of the safety of nuclear weapons are immersed, nothing approaching a compelling argument can be offered—only an argument from authority: "Trust us, the arsenals are safe." We are not reassured.

before them, a method is needed to rank priorities—which predicted catastrophes to believe, to which to give the most urgent attention, to which to devote the largest fiscal and intellectual resources. Such prioritization is necessarily a cold-blooded business, even though it is frequently admixed with shorter-term political considerations. It requires a manner of thinking abhorrent to many. But we invest a few pages in the hope that we can demonstrate that the potential costs of nuclear war—especially when we include nuclear winter—imply that its reliable prevention deserves by far the highest priority of all the entries on the policymaker's agenda. No recent lessening of tensions between the superpowers makes this imperative less urgent. Surely preserving the lives of most of its citizens is the minimum goal any nation can ask of its leaders.

The "cost" of a war, if we are required to think in such terms, might be measured by the attendant human deaths and suffering. For calculating risk there is no standard measure of the cost of a human life, nor can there be. Nevertheless, there are standard indemnities that are offered or paid by corporations or national governments when they are responsible for the deaths or serious injury of innocent civilians. The Union Carbide Corporation's offer following the Bhopal, India, disaster is a case in point. So are the reparations discussed (for non-Iranian victims only) after the U.S. Navy shot down Iran Air's Flight 665, a jumbo jet filled with civilian passengers, in July 1988. Or what insurance companies generally offer as out-of-court settlements to the next-of-kin of those killed in airplane crashes. In the same range are the reparations announced in August 1988 to American citizens of Japanese descent who were illegally interned—with authorization at the highest levels of the U.S. Government—in concentration camps for the duration of World War II. These offers are in the range of $10,000 to $100,000 per human being (ref. 6.2). With 100 million to a few billion nuclear winter–related fatalities, this works out to somewhere between one trillion and several hundred trillion dollars. The higher figure is also, very roughly, the current "replacement cost" of everything on Earth made by humans. The lost productivity of people killed by nuclear winter can be estimated from the Gross World Product as no less than $10 trillion a year, so that if—to make a very optimistic assumption—it required a century to reestablish the equivalent of the current global technical

# TWO KINDS OF CONSERVATIVE THINKING

In early 1939, Leo Szilard thought, on the basis of recent experiments, that a nuclear chain reaction and, therefore, an atomic bomb were possible. The nuclear pioneer Enrico Fermi thought this was only a "remote possibility," but on being pressed about what he meant by "remote" said, "Well, ten percent." "Ten percent is not a remote possibility," replied the physicist I. I. Rabi, "if it means that we may die of it. If I have pneumonia and the doctor tells me that there is a remote possibility that I might die [of it], and it's ten percent, I get excited."

"From the very beginning the line was drawn," Szilard concluded. "The difference between Fermi's position throughout this and mine was marked on the first day we talked about it. We both wanted to be conservative, but Fermi thought that the conservative thing was to play down the possibility that this may happen, and I thought the conservative thing was to assume that it would happen and take all the necessary precautions"—which, in Szilard's later years, after the "remote" possibility materialized, included verifiable limits on the number of nuclear weapons in the world (ref. 17.7, p. 54).

The tension between these two points of view, both considered "conservative" by their adherents, seems to have played a central role in the Cuban Missile Crisis of October 1962, the closest the United States and the Soviet Union have come to nuclear war. In a remarkable and most instructive retrospective of the events (James G. Blight and David A. Welch, *On the Brink: Americans and Soviets Reexamine the Cuban Missile Crisis* [New York: Hill and Wang, 1989]), the following proposition is put to Robert S. McNamara, U.S. Secretary of Defense at the time of the crisis:

> BLIGHT: On October 16, the President calls all of you into the first meeting of what would become the

> ExComm [Executive Committee to assist President Kennedy in the crisis]. People like Dillon, Nitze, Taylor, and Acheson say, in effect: "We've got to take the missiles out [by an attack on Cuban soil with several hundred U.S. aircraft] and, if we do, the *probability* of a drastic response from the Soviets is very tiny. So let's do it. Let's bomb them." But you seem to have said: "Wait a minute! A drastic response is *possible*. So let's be very careful about how we proceed." Is that about right?
>
> McNAMARA: Yes, exactly . . . I'm not interested only in probable risks. I'm interested in less than probable risks, if they may lead to disastrous consequences. That was what motivated me.

In a separate interview, C. Douglas Dillon, Secretary of the Treasury at the time, explained:

> I never thought the Soviets would use the missiles. I mean, if they had, they would have been committing suicide [because of U.S. strategic retaliation], and I never thought they'd do that . . . I think most of the difference between the hawks and doves had something to do with this[:] I think simple inexperience led to an inordinate fear of nuclear damage [from a Soviet nuclear attack on the U.S.], the fear of what *might* happen.

In fact the President, sharing McNamara's concerns, chose a naval blockade of Cuba rather than an air strike, and the crisis subsided.

civilization, the cumulative productivity loss would be well in excess of $1000 trillion, a quadrillion dollars. This number—a quadrillion dollars—is the order of magnitude of the cost in money of a full-scale nuclear winter.

Let us now compare this figure with the cost and risk of relatively slow global environmental changes—ozone layer depletion by CFCs and the increasing intensity of solar ultraviolet light at the surface of the Earth; or the growing abundance of infrared-absorbing greenhouse gases in the Earth's atmosphere and the consequent rise in average global surface temperature and sea level. Just before international agreements were made on phasing it out, manufacturing CFCs was a multibillion-dollar-per-year industry worldwide. At a time when the worst danger generally anticipated from ozone depletion was a few thousand incremental deaths per year from skin cancer over a period of about a century (ref. 6.3), it was still contemplated to spend billions—perhaps tens of billions—of dollars cumulatively in further research, in phasing out industrial production of CFCs, and in the development and production of new global industries for manufacture of substitutes for CFCs. This works out in the same rough range of "cost" per human life as for the indemnities mentioned above.

The slow warming of the Earth through the increasing greenhouse effect—in a way, the opposite of nuclear winter—will not (unlike nuclear war) kill many people directly (ref. 6.4). It is estimated eventually to produce severe dislocations in agriculture, and by the middle of the twenty-first century perhaps to convert such grain-growing regions as the American Midwest and the Soviet Ukraine into something approaching scrub deserts. But at the same time, climates more conducive to agriculture may for a time be generated further north—for example, in Canada and Siberia—and, provided the soils are fertile, it is undemonstrated that, on global average and over the long term, massive starvation on the scale of nuclear winter would ensue. As sea level rises from the volume expansion of seawater as it warms, and from melting glacial and polar ice—and especially after the slow deterioration or collapse of the West Antarctic ice sheet—a new category of economic consequence may emerge: the inundation of every coastal city on the planet, as well as such low-lying and unprotected nations as Bangladesh. The cost of massive emigration of agriculture, of building levees and dikes worldwide, of rescuing environmental refugees, and of relocating the principal coastal cities on Earth would be formidable. To prevent the greenhouse effect from continuing to increase requires—beyond energy conservation, banning of CFC production, and global reforestation—a massive conversion from fossil fuels to solar power, nuclear (preferably fusion) power, or other technologies that are not yet developed or are economically still noncompetitive (ref. 6.4). The research and development costs for a single such alternative technology—high-temperature fusion power, for example—are, at the very least, tens of billions of dollars, with worldwide conversion costing in the trillions. There are other reasons for taking these substantial steps to deal with greenhouse warming, and we are firmly in favor of them being taken. But we point out that compared with nuclear war, the cost of prevention, in relation to what is at stake, is very high.

A nuclear war would utterly destroy the societal infrastructure of the combatant nations and, through nuclear winter, very likely destroy the global civilization as well. The time scale would be very short—months to years—and there would be no opportunity, as there is with the foregoing, nonnuclear environmental catastrophes, to move by slow steps, to test out alternative forms of remedial action, and to contemplate optimal approaches at leisure.

If we were to assess the "cost" of a global nuclear war at $1000 trillion, and the likelihood of such a war even as low as 0.1% per year, it would then follow that we should be spending a trillion dollars a year to prevent nuclear war. This is roughly the combined military budgets of all nations—and if only we could be confident that this expenditure would with high reliability preclude nuclear war, we might consider it money well spent (ref. 6.5). However, there seems little grounds for such confidence.

A nation that took nuclear winter seriously might contemplate storing enough food for the surviving population for a period of at least a decade. But the disruption of national transportation systems and refrigeration by nuclear war greatly complicates the task, and even a partial attempt to stash such foodstuffs within walking distance of most survivors would be enormously expensive—although not as expensive as the global arms race and certainly not as expensive as nuclear war. Such a step, though, would carry serious penalties: It would suggest to potential adversaries at least a contingent intention to carry out a nuclear first strike, and it would be taken as an unfriendly act by noncombatant nations that might be destroyed by nuclear war and nuclear winter, even if no one intended to attack them directly.

In summary, we can say that the risks posed to the human species by nuclear winter at the most probable levels of severity now scientifically established are unacceptable by any standards that are currently applied to environmental issues—or even in comparison with the enormous prompt devastation of a nuclear war (where the direct damage is clearly unacceptable by similar standards). Its incremental cost, both in combatant and in noncombatant nations, whether measured in lives or money, is so enormous as to raise the stakes—which are already unacceptably high—of nuclear war significantly, and to justify heroic efforts to prevent nuclear war and nuclear winter from ever taking place (ref. 6.6).

CHAPTER 7

# TAMBORA AND FRANKENSTEIN:

## WHAT IT TAKES TO GENERATE NUCLEAR WINTER

Remember, thou hast made me more powerful than thyself. . . . If the multitude of mankind knew of my existence, they would do as you do, and arm themselves for my destruction. Shall I not hate them who abhor me? . . . I declared everlasting war against the species. . . . I vowed eternal hatred and vengeance to all mankind.

> —Victor Frankenstein's nameless and murderous monster, in exculpatory discourse with his maker. From Mary Wollstonecraft Shelley, *Frankenstein* (1816)

The bright sun was extinguish'd . . .
. . . and the icy earth
Swung blind and blackening in the moonless air;
Morn came and went—and came, and brought no day,
And men forgot their passions in the dread
Of this their desolation; and all hearts
Were chill'd into a selfish prayer for light.
. . . No love was left;
All earth was but one thought—and that was death,
Immediate and inglorious; and the pang
Of famine fed upon all entrails.

> —From George Gordon, Lord Byron, "Darkness" (1816). In *The Poetical Works of Byron*, Cambridge Edition (Boston: Houghton Mifflin, 1975), 189

PERHAPS THE MOST FAMOUS NOVEL OF TERROR, AS WELL as one of the earliest warnings that technology might be dangerous far beyond the well-meaning intentions of those who devise it, is Mary Wollstonecraft Shelley's *Frankenstein*. Its subtitle is *The Modern Prometheus*. In the summer of 1816 she was vacationing on a lake in the Swiss Alps with other literati, including Byron and Percy Bysshe Shelley, whom she was soon to marry. The early summer was "ungenial"—unseasonably cold and rainy. Much of Europe was marked by snowfalls and freakish weather. Cabin-bound, "in the evening we crowded around a blazing wood fire" and told ghost stories which "excited in us a playful desire of imitation." They set themselves a competition to produce a consummate tale of horror. *Frankenstein* was the winning (as well as the sole completed) entry. Mary Wollstonecraft was barely nineteen years old that summer.

The novel's early pages are full of foreboding weather: "frost and desolation"; "I . . . endured . . . cold [and] famine"; "encompassed as I am by frost and snow"; and the like. And the book ends with a haunting chase, over the Arctic wastes, of the monster by his creator; the narrator remarks, "The cold is excessive, and many of my unfortunate comrades have already found a grave amidst this scene of desolation." It seems possible that these images were elicited by that summer's ungenial weather.

George Gordon, Lord Byron, was inspired by the same strange weather to write his brooding poem "Darkness," which some—especially some of our Soviet colleagues—have taken as a kind of premonition of nuclear winter. What is odd is that the cold and dark that characterized the summer of 1816, and that produced untold hardship, hunger, and death across the Western world, may have really been something like a weak nuclear winter caused by a global pall of fine particles. The events of 1816 and after, to which we will return, help us to calibrate what it takes to generate a nuclear winter.

In physics, physiology, and many other areas of science, there is the useful idea of a threshold—the smallest input (stimulus) required to produce a measurable or perceptible output (response). There are, for example, levels of sound or light that we are unable to perceive, although at slightly higher levels we hear or see quite clearly. There is a threshold of kinetic energy below which collisions are elastic (e.g., billiard balls or air molecules bouncing harmlessly off one another) and above which they are inelastic (e.g., a steel wrecking ball hitting a building, or a crater-forming impact of a small world with the Moon). Many conventional high explosives as well as nuclear weapons require a detonator or equivalent—above a certain threshold input energy the explosion can occur, while below that level it cannot.

Some thresholds are of the "step function" variety, in which there is no effect for a steady increase in stimulus—until the threshold is reached, at which point the full effect is suddenly realized. The resulting graph looks like the side-view of a step on a flight of stairs. In other applications, the change in response to an increment in stimulus is never so abrupt or discontinuous; instead, a slow increase in stimulus has at first no effect, then reaches a transition region in which increasing stimulus produces increasing response—until the effect saturates, and thereafter an increase in stimulus again elicits (almost) no increased response. This behavior can be described as showing a logistic or sigmoidal threshold (named after the shape of the curve; sigma is the precursor of our letter "S"; the curve looks rather like the lowercase final sigma put at the end of Greek words). These two sorts of thresholds are indicated schematically in Figure 3.

The physics of the absorption of light by smoke shows a logistic and not a step function threshold. Here, the light-absorbing smoke is the stimulus, and the darkening of the Earth is the response. With increasing amounts of smoke in the atmosphere, sunlight is at first imperceptibly attenuated; then there is a broad transition regime in which the more smoke, the more the attenuation of sunlight; and finally there is a regime in which

FIGURE 3

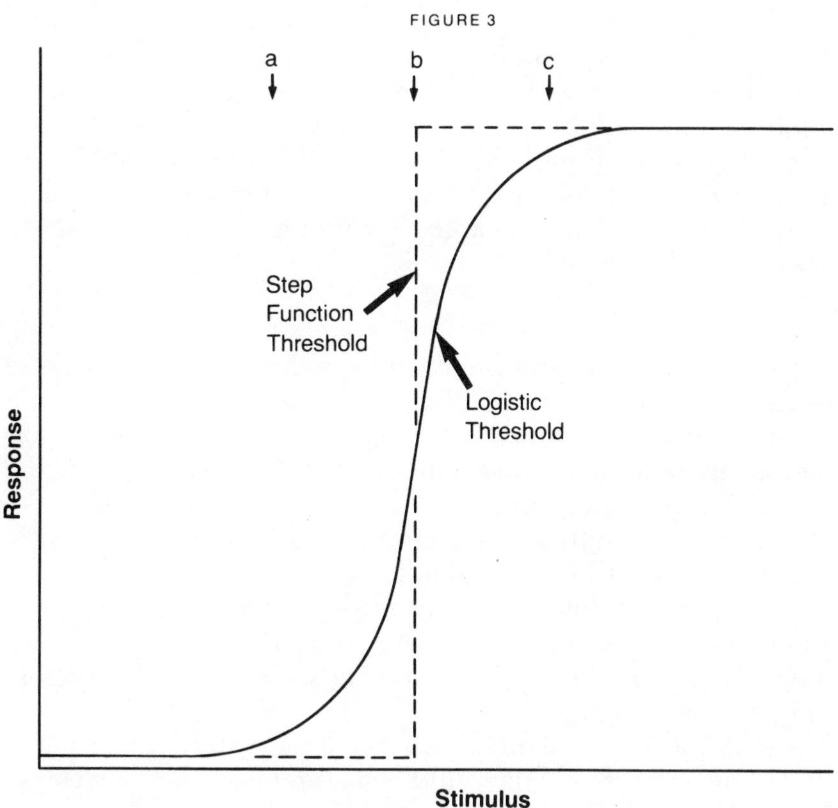


Two sorts of thresholds: a step-function threshold, in which the response occurs entirely at a unique value of the stimulus, and a logistic or sigmoidal threshold, in which there is a broad transition region in which the response follows the stimulus. The physics of smoke absorption and nuclear winter gives a logistic threshold (see Figure 4, later this chapter). In Figure 4, the response (temperature) *de*creases with the stimulus, so the resulting logistic curves are mirror images of the S-shaped curve shown here.

so much smoke is in the air that essentially no light penetrates to the ground, and increasing amounts of smoke have little additional effect. This logistic threshold for absorption of light translates into a logistic threshold for the dependence of surface temperature on quantity of smoke in the air (as shown in Figure 4, p. 105). Some confusion has been generated by those who understood the nuclear winter threshold to be a step function rather than a logistic.

However, the manifestations of nuclear winter would so depend on time and locale, and could span such a broad continuum of severities (depending, for example, on which targets are selected), that while the concept of "threshold" remains important, its definition is more ambiguous (ref. 7.1). In the S-shaped curve of Figure 3, where exactly is the threshold? At point a? Point b? Point c? Clearly it lies somewhere between a and c. But where? Is our choice a matter of physics, or politics, or emotional disposition?

The following general trends in the climatic consequences of nuclear war have been clear for some time:

(1) The severity of climatic impacts would, over an important range, tend to increase with the quantity of smoke (ref. 7.2) injected into the atmosphere;

(2) Inland continental regions would suffer more severe effects than coastlines and islands;

(3) The absolute temperature drops would be largest in summer and smallest in winter, although a springtime war might result in the most serious agricultural (and therefore human) consequences; and

(4) Superimposed on the average temperature changes predicted by current models would be natural extremes of weather adding significantly to the potential severity (ref. 7.3).

In spite of large spatial and temporal variabilities, we believe that a useful, if crude, threshold for disastrous effects may be defined when the weather is perturbed sufficiently to disrupt agricultural productivity over the great food-supplying regions of North America, Europe, and Asia. Beyond this threshold, severe disturbances could spread into Africa, South America, and Australia. We can calibrate any choice we make for a threshold amount of smoke by considering past temperature drops.

Temperatures have fallen and crops have failed after major volcanic eruptions. The connection was first understood in 1784 by the American polymath Benjamin Franklin, who proposed that the early frosts and severe weather of the winter of 1783–1784 were due to a "dry fog" that had been observed for months over the preceding summer. The fog, in Franklin's words, rendered the Sun's rays faint, and cooled the Earth. He speculated that if the "fog" were not due to meteoric dust, it must have been caused by a "vast quantity of smoke" that issued that summer from two Icelandic volcanos.

Perhaps the best-known recent case is the volcanic explosion on the island of Tambora, in what is now Indonesia, in April 1815. This event, one of the most violent in historical times, was heard in New Guinea, and in Sumatra, 2,000 kilometers away (refs. 7.4, 7.5). Soon after, it was at noon dark as a moonless night hundreds of kilometers away, in Java. Two weeks later, temperatures were below freezing in Madras, India (late April 1815). Months later snowfalls in Europe were described as "brown or flesh-colored," because of the volcanic ash in every snowflake. After the fine stratospheric debris from the explosion had spread worldwide, incoming solar radiation at the Earth's surface fell by an average of some 10%, and the average global temperature declined by about 1°C. The next year brought local temperatures that were the lowest in American meteorological history—but on average, only about 3°C (5°F) below normal. The fluctuations were more severe. That summer of 1816 was afterwards known in the folklore of New England as "eighteen-hundred-and-froze-to-death." There was snow in June, frost in July and August. The temperature at New Haven, Connecticut, turned out to be the coldest in almost two centuries.

Dead birds in large numbers dropped from the skies onto the streets of New York. The New England corn crop failed. Agriculture was imperiled in North Carolina: "The very cool and dry weather in spring and summer hurt our grain fields badly, and it was with sorrowful and troubled hearts that we gathered our second crop of hay and our corn crop, which were so scanty that we reaped only a third of what we usually get, and wondered how we could subsist until next year's harvest" (from a Moravian settlement in North Carolina [ref. 7.5]). The sugar crop was meager in the British West Indies, and the planters and merchants of the Caribbean island of St. Kitts pleaded for food from abroad to "secure them against the horrors of famine." With the scanty harvest of hay, farm animals were dying of starvation by the spring of 1817. Seed for new crops had almost all been eaten. Nova Scotia, many German states, and Morocco banned the export of staple foods. The grain crops (especially wheat, oats, and potatoes) failed in Western Europe, where 1816 was known as "the year without a summer."

It was also called "poverty winter" and "the year of the beggars." Famine stalked Ireland. This was the summer in which *Frankenstein* was written, and a Swiss cleric described 1816–1817 as years of "hunger, want, sickness, death, manufacturing unemployment, trade stagnation, and calamitous weather." There were food riots in England and France, famine in Italy, and dearth conditions in the Ottoman Empire. Nutritional diseases, such as pellagra and scurvy, became common. Starvation would have been much more widespread if not for massive grain imports from America and, especially, from Russia. Vagrancy, begging, theft, looting, and rioting were endemic. Poverty and hunger led naturally to violence. Roving bands in Ireland broke into homes looking for food. In England 1,500 people, carrying clubs studded with metal spikes and a flag emblazoned with the motto "Bread or Blood," roamed the streets, destroying homes as they went. Arson was rife. Juries were unwilling to convict looters. Governments used police and the army to suppress riots, and public works and military recruitment to provide subsistence for the large numbers of unemployed men. No such provision seems to have been made for women and children. People ate bread baked from straw and sawdust, and meat from dead and decomposing animals. An 1817 account from Württemberg:

> One saw wandering around in the towns and villages persons who looked like cadavers, and among them multitudes of children crying out for bread. Hunger and unnatural food produced wretched and chronic ill health among some, outbreaks of frenzy among others; those in the most desperate condition deemed themselves no

longer bound by the laws that are adopted for the protection of private property.

A medal struck to note the famine in Southern Germany read "Great is the distress. O Lord, have pity." Indeed, there was a resurgence of interest in religion (and, in Germany, in anti-Semitism). Vast numbers of people emigrated to America and Russia, and "a sort of stampede took place from cold, desolate, worn-out New England, to this land of promise"—Ohio.

It has been described as "the last great subsistence crisis in the Western world" (ref. 7.5). It occurred, it is true, only a few years after the end of the Napoleonic Wars, but the historians of these events attribute little of this human agony, even in Europe, to the wars. It was accompanied by and may have set the conditions for epidemics of typhus and plague. A cholera pandemic originated in Bengal in 1816–1817. Although the climate anomaly was mainly restricted to Eastern North America and Western Europe, it extended as far as the Eastern half of India, where a failure of the monsoon rains decimated the grain crop.

All this resulted from an average global temperature decline of about 1°C (almost 2°F). Clearly, small global temperature changes can have major, even if geographically localized, consequences (refs. 7.6, 7.7). There are several other cases in historical times strongly suggestive of "volcanic winter," at least one of which was considerably more severe than Tambora (see box).

A single night below freezing is enough to destroy the Asian rice crop. A 2 to 3°C average local temperature drop is sufficient to destroy all wheat production in Canada, and 3 to 4° all grain production. Crops in the Ukraine and the American Midwest would also be severely injured by a 3 to 4°C temperature drop (refs. 3.7, 3.11). A 5°C drop would bring us back to Ice Age global temperatures (ref. 7.8). A 10°C temperature decline with the accompanying darkening of the Sun (Figure 4) would devastate grassland ecosystems throughout the Northern Hemisphere (despite their much larger temperature resiliency than our pampered crops); here something like a step-function *bio-*

*logical* threshold seems to apply (ref. 7.9). A 10°C temperature drop from contemporary conditions also represents the coldest climate at the peak of the Wisconsin Ice Age. But land temperature declines of 5 to 10°C are calculated even for mild nuclear winters (Figures 4, 6), and much larger temperature declines may well occur (refs. 2.2, 3.14; Figures 4, 6).

This gives us a sense of where the threshold might be—the injection into the atmosphere of enough smoke to produce a few-degree (perhaps 1 to 4°C) temperature drop on land.

How much smoke is that? We need some measure of how much smoke there is overhead. A convenient gauge of the extent to which fine absorbing particles in the atmosphere block or attenuate sunlight is a number called the "optical depth" (ref. 7.10). Commonly used in meteorology and astronomy, it is measured or calculated for the case that the Sun is overhead. The farther the Sun is from the zenith, the longer is the slant path of sunlight through the atmosphere, and so the more absorption of sunlight along the path for a given amount of smoke in the air. The larger the optical depth of the cloud, the more sunlight it absorbs and the darker and colder the underlying ground or ocean can get. At an optical depth of zero, there is no effect; at an optical depth of 1, it is significantly darker; and at an optical depth of several, serious climatic change is underway. So it's possible to calibrate the severity of a nuclear winter by specifying the value of this key parameter—the optical depth of the (mainly) soot clouds that generate the cold and the dark.

We can gain some intuitive feeling of what different optical depths mean by looking at the historical record of volcanic explosions and their associated climatic effects (see box). But it's vital to bear in mind (see Figure 9, Appendix A) that a given optical depth for transparent sulfuric acid droplets from volcanic explosions (e.g., Tambora) has a much smaller climatic effect than the same optical depth for dark sooty particles, the principal agents of nuclear winter. In both cases, the *total* optical depth comes from light being absorbed *and* reflected (or "scattered") by the particle. For sulfuric acid droplets (or ice crystals, or silicate dust), scattering is more important than absorption. For soot, absorption is more important than scattering. When we discuss volcanic winter, the optical depths are for scattering alone. But in the following discussion of nuclear winter smoke, the optical depths we describe are those for absorption alone. Scattering would in general make things worse.

When the average absorption optical depth of the smoke is zero, it's a clear day; an optical depth of zero is just another way of saying there isn't a (soot) cloud in the sky—nor haze, nor smog. The absorption optical depth is generally written $\tau_a$, where the symbol is the Greek letter tau, and the subscript "a" reminds us we are considering absorption, not scattering. A value $\tau_a = 0.2$ corresponds to an average decrease in sunlight at the Earth's surface of about 30%; it would cause a few-degree average land temperature drop and requires roughly 6 million tons of sooty smoke to be uniformly distributed over one hemisphere of the Earth. It would be much worse than the events following the Tambora explosion. We might consider this a "threshold" at which global-scale climatic anomalies begin to threaten agriculture.

However, we humans are not yet very good at predicting climatic responses for such moderate changes in the intensity of sunlight reaching the ground. And we want to pick a threshold that makes sense to those who feel they need—even with the stakes so high—a considerable margin for error before the prospect of nuclear winter can begin to affect policy. Accordingly, we make a very cautious estimate of the average value of $\tau_a$ just at the threshold for major global effects: a value 5 times larger, corresponding to $\tau_a = 1.0$, and generated by about 30 million tons of sooty smoke.

In typical cases (cf. Figure 4) $\tau_a = 0.5$ will produce a 4 to 6°C temperature decline; and $\tau_a = 1.0$ will produce up to a 10° drop —easily enough to wipe out agriculture in the growing season. All these cases are for smoke at high altitudes. (In general, greater temperature declines are associated with larger quantities of smoke at higher altitudes for longer periods of time over the most extensive land masses.) A value $\tau_a = 2.0$ will produce an average surface temperature decline of up to 15°C, unless the smoke is all in the lowest part of the atmosphere, in which case the temperature decline would be only 3 to 4°C. For $\tau_a = 3.0$ the corresponding temperature drops are 20°C and 10°C. Since in most cases the smoke would be at high altitude (see Figure 2), a threshold optical depth $\tau_a = 1$ seems extremely conservative for major climatic effects—provided the smoke remains aloft for at least some weeks.

Equilibrium average land surface temperature as it depends on the smoke absorption optical depth, $\tau_a$ (ref. 7.10), for smoke injected into three specific altitude regions (in the lower troposphere, 0 to 4 km; in the upper troposphere, 4 to 8 km; and in the uppermost troposphere and lower stratosphere, 8 to 12 km). In these rough calculations, the smoke is held fixed in each layer. $\tau_a$ applies to the amount of smoke directly overhead ("vertical optical depth"). The optical depth scale at the bottom of the figure is logarithmic—ranging from an optical depth of 10 at the right-hand margin, to 1, to $10^{-1}$ (= 0.1), to $10^{-2}$ (= 0.01) at the left-hand margin. Such a scale is a way of compressing a wide range of optical depths into a figure of reasonable size and clarity.

The ratio of visible absorption and scattering (ref. 7.10) to thermal infrared absorption and scattering is assumed to be 10:1, typical of soots. Absorption is much more important than scattering. The temperatures shown are land averages; the temperature decreases over continents are about twice as large as for continents and oceans together.

The fraction of sunlight transmitted by the atmosphere is shown in the upper scale for the case that the Sun is overhead (see ref. 7.10). When it is nearer the horizon, or if daily averages are considered, the attenuation will be greater. The average attenuation can be read off the figure by considering an optical depth about 1.7 times larger than is appropriate for the Sun at the zenith.

If there is an amount of smoke in the right range ($\tau_a$ between roughly 0.03 and 0.1)—*and* if the smoke is in the middle atmosphere (below about 8 km)—then a small warming, rather than a cooling, results. But it amounts only to some 1 to 2°C—hardly a "nuclear summer."

This figure is based on one-dimensional radiative-convective model calculations by T. Ackerman, with continental interior temperature changes reduced by a factor of 1.5 to take account of the moderating influence of the oceans. Temperatures are given on the left-hand margin in degrees Kelvin (°K). The freezing point of pure water is 273°K = 0°C (= 32°F). Thus, 260°K = −13°C, 270°K = −3°C, 280°K = +7°C, etc. A Fahrenheit scale is given in the right-hand margin. (See also Appendix A.)

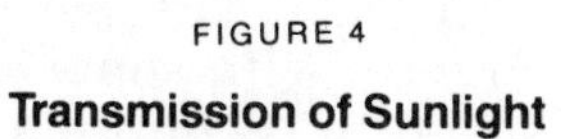

FIGURE 4

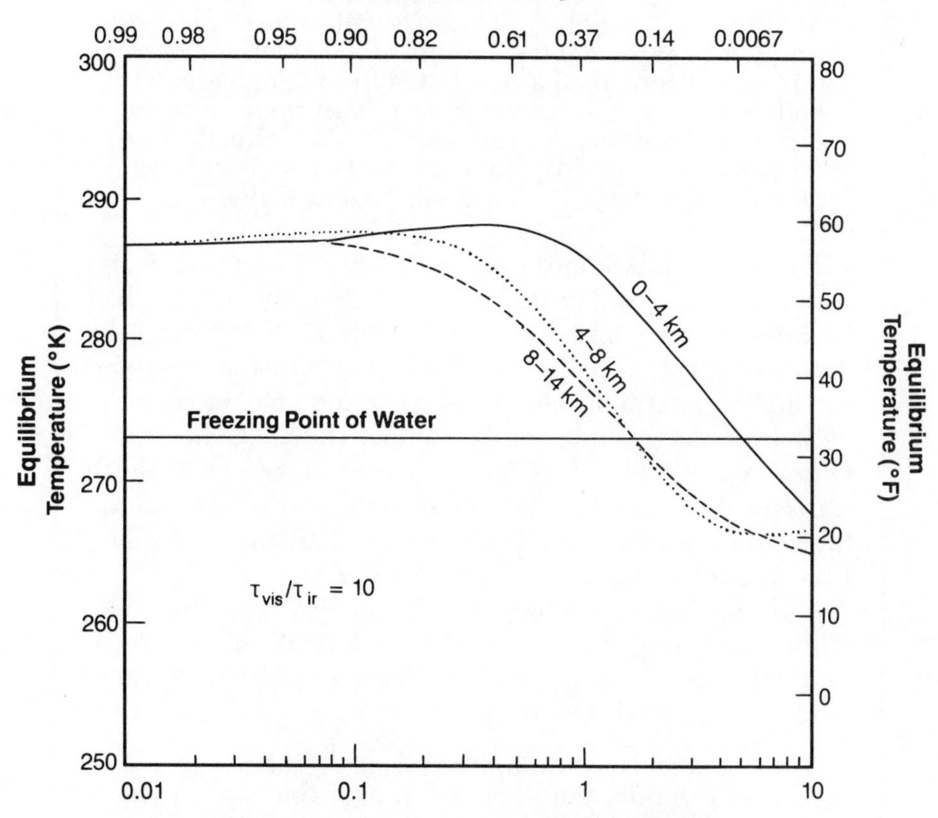
Transmission of Sunlight
0.99
0.98
0.95
0.90
0.82
0.61
0.37
0.14
0.0067
300
290
280
270
260
250
80
70
60
50
40
30
20
10
0
Equilibrium Temperature (°K)
Equilibrium Temperature (°F)
0–4 km
4–8 km
8–14 km
Freezing Point of Water
τvis/τir = 10
0.01
0.1
1
10
Absorption Optical Depth at Visible Wavelengths

# HIROSHIMA AND NUCLEAR WINTER

The Japanese city of Hiroshima was wiped out on August 6, 1945, by an approximately 13-kiloton-yield nuclear weapon. Some 200,000 men, women, and children were killed, many from lingering deaths. Col. Paul W. Tibbet, Jr., was the pilot of the *Enola Gay*, the B-29 that, for the first time in human history, dropped an atomic bomb on a city. He named the airplane for his mother. Here is his description of what he saw:

> What had been Hiroshima was going up in a mountain of smoke. . . . First I could see a mushroom of boiling dust—apparently with some debris in it—up to 20,000 feet. The boiling continued three or four minutes as I watched. Then a white cloud plumed upward from the center to some 40,000 feet. An angry dust cloud spread all around the city. There were fires on the fringes of the city, apparently burning as buildings crumbled and the gas mains broke. [Donald Porter Geddes, Gerald Wendt et al., eds., *The Atomic Age Opens* (New York: Pocket Books, August, 1945), 21.]

The Hiroshima bomb produced a firestorm that left it a flattened and desolate ruin; a large part of the city had literally gone up in smoke. (See Plate 3.) *The New York Times* (August 7, 1945) described Hiroshima as engulfed in "an impenetrable cloud" of dust and smoke. (For the view from the ground, see Michihiko Hachiya, *Hiroshima Diary*, W. Wells, ed. [Chapel Hill: University of North Carolina Press, 1955]; Masuji Ibuse's novel, *Black Rain*, translated by John Bester [New York: Bantam, 1985]; and John Hersey's *Hiroshima* [New York: Knopf, 1946].)

A young Japanese physician, T. Akizuki, gave this eyewitness account of the immediate aftermath of the Nagasaki explosion, two days later, also the result of a single nuclear weapon dropped by a B-29:

> The sky was dark as pitch, covered with dense clouds of smoke; under that blackness, over the earth, hung a yellow-brown fog. Gradually the veiled ground became visible and the view beyond rooted me to the

spot with horror. All the buildings I could see were on fire. . . . Trees on the nearby hills were smoking, as were the leaves of sweet potatoes in the fields. To say that everything burned is not enough. The sky was dark, the ground was scarlet, and in between hung clouds of yellowish smoke. Three kinds of colour—black, yellow, and scarlet—loomed ominously over the people, who ran about like so many ants seeking to escape . . . That ocean of fire, that sky of smoke! It seemed like the end of the world. [Akizuki, *Nagasaki 1945* (London: Quartet, 1981). See also Takashi Nagai, *We of Nagasaki*, I. Shirato and H. B. L. Silverman, eds. (New York: Duell, Sloan, and Pierce, 1951).]

Radioactive black rain fell on both Hiroshima and Nagasaki; in Hiroshima the rain was accompanied by a sudden chill, many survivors "shivering in midsummer" [*Hiroshima and Nagasaki: The Physical, Medical, and Social Effects of the Atomic Bombings*, Eisei Ishikawa and David L. Swain, trans. (New York: Basic Books, 1981), 92]. Nevertheless, no nuclear winter followed Hiroshima and Nagasaki because the burning of one or two medium-sized cities does not produce nearly as much hemispheric obscuration as a large volcanic explosion.

Today, the world nuclear arsenals contain nearly 60,000 nuclear weapons. Typical strategic weapons are 10 to 100 times more powerful than the Hiroshima or Nagasaki bombs. The aggregate explosive yield of the present world arsenals is the equivalent of a million Hiroshimas. But something like a thousand Hiroshimas, or a hundred big-city downtowns burning all at once, our calculations suggest, is all it takes to generate a nuclear winter far more severe than the events following the Tambora volcanic explosion.

In deriving this threshold, we do not mean to imply that any optical depth greater than 1.0 would cause nuclear winter, while any optical depth less than 1.0 would cause no effects at all. The logistic curve still applies. There is a transition region of generally worsening outcomes with increasing amounts of smoke, over the range of average widespread absorption optical depths now thought to be plausible after a nuclear war; i.e., from about 0.1 to 10 or so. We stress that the temperature declines associated with smaller values of $\tau_a$ may also work enormous and widespread damage. $\tau_a = 0.2$ would produce a climate much more severe than that of 1816–1817; $\tau_a = 0.2$ may be enough. Figure 4 shows roughly how continental land temperatures are predicted to vary with absorption optical depths in this range.

Defining a nuclear attack that might produce a nuclear winter reduces chiefly to estimating the amount of smoke generated. The smoke would quickly be distributed over the Northern Hemisphere (ref. 7.11). So the total quantity of smoke generated divided by the area it covers corresponds to some average value of $\tau_a$ (ref. 7.10). Most of the recent technical assessments find that an emission to the atmosphere of about 30 million tons of soot (ref. 3.13) would be sufficient to cause major planet-wide climatic perturbations, consistent with the definition of a rough optical depth threshold $\tau_a = 1$. We stress once more that this is a conservative estimate, corresponding to a threshold near point b, and not a, in Figure 3. (Also, compare Figures 3 and 4).

The calculation of what smoke optical depth is generated in a given nuclear war requires data on the abundance of combustible materials in the target zones, the quantity and darkness of smoke produced, the sizes of the smoke particles, and the rate of removal—for example, by rainfall—of the particles. The TTAPS team estimated, for its baseline nuclear war, $\tau_a = 1.4$ for urban smoke (plus an additional scattering optical depth from dust of about 2).* We have made a full analysis of the range of likely soot injections and absorption properties (ref. 3.14). The most likely values of $\tau_a$ in a major nuclear war are,

* Most recent three-dimensional computer simulations of nuclear winter for ease of computation consider only the smoke and ignore the high-altitude dust—which may contribute half the total scattering optical depth. The dust is produced by high-yield groundbursts against hardened targets (missile silos, underground command-and-control centers, etc.). U.S. nuclear forces are now reacquiring such high-yield weapons. High-altitude dust is an irreducible part of nuclear war with the present arsenals and should be considered carefully in future simulations. Multiburst phenomena—e.g., when the debris from one explosion is sucked up to high altitudes by a nearby explosion—need also to be examined.

we hold, between 0.5 and 3, although smaller and larger values are both possible. Our values (ref. 3.14) of possible absorption optical depths averaged over the Northern Hemisphere range from 0.3 to about 10 (depending strongly on targeting—i.e., on how the nuclear war is fought). This corresponds to masses of atmospheric soot between 10 million and roughly 300 million tons.

That's what it takes to generate nuclear winter.

Tens of millions of tons of soot.

Our technology has risen to the challenge.

# VOLCANIC WINTER

> Between 3 and 4 o'clock in the afternoon of the said 29th, it began to rain mud and ashes at Caysasay (12 miles from the volcano) and this rain lasted three days. The most terrifying circumstance was that the whole sky was shrouded in such darkness that we could not have seen the hand placed before the face had it not been for the sinister glare of the incessant lightnings. Nor could we use artificial lights, as [these were] extinguished by the wind and copious ashes, which penetrated everywhere. All was horror during those three days, which appeared rather like murky nights. [From an account of the events following the November 29, 1754, eruption of Mt. Taal in the Philippines, as described by an eyewitness, a Father Buencuchillo. Quoted in "Taal Volcano and Its Recent Destructive Eruption," by Dean Worcester, *National Geographic Magazine* 23 (4), April 1912, 320.]

In a major volcanic explosion, vast amounts of material are ejected—from huge boulders that fall nearby, to volcanic ash that is carried by the upper-level winds, to sulfur-rich gases that at very high stratospheric altitudes form tiny droplets of sulfuric acid. After the El Chichón volcanic explosions near the Yucatan Peninsula in March–April 1982, Earth satellites and ground-based laser-radar ("lidar") stations observed a cloud of fine sulfuric acid droplets form in the part of the stratosphere above the volcano, spread in a matter of a week all around the world in longitude, and then in a matter of months extend in latitude to cover most of the Northern Hemisphere. In seven weeks, 10 to 20% of these stratospheric aerosols had crossed the Equator into the Southern Hemisphere. (This resembles the pattern of dispersion expected for fine smoke particles injected into the high atmosphere by burning cities and petroleum repositories attacked in a nuclear war.) The estimated scattering optical depths were a few tenths at most. The following August saw record low temperatures in the American Northeast, with an August snowfall in Vermont and average temperatures in the low 40s on the Fahrenheit scale (several degrees above freezing on the Centigrade scale). Some scientists have proposed that the cold winter of 1984-1985 was due to the lingering effect of these fine particles.

A number of volcanos recorded in historical times have generated large amounts of fine particles in the stratosphere. But the amount of *stratospheric* aerosols—which is what counts here—doesn't depend only on the size of the explosion; sometimes smaller volcanic events produce more high-altitude aerosols than larger ones. For a few cases in which the scattering (not absorption) optical depth of the stratospheric particles could be estimated as around 1, the resulting hemispheric temperature declines were predicted to be a few tenths to 1° Centigrade. Where climatic records are available, this is just the temperature decline that was observed. This work, by James B. Pollack, Brian Toon, and Carl Sagan of the TTAPS team (performed in the middle 1970s), was another precursor study on the road to nuclear winter (see Appendix C). Our success with volcanos already suggests that science can now perform such calculations with fair accuracy. (Note again that an optical depth of around 1 for transparent particles will have much less of a climatic influence than an optical depth of around 1 for otherwise identical dark, sooty particles—see Figure 9 in Appendix A. Both sulfuric acid droplets and soot particles scatter visible light, but soot is much more absorbing.)

The Tambora, Indonesia, eruption (1815) that produced the famous "year without a summer" is estimated to have had a scattering optical depth of around 1.2, and to have taken two and a half years to fall to about 0.4. The celebrated Krakatoa, Indonesia, explosion (1883), which produced strangely beautiful sunsets all over the world, had a scattering optical depth of around 0.5 and a resulting global temperature decline of a few tenths of a degree C. (Extensive dust storms arising in the Sahara typically have scattering optical depths around 0.7; about a week later, when the dust has crossed the Atlantic Ocean—moving at lower speeds and lower altitudes than stratospheric volcanic aerosols—the optical depths are down near 0.2.)

The speed with which land cooling follows a volcanic explosion is another way in which volcanos help us to understand nuclear winter. In a study of the climate effects of eleven late-nineteenth- and twentieth-century volcanic events, "the most striking feature of our results is the speed of the climate system's response. . . . Our results provide empirical support for the short response time suggested by recent attempts to simulate the climatic effects of a nuclear exchange." The same study finds that global temperatures usually return to normal about two years after the explosion.

The largest eruption in the last 10,000 years whose effects have been quantitatively determined seems to have been Mt. Rabaul on the island of New Britain in the present Republic of Papua New Guinea. This explosion occurred in the year 536. Richard B. Stothers of NASA estimates that the scattering optical depth of the Rabaul explosion was about 2.5. If the sulfuric acid droplets it generated were pretty evenly distributed over the Northern Hemisphere, as was the case for El Chichón, significant darkening and cooling for a period of months or more is to be expected—from the same sort of analysis that predicts nuclear winter. In fact, state records in South China for the year 536 report snow and frost in July and August, the killing of the subsequent cereal crop, and a major famine the following fall. In some regions of South China, 70 to 80% of the population seem to have starved to death. The corresponding temperature decline was perhaps a few degrees Centigrade. A Mesopotamian record for the same period reads, "the Sun was dark and its darkness lasted for eighteen months; each day it shone for about four hours, and still this light was only a feeble shadow."

The Chinese historical records, the most comprehensive of any civilization on Earth until modern times, reveal a number of other examples. The explosion of Mount Hekla in Iceland, in about 1120 B.C., left in its wake, halfway around the planet, notations such as "It rained dust. . . . For ten days it rained ashes and the rain was gray. . . . It snowed in the sixth month [July, in the Chinese calendar]. The snow was deeper than a foot. . . . Frosts killed the five cereals. Fibre crops failed."

Mount Etna in Sicily erupted in 44 B.C. The next year Chinese chroniclers reported: "The sun was bluish-white and cast no shadow. At high noon there were shadows, but dim. . . . Frosts killed crops, widespread famine. Wheat crops damaged, no harvest in autumn." Sixty years later a Roman historian wrote about events in Italy at this time: "There was . . . obscuration of the Sun's rays. For during all that year its orb rose pale and without radiance. . . . And the fruits, imperfect and half-ripe, withered away and shrivelled up on account of the coldness of the atmosphere." This chronicler was Plutarch, who associated these events with a "divine ordering . . . after [Julius] Caesar's murder" on the Ides of March, 44 B.C.

The worldwide nature of these effects can be established from tree-ring dating, in which the cold produces frost-damaged or unusually narrow tree rings; and from sulfuric acid droplets (eventually fallen to Earth from the stratosphere) and volcanic ash, both preserved—laminated between layers of snow and ice—in the Arctic and Antarctic.

This varied evidence points to a consistent picture—major volcanic explosions injecting fine transparent particles into the stratosphere, where they are spread worldwide, persist for months or years, obscure the Sun when the scattering optical depths are larger than about 1, cool the Earth, and induce widespread crop failures as well as, in the most severe cases, mass famine. In nuclear winter, the fine particles are injected into the high atmosphere from many different locales, more or less simultaneously. The individual particles are much more absorbing. The resulting cold and dark can be much more severe.

Note that a climatic effects threshold at scattering optical depth 1 for *nonabsorbing* (transparent) particles suggests in a different way that our choice of nuclear winter threshold at $\tau_a = 1$ for *absorbing* sooty particles is very conservative: Smoke aside, a scattering optical depth around 1 is expected (ref. 2.2) from groundbursts against targets (e.g., missile silos) far from cities in a major nuclear war (ref. 7.12).

# CHAPTER 8

# TARGETING

And the combat ceased for want of combatants.

—Pierre Corneille, *Le Cid* (1636), Act IV, Scene III

AS LONG AS NUCLEAR WEAPONS ARE BELIEVED ESSENTIAL for national security, plans must be laid on what needs blowing up in other countries. The existence of nuclear weapons means that governments must engage in the plotting of mass murder. This description may sound harsh, but it is wholly accurate. The decisions about which facilities to destroy and which people to kill (the two are hardly distinguishable) is called "targeting." All targeting plans and "exercises," so far as we know—and certainly everything before 1983—have been performed innocent of any knowledge of nuclear winter. What would the climatic consequences have been if past targeting plans had actually been implemented? Or present plans?

The types and numbers of targets, and numbers and explosive yields of warheads in the strategic arsenals are summarized in a number of authoritative reports (ref. 8.1)—from which we conclude that even a small number of nuclear weapons (relative to the total in the world's arsenals or what is currently projected for the next few decades) could produce enough smoke to cross the optical threshold for nuclear winter. But the conclusion depends on targeting. Fewer than 100 big-city downtowns burning simultaneously may be adequate (refs. 2.2, 2.3). The smoke must rise to high altitudes if it is to do its maximum cooling, but we now know this will happen even if firestorms are not prevalent—in part because of the self-lofting of atmospheric soot heated by sunlight.

Targeting cities is an essential component of all NATO, French, and Soviet (e.g., ref. 8.2) as well as Chinese (ref. 8.3) plans for strategic war. National Security Decision Memorandum (NSDM)-242, signed by Richard Nixon on January 17, 1974—not long before he was forced to resign as President of the United States—gives as one goal of U.S. strategic targeting

## "DESIRED GROUND ZEROS"

The staff of the JSTPS (Joint Strategic Target Planning Staff) is divided into two major directorates. The first is the National Strategic Target List Directorate, which screens and selects targets to be struck and publishes the National Strategic Target List. This directorate then establishes aim points, referred to as DGZs (Desired Ground Zero[s]), and catalogs them in the National DGZ List. This directorate also evaluates overall target coverage and determines if the necessary damage expectancy goals required by national policy will be met.

The Single Integrated Operational Plan Directorate prepares the SIOP. This involves the application and timing of all committed weapons in the most effective manner possible against the aim points contained in the National DGZ List.

—Gen. Richard H. Ellis (U.S. Air Force, then Director, Strategic Target Planning), *Building a Plan for Peace: The Joint Strategic Target Planning Staff*, 1980 (issued to mark the twentieth anniversary of the founding of the JSTPS), p. 6.

In recent years the "D" in DGZ is almost always explained as "Designated," rather than "Desired." Perhaps the passion is going out of the targeting business. And the increasing computerization in this line of work has led to a new acronym, the National Strategic Target Data Base (NSTDB).

"the destruction of the political, economic and military resources critical to the enemy's post-war power, influence and ability to recover . . . as a major power" (ref. 8.4). Among other targets, that means cities. Admiral Noel Gayler (former Commander-in-Chief of all U.S. Forces in the Pacific; former Director, National Security Agency; and former Deputy Director, Joint Strategic Target Planning Staff), in testimony before the House-Senate Joint Economic Committee (ref. 8.5), answered:

> Would cities be struck? Almost certainly. The deterrent targets are embedded in them. Whatever the declaratory

> policy of either country, the weapons that go after leadership, control, military capability, industrial capability, or economic recovery will hit cities. Whatever our rhetoric, or theirs, in a general nuclear war cities will be struck, and they will burn.

It immediately follows that in any major nuclear war (or "central exchange") between the United States and the Soviet Union some level of nuclear winter is likely.

The petroleum stores of the warring nations and their suppliers are alone sufficient to cause major climatic disturbances —with the expenditure of only a few hundred small warheads or fewer (Figure 5) out of a global arsenal of nearly 60,000. This sensitivity follows because petroleum refining and storage facilities are highly localized, extremely vulnerable to nuclear detonations, produce enormous clouds of black oily smoke, and offer strategically critical targets to the war planners (refs. 8.6, 8.7).

Let's now compare the relatively small nuclear war—small at least by current standards—necessary to generate a nuclear winter with what was and is actually planned. Here, as elsewhere in this book, we mainly discuss U.S. war plans, not because they were necessarily more unconstrained than Soviet war plans—but because we have some knowledge of what they were. As a rough approximation, we can assume similar plans on both sides for comparable force levels.

In the period 1945–1949, before the first Soviet nuclear explosion, U.S. strategic targeting laid heavy emphasis on cities and petroleum refining and storage facilities. As early as October 1945, airborne nuclear attacks on 20 Soviet cities were envisioned; by December 1947 (plan codename *Charioteer*) the number had grown to 70 cities and "complete" destruction of the Soviet petroleum industry. In *Dropshot*, a targeting plan completed in late 1949, the number of cities had risen to 100 and "destruction of 75–85% of [the] petroleum industry, including storage facilities" was contemplated (refs. 5.8, 8.4, 8.8, 8.9). We can see how such plans both depended on and drove nuclear weapons production.

As Soviet nuclear strike forces and retaliatory capability developed, they also were targeted in U.S. war plans; but the attention devoted to cities and petroleum facilities seems not

FIGURE 5

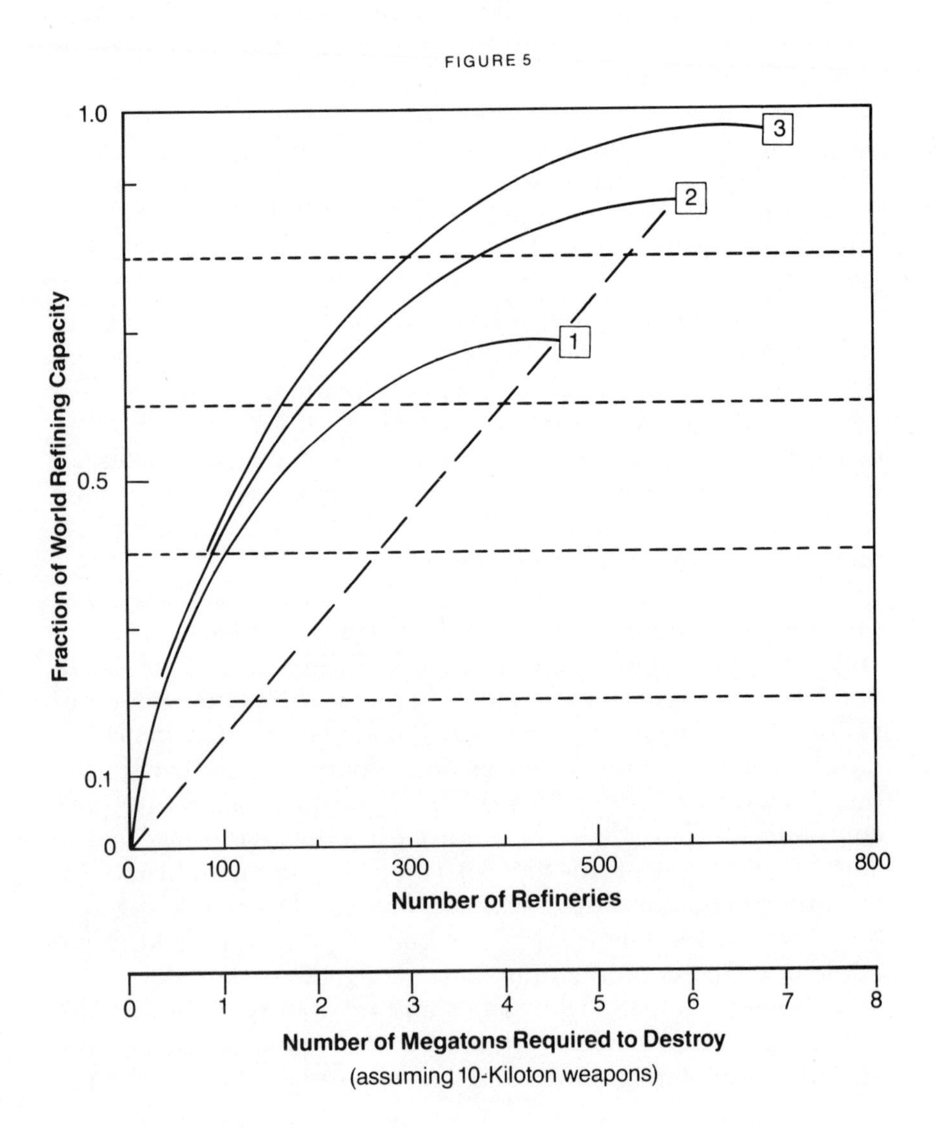
1.0
0.5
0.1
0
Fraction of World Refining Capacity
1
2
3
0
100
300
500
800
Number of Refineries
0
1
2
3
4
5
6
7
8
Number of Megatons Required to Destroy
(assuming 10-Kiloton weapons)

The fraction of world petroleum refining capacity as it depends on the cumulative number of refineries. Curves are given for three groups of countries: (1) the U.S., its NATO allies, the U.S.S.R., its (present and former) WTO allies, and China; (2) the countries of Group 1 plus the Middle East, Japan, Australia, and several other nations closely aligned with the superpowers; (3) all nations with significant refining capacity. For each group, the individual refineries are ordered by refining capacity, with the largest added first. A random summation of refining capacity would correspond to the long dashed line (for Groups 1 and 2). A second scale below the figure indicates the nuclear weapons megatonnage required to destroy the cumulative number of refineries; a 10-kiloton warhead (roughly equivalent in yield to the Hiroshima atomic bomb) would be sufficient to destroy and burn any refinery in the world today. [Data on refinery capacity taken from the *International Petroleum Encyclopedia* (Tulsa, Okla.: PennWell Publ. Co., 1986).]

An analysis of the effects of blast overpressure on storage tankage [based on S. Glasstone and P. J. Dolan, *The Effects of Nuclear Weapons* (Washington, D.C.: Department of Defense, 1977)] shows that a 10-kiloton airburst would rupture petroleum storage tanks over an area of about 2 to 15 square kilometers, depending on the size of the tanks and how full they are. Storage tanks are even more sensitive to low-yield bursts (around 1 kiloton) than refineries are; typical containers may be destroyed over an area of up to 10 square kilometers by a 1-kiloton airburst [Glasstone and Dolan, 1977].

Since 1- to 10-kiloton yields are characteristic of tactical nuclear weapons, these considerations underscore the danger that tactical as well as strategic weapons may bring about nuclear winter.

to have seriously waned—in part because of the huge growth in the number of nuclear weapons and their delivery systems. "Forces capable of destroying the Soviet urban society, if necessary" had "the highest priority" in the Kennedy Administration, and "an even higher priority" in the Nixon Administration; the capability of significantly retarding "the ability of the U.S.S.R. to recover from a nuclear exchange and regain the status of a 20th century military and industrial power" was considered "an important priority"—indeed it too was given "highest priority"—in the Carter Administration (ref. 8.10).

By 1980, the war plans devoted a large fraction of U.S. nuclear forces to the following categories of targets: petroleum refineries; ammunition factories; tank and armored personnel carrier factories; railway yards and repair facilities; and the coal, steel, aluminum, cement, and electric power industries (ref. 8.11). All these tend to be in or near cities. While there were only a few dozen targets in the late 1940s, by the early 1980s there were over 40,000 Soviet and other potential targets in the U.S. SIOP ("Single Integrated Operational Plan"—the detailed prescription for waging global nuclear war) (ref. 8.4; see also ref. 8.19).

There are always more targets than weapons. Target lists have grown in part to provide justification for still larger nuclear arsenals, and in part in the quixotic pursuit of "strategic superiority" by both sides.

It is useful to bear in mind that the U.S. and U.S.S.R. really have made contingency plans for a premeditated first strike—what in an earlier epoch was called a sneak attack. The "fully generated" U.S. SIOP of late 1970s vintage is said to include 4400 "Economic/Industrial" (E/I) targets and some 3600 E/I aimpoints. For "non-generated" (i.e, retaliatory) circumstances the numbers are 2300 and 1000 respectively (ref. 8.12; cf. ref. 8.19). E/I targets are still given priority, however anachronistically, in part because modern conventional weapons have such power and accuracy that unprecedented attrition in materiel (e.g., tanks, tactical aircraft) can be expected in a conventional war. Both sides might run out of equipment in a few weeks, and "winning" or "losing" the war might depend on further procurement—which in turn provides incentives for E/I targeting (ref. 8.13). Because of the proximity or collocation of strategic and E/I targets, the 200 largest cities in the U.S.S.R. are today effectively targeted by the U.S.; so are 80% of the roughly 900 Soviet cities with populations above 25,000 (refs. 8.14, 8.15).

A comparable proportion of American cities is doubtless included in Soviet targeting lists (ref. 8.16). There is now first-hand evidence that in the Cuban Missile Crisis of October, 1962, Soviet warheads were targeted "on cities like Washington and New York and on U.S. military installations and industrial centers" (ref. 8.17). Efim Slavsky, the Minister of Medium Machine Building—the cover agency for the Soviet nuclear weapons program—in July 1968 told Andrei Sakharov, "We've got to be strong, stronger than the capitalists—then there'll be peace. If the imperialists use nuclear weapons, we'll retaliate at once with everything we've got and destroy every target necessary to insure victory." Sakharov concluded from this, "So our response would be an immediate, all-out nuclear attack on enemy cities and industry, as well as on military targets." (ref. 8.18). In the most recent war to use "strategic" rockets (although not nuclear weapons), that between Iran and Iraq in the 1980s, few inhibitions were in evidence about countervalue targeting; cities were routinely attacked.

By the late 1980s U.S. war plans put comparatively less emphasis on Soviet economic recovery and more on Soviet war-sustaining industry. Not so much effort is said now to be given to destroy, for example, fertilizer plants, coal mines, and cement factories, but high priority is still directed to petroleum refineries, tank and ammunition factories, and rail repair facilities (ref. 8.19). Cities and petroleum depots continue to be so heavily targeted that the threat of nuclear winter is negligibly reduced by the latest U.S. targeting doctrines. But now that the likelihood of a European conventional war has hugely diminished, the perceived imperative for E/I war procurement targeting should, one might think, dramatically erode. This has not yet happened.

We conclude that a nuclear war even as early as the late 1940s to late 1950s would have carried a significant risk of generating nuclear winter—especially if it included Soviet conventional and nuclear attacks on Western European cities and

# AERIAL BOMBING AND NUCLEAR TESTING: ARE THEY CONSISTENT WITH NUCLEAR WINTER?

On July 27, 1943, British bombers dropped about 1 kiloton of high explosives plus 1 kiloton of incendiaries on Hamburg. Most buildings in a 5-square-mile area were in flames twenty minutes after the first wave of bombers had passed. The resulting firestorm, with winds of hurricane force, took two or three hours to reach maximum intensity. Then people were sucked off the streets into a pillar of fire. According to some reports, the smoke plume, including the ashes of the immolated, reached almost to the stratosphere [data supplied by David L. Auton, Defense Nuclear Agency, private communication, 1986]. An R.A.F. pilot commented, "You really don't see the great sheets of flame when you pass over the city . . . because of the smoke." (Martin Caidin, *The Night Hamburg Died* [New York: Ballantine, 1980].)

There were also World War II firestorms in Dresden and Darmstadt and Tokyo, to say nothing of Hiroshima and Nagasaki. Shouldn't the historical records show a slight lowering of the global temperature in the middle 1940s? The answer is that, despite all that devastation, not nearly enough smoke was put up into the atmosphere all at once to produce a significant temperature decline.

What about the years between 1945 and 1963, when aboveground nuclear testing was performed on a very large scale? Shouldn't those hundreds of nuclear explosions have produced enough debris to darken and cool the Earth? Again, the answer is no. The explosions were spread over an eighteen-year period; more important, and happily, they were not detonated on or above cities, oil refineries, or even forests. These explosions occurred in deserts, island atolls, the Arctic tundra, and in space. The total optical depth for any significant area of the Earth was always far less than 1 due to nuclear explosions. There should not have been any climatic anomalies from this cause, and there do not seem to have been any.

But, especially after witnessing high-yield nuclear explosions, it was easy to imagine a connection with the effects of large volcanic explosions that do cause climatic anomalies. An American military observer at the "Mike" Pacific test of 1952 recorded

> an amber glow along the entire horizon. It was the most artificial thing I have ever seen and sensed in my life. We had displaced many millions of tons of coral debris that had been lifted up to forty and fifty thousand feet by the blast. [Tom Stonier, *Nuclear Disaster* (Cleveland: World Publishing, 1964), 137.]

A few scientists have attempted to connect, say, the cool summer of 1954 with the nuclear weapons tests at Bikini atoll that same year: e.g., A. Arakawa et al., "Climatic Abnormalities as Related to the Explosions of Volcano and Hydrogen-bomb," *Geophysical Magazine 26*, 1955, 231–255. But the evidence for a causal connection is unconvincing [G. Sutton, "Thermonuclear Explosions and the Weather," *Nature 175*, 1955, 319–321; P. J. R. Shaw, "Nuclear Explosions and the Weather," *Australian Meteorological Magazine 36*, 1962, 39–40; B. J. Mason, "Man's Influence on the Weather," *Advancement of Science 12*, 1956, 498–504], and nuclear winter theory predicts a null effect. A poll of some 80 leading meteorologists, conducted by the U.S. Weather Bureau, turned up not one expert who thought that nuclear testing might affect the weather beyond the test site itself (L. Machta and D. L. Harris, "Effects of Atomic Explosions on Weather," *Science 121*, 1955, 75–80: "There appear to be many orders of magnitude separating the amount of dust required to produce any significant reductions in the worldwide incoming [solar] radiation and that produced by the Nevada [test site] explosions.")

However, modern three-dimensional general circulation models of the atmosphere, used in the study of nuclear winter, are greatly indebted to the pioneering research by the same A. Arakawa.

So historical records of obscuration and temperature declines (or their absence) for large forest fires, for volcanic explosions, for firebombing of cities, and for nuclear testing all are consistent with the predictions of nuclear winter.

petroleum facilities, which it is reasonable to infer might have been the case. But nuclear winter had not yet been discovered, and could have played no role in deterring nuclear war. This example illustrates the obvious proposition that it is vital to understand fully the local, regional and global consequences of specific targeting plans in the design of policy and doctrine—and underscores the dangers flowing from the fact that nuclear winter was overlooked for so long.

One of the most worrisome aspects of nuclear winter is that it could be initiated by only a small fraction of the strategic arsenals of the United States or the Soviet Union, and that it seems to lie within the capability of other nuclear-armed nations, such as Britain, France, and China. As intermediate-range ballistic missiles and nuclear weapons proliferate, other nuclear-capable nations may eventually have to be added to the list. Israel and Iraq come to mind.

In any general nuclear war—even a so-called "counterforce" exchange in which targets are intentionally restricted to military facilities, and cities are consciously avoided—urban damage could be extensive (refs. 8.2, 8.20, 8.21), and at least threshold levels of smoke could result. In a combined counterforce/countervalue central exchange—in which cities are attacked within hours or days of the attack on strategic forces—smoke optical depths, sunlight attenuation, and temperature declines much more severe than threshold values may occur. It is impossible to quantify such probabilities exactly, but they are worrisomely large.

There has been, as we have mentioned, a curious tendency to dismiss the most severe end of the spectrum of nuclear winter effects as "worst case" scenarios. A modest nuclear winter is bad enough. But it would be foolish and dangerous—even a dereliction of duty—to ignore the potential bad news. In any other area of military planning, policies responsive to the worst case would be considered simple prudence (cf. ref. 8.22). Where the stakes are so much greater, we cannot demand less of the political and military establishments.

CHAPTER 9

# WHAT DOES IT TAKE TO PREVENT NUCLEAR WINTER?

As I walked through the wilderness of this world . . . I saw a Man clothed with rags, standing in a certain place . . . , a Book in his hand, and a great Burden upon his back. I looked, and saw him open the Book, and read therein; and as he read, he wept and trembled; and not being able longer to contain, he brake out with a lamentable cry, saying . . . "I am for certain informed, that this our City will be burned with fire from Heaven; in which fearful overthrow, both myself, with thee my wife, and you my sweet babes, shall miserably come to ruin, except (the which yet I see not) some Way of escape may be found, whereby we may be delivered. . . ." And as he read, he burst out . . . crying, "What shall I do to be saved?"

—John Bunyan, *Pilgrim's Progress* (1678)

HOW CAN WE ENSURE, WITH REASONABLE CONFIDENCE, that even threshold levels of smoke would not be generated in a nuclear war? The simple answer is "Don't have a nuclear war." But we have mentioned the dangers of crisis instability and the unauthorized or accidental use of nuclear weapons and note below the likelihood, sooner or later, of madmen in high office in the nuclear-armed nations. If we cannot guarantee that nuclear *war* is impossible, what can we do to make nuclear *winter* impossible, or at least highly improbable? The only ways would seem to be:

(i) remove from the target lists all industrial/economic targets, in addition to all "military" targets that might generate substantial collateral damage in urban and industrial regions;

(ii) increase warhead accuracies, reduce warhead yields and specify detonation heights (e.g., subsurface) to restrict collateral damage and fires;

(iii) plan to carry out a nuclear exchange with few enough weapons, or at a slow enough pace, or under sufficiently "favorable" meteorological conditions to minimize smoke emission and accumulation in the atmosphere;

(iv) deploy an effectively impermeable defensive shield not only over the entire territory of the United States and the Soviet Union, but also over Europe, China, Japan and eventually the whole planet; or

(v) reduce the nuclear arsenals to levels at which threshold quantities of smoke cannot be generated, no matter how a nuclear war is "fought" or who is in charge of the nuclear-armed nations.

Let's consider these five options in turn. Option (i), even if negotiated and agreed to by the superpowers, could obviously never be verified. Each U.S. Minuteman II and Minuteman III missile reentry vehicle is programmed with 4 to 8 alternative targets; those equipped with a command data buffer system can choose among the alternatives instantaneously, and each missile can have all targets completely reprogrammed in about half an hour (ref. 8.4) or less. Improvements in retargeting capability are doubtless continuing on both sides. Furtive electronic signals directed to computer chips could undo any comprehensive targeting agreement, no matter how well verified (ref. 9.1).

Option (ii) is the "technical fix" to nuclear winter which, it was predicted (refs. 2.3, 9.2), would be favored by advocates of nuclear warfighting and "limited" nuclear war options. On its face, this option may seem rational or at least practical. But its successful application would demand that a number of stringent conditions be met: (a) the numerous new weapons technologies required would have to be developed and successfully tested; (b) the existing strategic and tactical nuclear forces would have to be wholly reconstituted at stupefying expense;* (c) all of the nuclear powers would have to move simultaneously in this direction; (d) a foolproof means of verifying or guaranteeing limits on warhead explosive yield would have to be invented; (e) a long and dangerous transition period—in which such "usable" warheads would coexist with the present, much more dangerous arsenal of larger warheads—would have to be safely negotiated; and (f) a new mode of stability would have to be found under which the reduced threat of massive nuclear retaliation would not increase the possibility of a war utilizing such "safe" nuclear weapons. Indeed, to the extent that Option (ii) would make nuclear warfare look more like conventional warfare, the risk of nuclear confrontation and conflict could increase.

Option (iii)—which depends on the execution of a highly choreographed, controlled, and contained nuclear war—is untenable. There are both technical and psychological reasons that a small nuclear war is likely to escalate quickly to a "central exchange" (ref. 9.3). Even if on the day the war begins there were meteorological conditions more advantageous to one side than another (in itself a very dubious proposition), protracted war would involve long periods of time in which weather and

* Simply *adding* such weapons to the existing arsenals does not reduce the dangers of nuclear winter—unless only such weapons are targeted on cities. But this then reduces to the unverifiable Option (i).

climate had entered unprecedented and therefore unpredictable regimes. The uncertainty and risk of this option seem very high.

Option (iv) is of course similar to (although even more difficult than) the objective of an impermeable strategic defense that would protect the civilian population of the United States. Such an SDI is now widely acknowledged to be impossible, at least in the foreseeable future (ref. 17.5). Some relevant aspects of much less ambitious strategic defense systems are discussed below.

Option (v) is tantamount to minimum deterrence, or "sufficiency." It envisions a decrease in the world tactical and strategic arsenals such that, even if the worst happened—with every available nuclear weapon detonated, and city targeting emphasized—nuclear winter would still be unlikely. It requires a reduction in the arsenals to below 1 percent of present levels, as outlined later in this book.

Option (v) encounters difficulties on at least three counts. Until the late 1980s it seemed politically hopeless to imagine even *small* (a few to 50%) reductions in the strategic and "theater" nuclear arsenals, let alone the enormous reductions required to preclude nuclear winter. However, the Intermediate-Range Nuclear Forces (INF) Treaty, signed and ratified by the United States and the Soviet Union, demonstrates that reductions are possible—although in the INF case they constitute only something like 3% of the warheads in the global arsenals (which, moreover, are not destroyed). George Kennan's celebrated proposal (ref. 9.4) to halve the strategic arsenals was at the time (1981) widely considered to be "utopian." However, at the Reykjavik Summit the leaders of the United States and the Soviet Union at least fleetingly entertained the notion of much more massive decreases in the nuclear arsenals (ref. 9.5). Serious negotiations on a Strategic Arms Reduction Treaty (START) have now taken place, and, while a number of policy and doctrinal issues as well as technical difficulties remain (ref. 9.6), there are many on both sides who consider a 30 to 50% reduction in strategic arsenals feasible in the relatively near future. The INF Treaty embodies, and START would necessarily embrace, intrusive verification procedures with inspectors from each nation given unprecedented access to the military facilities and weapons systems of the other. These developments represent revolutionary changes in the usual way of doing business in nuclear-armed nation-states, and suggest that Option (v) is not nearly so out of reach as had once been thought.

However, the United States and the Soviet Union reducing their strategic arsenals by a factor of two is vastly easier than reducing their total nuclear arsenals by a factor of 100. Even if the one is feasible, it by no means follows that the other is as well (cf. refs. 9.7, 9.8). [One pre-publication technical reviewer of this book in a position to know the attitudes of strategic policymakers estimates that if, in the United States of the late 1980s, a choice had to be made, the war-fighting Option (ii) would have been much more easily implemented than the minimum deterrence Option (v).]

The fate of the smaller nuclear forces—e.g., those of Britain, France, and China—poses special difficulties, particularly to the extent that these forces do not serve merely military ends, but are also means of accomplishing or resisting coercion, as well as serving as tokens of world power status.* But these nations are not monolithic or immune to superpower influence and example. In Britain the Labour Party had, until 1989, committed itself to unilateral nuclear disarmament in tandem with a build-up of conventional forces (ref. 9.9); in 1989 it foreswore unilateralism, committing itself to nuclear disarmament on a longer time scale as part of a worldwide process involving, especially, the United States and the Soviet Union (ref. 9.10). The Social Democratic and Alliance Parties were at least at one time favorably disposed to a proportional build-down of British nuclear forces if there is a major reduction in the U.S. and Soviet strategic arsenals (ref. 9.11). China has, from time to time, announced its willingness "to negotiate the general reduction of nuclear weapons by all nuclear weapons states" as soon as the U.S. and U.S.S.R. reach the 50% mark in weapons and delivery systems (and cease weapons testing) (ref. 9.12). There is some reason to think that France would be amenable to a very small strategic arsenal in a superpower minimum deterrence regime (ref. 9.13).

* If their economies continue to decline, nuclear weapons may increasingly serve this latter function for the United States and the Soviet Union as well.

An additional objection to Option (v) is the contention that arsenals so small as to be incapable of inducing a nuclear winter would necessarily provide a much weaker deterrent. However, even a few hundred strategic weapons are clearly enough to obliterate the United States or the Soviet Union as functioning economic and political entities, and just such a deterrent to nuclear attack was widely considered more than adequate during the Kennedy and Johnson administrations (ref. 9.14). We argue below that such force levels, properly configured, could provide *greater* strategic stability than the present arsenals, and constitute a robust deterrent to conventional attack. But this is not the prevailing wisdom.

# VLADIMIR ALEXANDROV: THE FIRST CASUALTY OF NUCLEAR WINTER?

Vladimir Alexandrov was the Soviet Union's leading expert on nuclear winter. But only two years were to elapse between the time he first heard of the possibility of nuclear winter and his mysterious and still unexplained disappearance in Madrid—seemingly vanishing off the face of the Earth. He was forty-seven years old.

The two nations that naturally should be most concerned about nuclear winter are the two nations most likely to bring it about—the United States and the Soviet Union. Early in 1983 we were planning (see Appendix C) a three-day closed scientific meeting to evaluate the claims of nuclear winter. Experts in many relevant fields were being invited from the United States and a few other nations, but the absence of any Soviet scientists from our list seemed a glaring omission. The U.S.S.R. has excellent scientists who might be able to contribute to our understanding of the subject, we thought; and, even more important, how could whatever policy implications that derive from nuclear winter be taken seriously by the Soviets if they had not performed an independent confirmation of our findings?

So one of us (C.S.) met in Washington with Y. P. Velikhov, Vice President of the Soviet Academy of Sciences, to urge, on the very short notice of two or three months, some Soviet participation. We knew this was a difficult request, because according to then standard Soviet practice something like a year's processing time was required by the KGB before a scientist was permitted to travel abroad. Two Soviet scientists who had studied related matters in the atmospheric sciences were suggested, but Velikhov said no, he had another candidate, V. V. Alexandrov, a specialist in computer modeling of the Earth's atmosphere. At the time, we had never heard of Alexandrov. But other American scientists had. Alexandrov had done research at such institutions as the National Center for Atmospheric Research in Boulder, Colorado; spoke excellent English; had lived in the United States and had an Oregon driver's license; and had adapted the Oregon State University computer model of the three-dimensional circulation of the Earth's atmosphere to a Soviet computer in Moscow. Alexandrov was chief of the Climate Research Laboratory in the Computing Center of the Soviet Academy of Sciences. He seemed a good candidate, but we were skeptical that Velikhov could deliver.

Confounding all expectations of Soviet bureaucratic inertia, Alexandrov arrived in Cambridge, Massachusetts, for our meeting in April. Furthermore, he arrived unencumbered by the usual plainclothes KGB officer whose job it was to guide the scientist away from potentially compromising situations and to prevent him from indulging any temptation to defect. Except for the most senior scientists, this was also remarkable, although not unprecedented. And there were other things about him that were unusual. He was far more stylishly dressed than the usual visiting Soviet scientist of those days. He spoke with startling candor of his love for his wife and children, and with particular tenderness for his daughter, a young ballerina who suffered from asthma.

At the meeting, Alexandrov seemed affable and competent. He listened carefully and took notes. His principal comment in the general discussion was to warn of the potential unreliability of three-dimensional general circulation models for this problem. No such models had yet been run on nuclear winter, but they were the obvious next step. Such computer models generally "tune"—or arbitrarily adjust—certain free physical parameters in order to get results that match the present Earth's atmosphere. This might be suitable for examining the contemporary environment of the Earth, and small perturbations from it, but it could lead to grave difficulties, he warned, when applied to what he called the "shockingly different" atmospheric conditions of nuclear winter. Other scientists replied that in simulating the atmospheres of other planets and in reproducing the changing seasons on the Earth such models do quite well. If proper precautions were taken, they said, three-dimensional models might yield useful results. Alexandrov agreed.

Afterwards we urged him to run his own computer model in Moscow to see if he got the same results as our TTAPS study. It would be especially interesting, we thought, if some Soviet results could be available in time for the first public discussion of nuclear winter, scheduled for Washington later that year. Alexandrov confessed that he would dearly love to run the first three-dimensional nuclear winter model and test our results, but computer facilities were so limited in Moscow that the chance of getting adequate time to run the models in only half a year was essentially zero.

But when the Conference on the Long-term Biological Consequences of Nuclear War opened in Washington on October 31, 1983, Alexandrov was there, his preliminary results neatly bound in a blue-covered, English-language pamphlet. "The work I shall present was inspired," he said, by his participation in the earlier Cambridge meeting. For its baseline nuclear war, the TTAPS one-dimensional model had been able to predict temperatures (changing with time) for continental interiors and for oceans (and we had made a rough estimate for coastal regions). But Alexandrov presented world maps with predicted temperature contours laid out. Some of the temperatures were colder than we had predicted—a drop of 40°C or more in eastern Canada, Scandinavia, eastern Siberia, and even the Indian subcontinent forty days after the war.

Alexandrov was gratified that his results were in general agreement with those from the other three-dimensional general circulation model reported at the same meeting—that of Stephen Schneider and his collaborators at the National Center for Atmospheric Research in Boulder, Colorado. The Moscow and Boulder groups then went on to publish a joint paper in the Swedish journal *Ambio* on their mutually compatible findings. Neither of these models independently derived how much smoke would be put into the air, how absorbing the smoke would be, or how fast it would be removed from the atmosphere; for that they relied on the TTAPS estimates. Nor did they run any but a small fraction of the nearly fifty different cases discussed by TTAPS. Nevertheless, the first three-dimensional models provided an important confirmation of nuclear winter theory.

There was one puzzling result that Alexandrov mentioned in his Washington talk: As the soot and dust fell out of the atmosphere many months after the war, his computer predicted a pulse of intense heating near the ground; at the Tibetan plateau the temperatures seemed to rise by as much as 20°C (36°F): "This would cause the mountain snow and mountain glaciers to melt, and would probably result in floods of continental size—I repeat, for emphasis, of continental size." None of the other models, one-dimensional or three-dimensional, found anything like this, and in private discussion then and later Alexandrov was unable to give a physical explanation of how it came about: "It's what the computer tells us," he would say. Perhaps the very problem about which Alexandrov had warned in Cambridge was plaguing at least this aspect of his own calculations.

After the first announcement of his nuclear winter results, Alexandrov was much in demand, both in scientific and in political forums. He appeared in a special symposium in the U.S. Senate convened by Senators Edward Kennedy and Mark Hatfield; in a separate and highly unusual event, he was summoned from the audience to testify in a hearing in the U.S. Congress; in conjunction with an assessment of nuclear winter by the Pontifical Academy of Sciences, he was called, along with Carl Sagan, to present the new findings to Pope John Paul II in the Vatican. He wrote many articles and contributed to a number of books. He was not shy about drawing political conclusions—about the need for improved relations between the United States and the Soviet Union, and about the necessity to make massive reductions in the world's nuclear arsenals.

The machine that Alexandrov was using at the Soviet Academy of Sciences was about ten times faster, but had even less memory, than a run-of-the-mill IBM personal computer of the time. This was pitifully little with which to pursue the nuclear winter calculations, and he longed to use a more capable machine. But by January 1985 the United States had withdrawn permission for Alexandrov to have any access to American supercomputers used for predicting weather and climate. His visa displayed the handwritten admonition "Not permitted direct or indirect access to the supercomputers in the United States." Evidently the government was worried that Alexandrov had another item on his agenda: stealing U.S. computer secrets for the Soviet military—although, as nearly as we can determine, every American scientist whose laboratory Alexandrov visited vouches that he had neither the opportunity nor the inclination. At about the same time, Alexandrov's request to use the (by Soviet standards) much more advanced computers of the Institute for Cosmic Research of the Soviet Academy of Sciences in Moscow was turned down. Also, a news item in *Science* magazine (R. Jeffrey Smith, "Soviets Offer Little Help," 225, July 6, 1984, 31) quoted the opinions of some American scientists, including one of us (R.T.), about the slowly moving pace, low-level effort, and derivative nature of Soviet nuclear winter research. Alexandrov's rebuttal, submitted to *Science*, was considered by the editors to be inadequate, and was never published.

On March 31, 1985, he was in Madrid, on his way back to the Soviet Union (with a planned stopover in Italy), after attending a meeting in Córdoba, Spain, of communities, worldwide, that had declared themselves nuclear-free zones. This is the last time that anyone, on the public record, saw Alexandrov alive.

There are many conflicting accounts of his last days: He was depressed and habitually drunk (although many of us can testify that he was not given to heavy drinking). The Córdoba authorities had called the Soviet Embassy in Madrid to complain that Alexandrov was behaving bizarrely (although the Soviets claim that when they called the Mayor's office back, the Córdoba authorities denied that any such call had been placed). He was seen being hustled into an automobile by several burly men; he was dropped off at his hotel alone. His wallet (with money intact) was found in his room; his wallet, passport, and plane tickets were found in a trash can nearby. He told the driver who took him from Córdoba to Madrid to go straight to the Madrid airport; the driver took him to the Soviet Embassy. He was driven from Córdoba to Madrid in a taxi; he was driven by "Municipal Council chauffeurs"; he was driven by "townspeople." His lecture at Córdoba was well received; his lecture was perfunctory, distracted, and disappointing.

In the months after Alexandrov's disappearance, there were no stories in the Spanish press, no advertisements or "Missing" placards sponsored by Soviet authorities. The Soviets say that the Spanish government was uncooperative. Spanish government sources say the same thing about the Soviets. The Mayor of Córdoba at that time was a Communist, but the Spanish Communist Party takes delight in demonstrating its independence of the Soviet Union. It was more than three months after his disappearance before a government-to-government request was made by the U.S.S.R. for the Spanish police to search for Alexandrov.

A strange fog of confusion hangs over the actual events. All we are sure of is that Alexandrov disappeared and that his body has never been found.

A few days later, FBI agents began calling on American scientists who knew Alexandrov to inquire how "defectable" he might be. About a week later, KGB agents called upon Alexandrov's Soviet colleagues, asking exactly the same question. Most who knew him agree with Hugh W. Ellsaesser of Lawrence Livermore National Laboratory that he was not likely to be a defector: "He was too much of a family man. He loved his wife and daughter too much."

In response to an inquiry, U.S. Government officials stated that no agency of the U.S. Government had Alexandrov, and that they had no idea what had happened to him. Perhaps the Soviets . . . , they suggested. Soviet officials claimed total ignorance of Alexandrov's whereabouts. "He was on his way back to Moscow in another week," one of them told us. "Vladimir had a promising career, and anyway the Soviet Union hasn't assassinated one of its people abroad since Trotsky." Perhaps the Americans . . . , they suggested.

The right-wing American columnist Ralph de Toledano first intimated that Alexandrov was in CIA hands, and then later that he was assassinated by the KGB—because he was about to reveal scientific misgivings about nuclear winter. In a still later column, de Toledano asserts "that the CIA did kidnap him, discovered that he was an alcoholic and not a potential defector, and gave him back to the KGB." Iona Andronov, a special correspondent for *Literaturnaya Gazeta*, suggested that it was the Americans, not the Soviets, whom Alexandrov was annoying with his accounts of the perils of nuclear winter, that the Americans were the only ones with a motive to snatch him, and that Alexandrov had been abducted so that he would be compelled "to sign a request for political asylum, to appear in the Western press with attacks on his mother country, and at the same time forfeit the last hope of returning home with an untarnished name." The implication is that the abduction was bungled. (Andronov and de Toledano have interviewed each other—each apparently coming away with the suspicion that the other works for his respective nation's espionage service.) Some American officials, oddly, suggested that Alexandrov had been abducted by the British or French secret services. A high Soviet official suggested that a particular American-based right-wing organization of demonstrated craziness had abducted and killed him. We are not sure that much credence can be given to any of these suggestions.

Just a few months after Alexandrov's disappearance, an elegantly produced, semi-technical, semi-popular account of nuclear winter, called *The Night After . . . : Climatic and Biological Consequences of a Nuclear War*, was put out by the Mir publishing house in Moscow. In it, many aspects of nuclear winter are discussed, as well as a number of items of peripheral relevance. Startlingly, there is no mention of Vladimir Alexandrov at all. Some of his scientific papers are referred to, but—unlike other referenced articles in the same book—without any hint of who the author might be. "They told us nothing about it," one of the contributors to the book confided to us. "They just removed all the references to Vladimir while the book was in press." Alexandrov was rendered a nonperson in other publications as well, and his salary was no longer provided to his seriously ill wife. But shortly thereafter, the salary was restored and Alexandrov could once again be referred to in Soviet scientific literature. All of this looks as if the Soviets had first decided that Alexandrov had defected to the West, and then changed their minds. But here, too, there are other possibilities.

For a while, some Soviet scientists were calling Alexandrov "the first casualty of nuclear winter." Because his body has never been found, it seems unlikely that he was the victim of a random, nonpolitical, criminal attack. It seems to follow, then, that someone, somewhere found Alexandrov a sufficient threat to abduct and perhaps to kill him. Now that nuclear winter is widely discussed, now that large computers from many nations are pouring into the Soviet Union, now that there is palpable movement toward nuclear arms reductions, it is hard to see what killing Alexandrov could have accomplished.

We remain perplexed and troubled about what his fate may have been (ref. 9.15).

# CHAPTER 10

# DETERRING OURSELVES

The supreme act of war is to subdue the enemy without fighting.

—Sun Tzu, *The Art of War*, 6th century B.C.

IMAGINE TWO NUCLEAR-ARMED NATIONS, A AND B. SUPPOSE that, for whatever reason, A launches a massive nuclear first strike on B and that B, for whatever reason, does not retaliate. Great clouds of smoke and soot rise over B, and in roughly a week are carried by the winds halfway around the planet to darken the skies of A. Temperatures plummet, radioactive debris falls, epidemics rage, and (later) deadly ultraviolet light reaches the surface. Country A has contrived by elaborate indirection to destroy itself. Countries interested in committing national suicide have more straightforward methods available to them. If this is a plausible scenario, and if both A and B recognize its implications, both are deterred from starting such a war. The more likely the scenario, the stronger the deterrent.

Preparations to make a first strike, and concerns about the adversary's capabilities for first strike, have powerfully spurred mutual anxieties, leading to the present disposition of forces and to the nuclear arms race itself. But if a massive nuclear first strike, intended to destroy much of the retaliatory forces of the adversary, could boomerang and destroy the attacker without any effort on the part of the victim, how credible is it that such an attack would ever be launched?

Nuclear winter deters the use—both for warfare and for intimidation—of the vast bulk of the present strategic arsenals. It goes some additional distance toward making it "as nearly certain as possible that the aggressor who uses the bomb will have it used against him"—the early formulation of nuclear deterrence by the pioneer in the field, Bernard Brodie (ref. 10.1). This is the principal and novel way in which nuclear winter strengthens deterrence. The incidence of armed robbery would decline dramatically if firearms routinely blew up in the faces of those who use them (and if the fact that they blow up were widely known).

It is (almost) inconceivable that any sane national leader would count on immunity from retaliation after launching a first strike, nuclear winter or no nuclear winter. The present deployment of global nuclear forces permits a nation to visit a devastating counterattack, even after absorbing a massive first strike. Mobile missiles on trucks or railway cars may evade attack. Some intercontinental ballistic missiles (ICBMs) in hardened underground silos will survive imperfect targeting. Strategic aircraft aloft will be unaffected by nuclear attack on their bases. Submarine-launched ballistic missiles (SLBMs) waiting in the ocean depths are effectively invulnerable. A single surviving modern ballistic missile submarine on either side could obliterate any nation on Earth. The widely acknowledged dissuading influence of such small invulnerable retaliatory forces is a kind of minimum deterrence. It already exists without nuclear winter (ref. 10.2).

Some analysts believe that the deterrence provided by nuclear winter does no more than strengthen the existing deterrence provided by the prospect of blast, radiation, fires, and prompt fallout at hundreds or thousands of targets; and that a leader insane enough to be undeterred by the prompt effects of nuclear retaliation will not be slowed down much by the further prospect of nuclear winter. However, this traditional style of deterrence—deterrence out of fear of the consequences of nuclear retaliation—does depend on the state of mind of the leaders of the adversary forces. And leaders can sometimes be irresolute, or confused, or immobilized with fear, especially when sleep-deprived in times of crisis (ref. 5.14).

That the civilian leader—perhaps out of fainthearted humanitarian motives or a failure of nerve—might decide not to order nuclear retaliation after a nuclear attack is a dagger aimed at the heart of deterrence. Strategists sometimes worry about this (ref. 10.3). Gen. Maxwell Taylor, Chairman of the Joint Chiefs of Staff in the Kennedy Administration, reassuringly noted that deterrence depends on "a strong President unlikely to flinch from his responsibility" (ref. 10.4). But presidential candidates cannot usually be tested beforehand to see whether they are able to commit mass murder under trying circumstances. Cour-

## "THE DILEMMA WE FACE"

One of the shocking implications of nuclear winter is that we can destroy ourselves by attacking our enemy. Some experts have long maintained that there is no military usefulness to nuclear weapons, or to some types of nuclear weapons. Now we learn that, by employing the nuclear option, we may be shooting ourselves in the head. If the nuclear winter study is correct, we have thousands of weapons in our own arsenal that are pointed at ourselves. Of course, the same situation exists for the Soviet Union. The dilemma we face is that the strategic option we have adopted to assure deterrence could also assure our self-destruction.

From "Opening Statement of Senator William Proxmire, Vice Chairman," in *The Consequences of Nuclear War: Hearings Before the Subcommittee on International Trade, Finance, and Security Economics of the Joint Economic Committee, Congress of the United States*, 98th Congress, 2nd Session, July 11 and 12, 1984 (Washington, D.C.: U.S. Government Printing Office, 1986), 99.

age—if that's what "unlikely to flinch" is meant to convey—is an unpredictable character trait. You may never know, yourself, whether you have it until the time comes. What if a leader finds himself unable to press the button? What if, pre-crisis, he (or she) is suspected of an inclination to flinch? And is the obdurate readiness to launch thousands of nuclear warheads a character trait we *want* in our national leaders?

If nuclear winter precludes "winning," never mind what the other side does, and if retaliation is to be visited even if the adversary is timid or scrupulous, then any conceivable advantage from striking first dissipates, and deterrence becomes much more reliable. This is a nonstandard kind of deterrence, one that short-circuits concerns about leadership psychology. Of course, it could not, with anything like the present arsenals, reliably deter madmen at the helms of nuclear-armed nations. For that, other measures are needed.

. . .

There are ways in which nuclear winter makes it harder to fight a nuclear war. The classic method of "damage assessment" is to obtain a high-altitude image of your target. If there's a smoking crater where a city used to be, your weapon did its job. (Congratulations, doubtless, will be in order.) But if the entire region surrounding the target is covered by opaque smoke, you may be unsure of which targets you've hit, or if you've inadvertently hit the wrong target (important especially to those trying to fight a "counterforce" war). What about jet engine performance in the smoke and dust clouds (ref. 10.5)—needed for military operations, for intelligence, and for controlling the war from airborne command posts? What about orderly advances or withdrawals of infantry, or bomber and cruise missile navigation, in near darkness? Tank performance at low temperatures? The increased and indiscriminate injuries and deaths in the armed services from nuclear winter effects? Civilian and military morale? The U.S. Department of Defense has attempted to find ways to fight a nuclear war and "prevail" even in a nuclear winter environment. It isn't easy.

There is already, without nuclear winter, concern that command and control of nuclear forces will deteriorate rapidly in the course of a nuclear exchange (ref. 9.3). This provides a powerful incentive to launch nuclear weapons early, all nuclear weapons—even those that might be considered part of an invulnerable retaliatory force (ref. 10.6). This makes nuclear winter more likely. For reasons of this sort, nuclear winter provides additional motivations for improvement of command and control, and for restraint in launching nuclear war. (See below.)

A temptation might arise, akin to that used to justify the so-called window of vulnerability of the Reagan Administration, to launch a partial attack on the adversary's strategic forces, sparing many cultural and economic targets, and cities—retaliation being discouraged by holding those targets as ransom to the attacker's withheld and invulnerable weapons. Some future national leader might convince himself that massive preemption of this sort was the least unsatisfactory response to a projected attack; or that new technologies permit a much more thoroughgoing counterforce first strike than the adversary believes possible, and that a first strike is needed before the adversary understands its own vulnerability. But nuclear winter frees leaders from the necessity of making such calculations. The outcome of nuclear war then depends less on the wisdom, courage, sobriety, sanity, or other conceivable virtues of national leaders.

Nuclear winter enhances self-deterrence by increasing the ultimate risk to the aggressor through the likelihood of retaliation—not by the adversary, but by the Earth's climate system. Knowledge of nuclear winter must therefore act at least incrementally—but perhaps decisively—as a stabilizing force.

Concern has been expressed in the United States about an "asymmetry of perception" about nuclear winter: concern, that is, that Americans might take nuclear winter seriously while Soviets might only *pretend* to take it seriously, leading to unbalanced incentives for discontinuing the nuclear arms race. Even if this were plausible, it is hard to see how it would decrease the security of the United States, since each side is able to inflict unacceptable damage on the other with substantial, indeed enormous, margin. However, the laws of nature are the same in Moscow as in Washington, and there is abundant evidence that nuclear winter is taken seriously in the Soviet Union (see Chapter 13). If anything, official American reaction about nuclear winter, at least for public consumption, appears to be more restrained than official Soviet reaction. Curiously, this may derive, in part, from the fact that nuclear winter implies a major increase in the number of U.S. civilian casualties in a nuclear war. Colin Gray and Keith Payne—the former an influential U.S. nuclear strategist of the school of Herman Kahn—argue that if only 20 million American casualties are anticipated in a nuclear war, then "U.S. strategic threats" are "more credible," while if American casualties are, say, 100 million, then "a U.S. President cannot credibly threaten" nuclear war to achieve his policy objectives (see box). This latter case is considered an unfortunate constraint on essential presidential prerogatives. Nuclear winter makes it highly unlikely that American casualties can be kept as low as 20 million in a central exchange; indeed, it makes it far more likely that nearly all Americans will die. Nuclear winter therefore challenges the coercive (nondeterrent) use of nuclear weapons in U.S. (as in Soviet) foreign policy (ref. 10.7).

There is also an asymmetry of *effects*, because of the more marginal nature of Soviet agriculture. From the propagation of the most severe climatic effects (as appears in computer simulations of the three-dimensional general circulation of the atmosphere—essentially, movies made of weather maps calculated for the postwar world), it seems likely that for anything beyond a mild nuclear winter, the Soviet Union would suffer considerably more than the United States (see, e.g., refs. 3.7, 3.11, 3.13, 3.16, and Plates 11–18). In addition, the Soviet Union is heavily dependent on American grain, but not vice versa; even if—contrary to all the evidence—a nuclear winter, following a Soviet attack, could be restricted to North America, mass starvation in the Soviet Union might still be a consequence. These facts may have contributed to the seriousness with which the Soviet Union seems to take nuclear winter.

These facts, you might also think, should—even with the residual uncertainties—by now have influenced strategic policy and doctrine. But there's a curious reticence to admit it. The Australian climatologist Barrie Pittock writes:

> It is my belief that the military elites in Washington and Moscow have not so far been willing to admit publicly that the new understanding of the environmental effects of nuclear war has brought about any significant change in their thinking, because such an admission would throw into question the rational basis for their long-held military strategies. To admit that the large-scale use of nuclear weapons is suicidal would be to confess that reliance on a policy of nuclear deterrence is no longer credible. The very act of admitting this would change the strategic situation in ways which might be dangerous to the strategic interests of either side (ref. 10.8).

What do you do if the basis of deterrence has suddenly shifted beneath your feet, and you cannot bring yourself to admit it in public? Do you ignore or deny the shift, or quietly let it influence policy? Or all of the above?

# BESIDES DETERRING WAR, WHAT ARE NUCLEAR WEAPONS FOR?

The standard justifications for having nuclear weapons and their means of delivery is to deter the potential adversary from any use, or threatened use, of its nuclear arsenal. This sounds inoffensive enough. The reality, though, is a little different. There is, at least at the beginning, a fundamental asymmetry between the United States and the Soviet Union on nuclear weapons policy: The United States has pursued nuclear weapons development and deployment as a relatively inexpensive hedge against aggression by what were viewed as "overwhelmingly" superior Soviet conventional forces, and also as a means, both explicit and implicit, of preventing intimidation by or of extracting concessions from other nations. This latter coercive function has been made clear by analysts of quite different ideological stripes (e.g., Colin S. Gray and K. Payne, "Victory Is Possible," *Foreign Policy* 39, Summer, 1980, 14–27; Daniel Ellsberg, "How We Use Our Nuclear Arsenals," in D. U. Gregory, ed., *The Nuclear Predicament: A Sourcebook* [New York: St. Martin's, 1986], 90–96).

American strategic forces do not exist merely for deterrence, Gray and Payne write, but also "to support U.S. foreign policy" through nuclear intimidation. "[The] West needs to devise ways in which it can employ strategic nuclear forces coercively, while minimizing the potentially paralyzing impact of self-deterrence. . . . A condition of parity or essential equivalence is incompatible with extended deterrent duties." This doctrine, if adopted by both sides, leads, of course, to arms races. That Gray and Payne's position has been a mainstream American view is clear from President Eisenhower's memoirs:

> [It] would be impossible for the United States to maintain the military commitments which it now sus-

tains around the world (without turning into a garrison state) did we not possess atomic weapons and the will to use them when necessary. [Dwight Eisenhower, *Mandate for Change*, Volume I (New York: Doubleday, 1963), 180.]

The United States was the first nation to develop nuclear weapons, and the only nation to use them to destroy cities. It has been responsible for most technical developments in delivery systems—including the invention of MIRVs (Multiple Independently-Targeted Reentry Vehicles), which has greatly increased both the destructiveness and the crisis instability of the nuclear arsenals of both superpowers.

In a common view of American policy analysts, major reductions in nuclear arms would lead to a world in which conventional armaments play a much larger role, and in which the massive land armies of the Soviet Union, China, India, the Koreas, Iraq, and Vietnam would be considerably more influential than they are today. Despite occasional public commitments to major cuts in the nuclear arsenals—which, because they are incorporated into the 1963 Limited Test Ban and the 1968 Nuclear Non-proliferation Treaties (ref. 20.4), are in fact the law of the land—many U.S. strategists remain deeply suspicious of massive mutual reductions in the nuclear arsenals, even if they are equitable and verifiable. They want vast U.S. nuclear arsenals to deter conventional forces of other nations and to give U.S. foreign policy a free hand. This is called "extended" deterrence.

An opposing point of view might stress the devastating destructive power of even a greatly reduced U.S. nuclear arsenal, as well as pervasive public misgivings about extended deterrence: In 1985, at the height of Reagan Administration characterizations of the U.S.S.R. as an "evil empire," fully three-quarters of the American people did not believe U.S. nuclear weapons should be used even if the Soviet Union were "over-running Europe with conventional, non-nuclear forces" (*A National Study of Attitudes Toward Nuclear Weapons and Arms Control* [Boston: Marttila and Kiley, September 1985]). By 1988—

before the sequence of revolutions in Eastern Europe—only 8% of the American people believed nuclear weapons should be used to counter a Soviet conventional invasion of Western Europe [*Americans Talk Security*, National Survey No. 6 (Boston: Marttila and Kiley, June 1988)]. To whatever extent the government is responsive to the will of the people, such views, held by so many, tend to undermine extended deterrence. The debate reminds us that there's a connection between nuclear and conventional arms reductions.

The Soviet Union has pursued nuclear weapons to neutralize the possibility of U.S. intimidation (and, more generally, U.S. power and influence), to counterdeter Western nuclear weapons and so disinhibit the use of Soviet conventional forces, as an additional entrée to acknowledged superpower status, and for purposes of coercing and intimidating other nations (cf. D. Holloway, *The Soviet Union and the Arms Race* [New Haven: Yale University Press, 1984]). A less charitable view holds that a principal function of the nuclear arsenals of each nation is to permit it to intervene in force in other (including developing) nations without having to worry that the other superpower might try to dissuade (cf. ref. 15.1).

A comparison of Soviet and U.S. national security objectives—more balanced than those given in previous Administrations, but still richly asymmetrical—has been provided by the Joint Chiefs of Staff:

> The national security objectives of the Soviet Union are to strengthen the Soviet political system, preserve rule by the Communist Party, extend and enhance Soviet influence throughout the world, defend the Soviet homeland, and maintain dominance over the land and sea areas adjacent to Soviet borders. . . . In the past, the Soviets have used this military might to advance their interests by intimidation, coercive diplomacy, or the direct use of force. The Soviet withdrawal from Afghanistan and recent willingness to cooperate in resolving regional conflicts reflect their reevaluation of this approach. . . .
>
> US peacetime military strategy is designed to:
>
> (1) Safeguard the United States and its allies and interests by deterring aggression and coercion across

> the conflict spectrum; and should deterrence fail, by defeating armed aggression and ending hostilities on terms favorable to the United States and its allies.
>
> (2) Encourage and assist our allies and friends in defending themselves against aggression, coercion, subversion, insurgency, and terrorism.
>
> (3) Ensure access to critical resources, markets, the oceans, and space for the United States, its allies, and friends.
>
> [The Joint Chiefs of Staff, *1989 Joint Military Net Assessment* (Washington, D.C.: Department of Defense, 1989), Executive Summary, ES-2, ES-3.]

The Joint Chiefs' item (3) clarifies the meaning of extended deterrence.

Still more simplistic and unbalanced comparisons have been provided by Soviet military spokesmen. For example, consider the following remarks, typical of their era, by a former Soviet Defense Minister:

> Of the modern military doctrines, that of the United States of America should be mentioned above all. Its main idea is to confirm the U.S. world hegemony. . . . Soviet military doctrine acts in sharp contrast with the military doctrines of capitalist states. It is a system of scientifically founded views on the essence, character and methods of waging a war which might be imposed on the Soviet Union. . . . The ideologists of the bourgeoisie. . . . have created and intensively maintain the myth of the so-called "Soviet threat" to peace. They spread false versions to the effect that the sources of wars in the modern era allegedly lie not in the aggressive nature of imperialism, but in the ideology of communism, and allegedly in the attempt by the Soviet State to "export revolution." [Marshall A. A. Grechko, *The Armed Forces of the Soviet State,* 2nd ed. (Moscow, 1975). The English translation (Washington, D.C.: U.S. Government Printing Office, 1978, Document 0-254-358) carries this unsurprising caveat: "The translation and publication of *The Armed Forces of the Soviet State* does not constitute approval by any U.S. Government organization of the inferences, findings and conclusions contained herein."]

Or, to take another statement by Soviet Defense Ministry analysts,

> While socialism is capable of bringing the accomplishments of scientific and technological progress into accord with the needs of social progress, capitalism is not yet up to it. The social antagonisms tearing it apart, the cult of force, the spirit of profit and the orientation towards confrontation prevailing therein contain objective prerequisites for using the achievements of the scientific-technological revolution in ways catastrophic for mankind. [Boris Kanevsky and Pyotr Shabardin, "The Correlation of Politics, War and a Nuclear Catastrophe," *International Affairs*, February 1988, 95–104.]

Consider also these remarks by two leading Soviet scientists:

> There are two directly opposed trends in the present-day world. The most aggressive imperialist circles are stockpiling weapons and making systematic efforts to accustom public opinion to the idea of the permissibility of the use of these weapons and the acceptability of nuclear war. This dangerous course is countered by the Soviet Union, the socialist community countries, and the forces of peace and reason throughout the planet. [V. Goldansky and S. Kapitsa, "Scientists on Global Consequences of Nuclear War," *Izvestiya*, July 25, 1984, 5.]

They neglected to mention that the Soviet nuclear weapons stockpile was then, as now, roughly as large as the American stockpile, and that downplaying the dangers of nuclear war had been a staple of Soviet military propaganda and field manuals for most of the Cold War (cf. refs. 5.16, 13.30). One of the authors of the *Izvestiya* article recalls the paragraph in question as inserted by the censors on the eve of glasnost (Vitaly Goldansky, private communication, January 17, 1990). Both Goldansky and Kapitsa have many times been outspoken and courageous in their public policy statements.

CHAPTER 11

# CONSEQUENCES OF EXECUTION

The beacon-fires of war
Have been lit . . .

The nation
Has been destroyed.

—From "Spring Prospect," by Du Fu (Tang Dynasty, 757), in Greg Whincup, ed. and trans., *The Heart of Chinese Poetry* (New York: Anchor Press/Doubleday, 1987), 65

THERE IS A KIND OF SAFE IN AN AIR FORCE BASE NEAR Omaha, Nebraska, in which is kept a document, regularly updated, marked with the highest security classification. Perhaps there is a copy in Washington as well. Its title reads, "Consequences of Execution." It is a detailed estimate of what would happen if any one of a variety of nuclear war plans were carried out. It exists so the National Command Authority—the President, his or her designees and successors—can make an informed judgment should the fateful hour ever come. It is one of the most important documents on Earth. We do not, of course, know what information that book contains. (We are not even sure that it exists—although we have been reliably told that it does.) There should be a comparable book sitting in a safe in Moscow—and much thinner books in London and Paris and Beijing and, perhaps, elsewhere. What is in these books? Is nuclear winter taken into account? If not, is the "National Command Authority" being given all the relevant information? How can decisions about nuclear war be made without understanding the range of possible consequences?

The combatant nations in a nuclear war can expect—apart from the cold and dark of nuclear winter—to be devastated by blast, fire, prompt radiation, pyrotoxins, radioactive fallout, and high ultraviolet intensities. Even in the most restricted counterforce scenarios, where urban targets are meticulously avoided, early radioactive fallout would exact an enormous toll (ref. 11.1). A more general strike on the hundred largest cities would promptly kill tens of millions (ref. 5.10) and destroy the economic infrastructure of the nation under attack. Even in the worst such scenarios, however, many tens of millions would initially survive, although under extraordinary physical and mental stress. Enormous numbers of lingering deaths would be expected (ref. 11.1). These are some of the consequences of execution.

The warring nations would be among the richest in the world, generally with ample stores of grain and other raw and processed foods. Many would enjoy self-sufficiency in mechanized agriculture. But given low temperatures and light levels; large areas of agricultural land contaminated by radioactivity and toxic gases, and later irradiated by ultraviolet light; the interruption or destruction of critical subsidies of fuel, fertilizer, seed, irrigation, herbicides, pesticides, and the facilities to harvest, store, process, and deliver foods; and with plagues of insects—agricultural yields should plummet sharply for long periods or disappear altogether (ref. 3.11). In the combatant nations, most of the survivors of the first few days of the war would die, chiefly from starvation. This also is among the consequences of execution.

Birds and mammals are more vulnerable to the cold, dark, and radiation than are the insects they prey upon, and plagues of insects, perhaps of biblical proportions, may ensue. Insects are carriers of disease microorganisms which would be spreading just when hospitals are destroyed, physicians killed in large numbers, medicines rendered unavailable, and the immune systems of the survivors weakened by unprecedented physical and emotional stresses. This is an example of "synergism," the adverse multiplicative interaction of the consequences of execution.

Despite these grim prospects, some optimistic forecasters have projected the recovery of the gross national product and quality of life in America within a few decades (ref. 11.2). Presumably, they imagine such notions of national recovery to apply, in varying degrees, to other parts of the world as well. It has also been suggested officially that residual agriculture could almost immediately support the survivors of a nuclear attack on the U.S. and that recovery would be swift (ref. 11.3). We are deeply skeptical. Nuclear winter casts a long shadow on predictions of recovery for the combatant nations. Even the mild end of the spectrum of climatic anomalies predicted by current nuclear winter models implies widespread crop failures in the year or two following a nuclear war (ref. 3.11)—as occurs on a smaller scale in "volcanic winter" after a major volcanic explosion. Little, if any, agriculture in the Northern

# NUCLEAR WINTER AND FAMINE

Perhaps the most profound influence of nuclear winter on human beings would be, successively, hunger, malnutrition, and starvation. Here is a stark account by an expert on nutrition:

> Historically, famines have been of two kinds: those in which there was an absolute shortage of food and those in which people simply did not have money to buy food and it was not distributed to them. Nuclear war would precipitate both kinds simultaneously. We have no precedent for the climatic effects of a nuclear exchange, but we have an abundance of grim historical evidence of what would happen to the nutrition and health of human populations. Money to buy food would not last long for people whose jobs had been eliminated by the conflict, in a country where government and infrastructure had been destroyed and money had probably lost meaning. As has happened so often in the past, most recently in Ethiopia and the Sahel but several times this century in India, the regional movements of food necessary to alleviate local famines would not occur.
>
> Famines caused by natural disasters and wars have plagued mankind throughout history, and the consequences are only too familiar. Hungry, desperate people do not remain orderly and disciplined. In the 18th-century famines in Europe, storehouses, markets, and even granaries were plundered, and the riots often could not be controlled, even by large numbers of troops.
>
> The evidence of the 20th century is even more shocking because it derives from events that have occurred within recent memory and is so well documented. Herbert Hoover, in his three-volume *An American Epic* [Chicago: Regnery, 1959–1961], describes an absolute famine affecting 25 million people in the Volga valley and Ukraine of Russia in 1921 [and caused, at least in part, by the Russian Civil War]. Death for the whole population of these areas was estimated to be only a few months away. Malnutrition and starvation were evident everywhere, and the dead were seen lying in the streets and on roads leading into towns, where they soon became prey to dogs

and birds. The naked dead were piled together to be transported later to the cemetery where great pits, approximately 3 m[eters] deep, could accommodate several hundred corpses. In the city of Orenburg alone, 800 deaths a day were reported for a time.

By January 1921, the bodies of those who had died were too numerous to bury and were piled in heaps in buildings. They were often stolen, and the flesh was boiled for food. Typhus and typhoid fevers were epidemic, and dysentery was prevalent, with a case-fatality rate of up to 50 per cent in children. Bread was made from leaves, the bark of birches and elms, sawdust, nut shells, rhubarb, rushes, peanuts, straw, potato peels, cabbage, beet leaves, and even horse manure. Dead animals were luxuries. By the summer of 1921, survivors were fighting for life by eating dogs, rats, roots, skins, bones, and all manner of refuse. Men lost their reason and became cannibals. A variety of infectious diseases aggravated the dreadful suffering of the famished. A contemporary account describes them as "Seeking food, the exhausted, sick, and naked starved people dragged themselves hither and thither seeking the larger towns and villages in the hope of finding food there. One met, at every step, living skeletons, scarcely able to move, or already completely exhausted, and dying where they lay." [Hoover, *ibid.*]

This occurred in the present century as the result of wartime devastation that was trivial by comparison to a nuclear holocaust and was uncomplicated by the effects of radiation. Can there be any doubt of similar scenes in North America and Europe among any concentration of survivors of a nuclear exchange with the kind of destruction certain to occur? At the end of World Wars I and II, the large shipments of food from the United States ultimately saved millions of lives.* With nuclear war involving both Europe and North America there would be no source for such a rescue. [Nevin C. Scrimshaw, "Food, Nutrition, and Nuclear War," *New England Journal of Medicine 311*, 1984, 272–276.]

---

* [The famine of 1921, described here, was ended with food supplied by the people of the United States, in an effort led by future President Hoover; it saved the lives of millions of Soviet citizens.]

mid-latitude target zone is likely to succeed in the first year, and production for at least several more years would probably be frustrated by unpredictable and anomalous weather (ref. 11.4). The prognosis for a population afflicted with widespread and profound injuries; unprecedented exposure to radioactivity and pyrotoxins; severe problems of sanitation and disease; societal and psychological trauma; lack of food, potable water, medicine, and medical care; bizarre and extreme weather variations—including intervals of deep cold, violent storms, drought, and eventually, increases in the intensity of searing solar ultraviolet radiation—would seem to be unfavorable.

Civil defense preparations in nations at risk have focused on the temporary protection of citizens from blast and fallout, or their resettlement in areas unaffected by the war. However, the U.S. and many other countries have realized that the construction and maintenance of effective shelters for the entire threatened population would be far too costly to undertake seriously (ref. 11.5). Soviet planners have pursued shelters for other than the civilian and military leadership mainly as a means of reassurance of the potential victims—i.e., for political reasons (see box). Moreover, even the most sophisticated mass shelters would be designed only for short-term occupancy (perhaps several weeks). There are no civil defense plans for the bulk of the surviving population in a severely degraded environment—nor could there be. To avoid the awkward admission that large-scale civilian shelters are worthless, given the probable post-war environment, American civil defense planners have chosen to ignore nuclear winter altogether; there is no hint even of the possibility of nuclear winter in the planning documents of the government entity with the sunniest disposition, the Federal Emergency Management Agency (ref. 11.6).

Ironically, a nationwide shelter system designed without regard for long-term effects might increase the post-shelter demands on agricultural and medical systems (the latter, because many of the shelter survivors would have received debilitating but sublethal radiation doses, acquired diminished immune system capacity and/or developed serious illnesses while in confinement). In the light of nuclear winter, standard civil defense shelters can be seen as briefly postponing, but hardly preventing, the deaths of large numbers of those obtaining shelter after a central exchange. And nuclear winter makes "cri-

# THE SOVIET SHELTER SYSTEM

Soviet doctrine for many years has stressed the importance of shelters—for the leadership, for the military, and for the civilian population. Great emphasis is laid on shelters in Marshal Sokolovskii's *Soviet Military Strategy*, a book that had an enormous impact on American strategists (V. D. Sokolovskii, ed., *Voennaya Strategiya* [Moscow: Voenizdat, 1962, and subsequent editions]).

A sober American assessment:

> The Soviet war management concept goes beyond sheltering a small number of key leaders and has been designed to support leadership continuity at all levels . . .
>
> The central component of the Soviet leadership continuity program is a comprehensive and redundant system of hardened command posts and wartime communications facilities, mobile command posts, and exurban relocation sites. Command posts now consist of near-surface bunkers as well as complexes deep underground. Some facilities in the Moscow area are hundreds of meters deep and can accommodate thousands of people. Similar smaller facilities exist beneath other major Soviet cities. Near-surface bunkers also have been built for military, Communist Party, and government authorities throughout the Soviet Union. [The Joint Chiefs of Staff, *1989 Joint Military Net Assessment* (Washington, D.C.: Department of Defense, 1989), 7-1.]

The Department of Defense estimates 1,500 hardened shelters for 175,000 Communist Party and government officials (U.S. Department of Defense, *Soviet Military Power* [Washington, D.C.: U.S. Government Printing Office, 1987], 52). Presumably, places will have to be found for officials of other political parties as well, if democratization proceeds much further in the U.S.S.R.

A perhaps less sober assessment has been made by a former Chief of U.S. Air Force Intelligence, Maj. Gen. George J. Keegan, Jr., who believes that the Soviet Union has built some 75 underground shelters "in every city in every military district" at a total implied cost of some $15 trillion. (William E. Burrows, *Deep Black: Space Espionage and National Security* [New York: Berkley Books, 1988], 5–6; cf. ref. 4.5, especially its Chapters 5 and 8.] This sum is larger than the entire amount spent by the United States on the Cold War since 1945. The real ruble figure for Soviet shelters is likely to be much less.

The scale of Soviet preparations to ensure the survival of the leadership elite has been mainly responsible for the continuing development by the United States of burrowing or "Earth-penetrating" nuclear warheads which, upon arrival in the U.S.S.R., dig down toward the leadership shelters before detonating. Both the MX and the Trident 2 missiles are, apparently, to be equipped with such warheads (cf. Edgar Ulsamer, "Missiles and Targets," *Air Force Magazine*, July 1987, 70; James W. Canan, "The Dangerous Lull in Strategic Modernization," *Air Force Magazine*, October 1988, 74; Warren Strobel, "U.S. to Make Nuclear Bomb That Burrows," *Washington Times*, September 12, 1988, 1). The motivation seems to be approximately the same as Premier Khrushchev's decision, despite the protests of Andrei Sakharov, to develop and test a nuclear weapon of 50 to 100 megaton yield—able to destroy American underground command posts. Indeed, the U.S. is also returning high-yield (9 megatons and possibly more) gravity bombs to its arsenals to be able, on an interim basis, to threaten underground Soviet leadership shelters until the burrowing weaponry is ready.

There is no doubt that extensive *civilian* shelters have also been built in the Soviet Union and stocked with food and other supplies. But even apart from nuclear winter, they would have little effect on the outcome of a nuclear war. According to former CIA Director William Colby, they were built for reasons of morale. [Burrows, *Deep Black*, pp. 14, 15.] When one of us (C.S.) asked a high-ranking Soviet officer why there was so much emphasis on shelters in the U.S.S.R. when their utility was marginal, he replied, "How can we tell our people that we can do nothing to protect them from a nuclear war?" (The same political reasoning was in part behind U.S. proposals for Star Wars.) The shelters play much more a political role in the absence of nuclear war than a strategic role in the event of nuclear war. This is sometimes thought of as prudent restraint before a skittish public given to fits of panic:

> Knowledge of the destructive effects of modern weapons must not morally disarm our people before the aggressor. While telling about the destructive characteristics of modern weapons, the propagandists should be guided by V. I. Lenin's teaching to the effect that our propaganda, aimed at raising discipline and at strengthening military preparation, must not overstep those limits where we, ourselves, contribute to panic. ["Concerning the Program for Preparing the Leadership Element: Civil Defense Propaganda," *Voyennyye Znaniya* (*Military Knowledge*), No. 7, July 1984, 23.]

A Soviet comment, by Yevgeny Velikhov, the man responsible for the clean-up after the Chernobyl nuclear power plant disaster:

> What good was civil defense at Chernobyl, where we had to mobilize the entire country to clean up a relatively small nuclear mess? It is absolutely crazy to think that any kind of civil defense would have any significance in nuclear war. Civil defense cannot work and we do not have it. That is what I tell our military people. [Interview with Velikhov, "Chernobyl Remains on Our Mind," in Stephen F. Cohen and Katrina van den Heuvel, eds., *Voices of Glasnost: Interviews with Gorbachev's Reformers* (New York and London: W. W. Norton & Company, 1989), 171.]

Even if Soviet shelters were equipped with sufficient food, water, air filtration systems, etc., for the initial months or years of nuclear winter, the world awaiting the occupants overhead would be most unpromising—dwarfing the famine world of 1921. And there would be no mountains of food donated from abroad. To whatever extent a "shelter gap" is a source of concern for American military planners, nuclear winter should help relieve that anxiety.

sis relocation" of urban refugees to a supportive and nurturing countryside a forlorn hope.

To summarize, in combatant nations, nuclear winter profoundly threatens the surviving population and poses severe challenges to societal and economic recovery after nuclear war. We hope this fact is noted in the locked books in Omaha and Moscow (ref. 11.7).

CHAPTER 12

# NUCLEAR WINTER IN NATIONS MINDING THEIR OWN BUSINESS

Whatever right a country may have to preserve its own form of government in the face of foreign opposition, it cannot, with any justice, claim the right to exterminate many millions in countries which wish to keep out of the quarrel. How can it be maintained that, because many of us dislike Communism, we have a right to inflict death on innumerable inhabitants of India and Africa?

—Bertrand Russell, *Has Man a Future?* (New York: Simon and Schuster, 1962), 43

Our sympathy is cold to the relation of distant misery.

—Edward Gibbon, *The Decline and Fall of the Roman Empire* (1781), Chapter 49

WHAT ARE THE CONSEQUENCES OF NUCLEAR WINTER IN nations minding their own business—nations unaligned with nuclear-armed treaty organizations, nations that play no part in the quarrel between the United States and the Soviet Union, nations far away that wish to be left alone?

Recovery after a nuclear war is, in the minds of some strategists, tantamount to "winning"—provided the adversary recovers more slowly, or preferably not at all. But if the "winner" could not reclaim its power in a timely manner, other nations, less completely obliterated, would quickly fill the economic and military vacuum, presumably obviating the winner's reason for fighting the war in the first place. What seems to follow is that in the chaos and routine mass murder of nuclear war, military and economic targets may be struck in noncombatant nations so they cannot become postwar rivals of the combatants. Also, old scores might be swiftly settled. The justification for such targeting is eased by the fact that ports and airfields far from the combatants potentially hold strategic significance for the reloading, refueling, and refurbishing of aircraft, submarines, and surface ships—especially in a protracted war, if one were possible.

Nuclear weapons states are unlikely to admit to targeting noncombatant nations, but it flows readily enough from what passes for the "logic" of strategic operations. Even a few nuclear bursts within a nation's borders could produce widespread devastation and unparalleled hardship (ref. 12.1). Prudent planners in noncombatant nations, aligned and nonaligned, would be wise, we think, to consider such possibilities.

In a massive nuclear conflict, noncombatant nations outside the two primary alliances might expect:

(a) potential nuclear detonations at key military or economic facilities within or near their borders, including airfields, ports (especially those equipped for submarines), communications and manufacturing sites, petroleum facilities, and other economic targets;

(b) cessation of trade with combatants and other supplier nations, particularly in food, medicine, fuels, fertilizer, seeds, and manufactured goods;

(c) influx of refugees, many suffering from serious malnutrition, injuries, and illnesses; and desperate pleas for aid from neighboring nations; and

(d) long-term environmental disturbances, perhaps of unprecedented severity.

Most of the nonaligned or weakly aligned nations of Africa, South America, and Asia apparently do not anticipate such nuclear ministrations at the hands of the superpowers. Others, such as New Zealand, have sought to minimize the risk of direct nuclear attack by reducing the number of potential targets within their borders (cf. ref. 13.4).

The SCOPE biological analysis (ref. 3.11) treats the effects of worldwide disruptions in agricultural trade. Hundreds of millions of people are found to be in jeopardy of starvation, even without major climatic perturbations. The projected 1990s food import/export deficits for many of the developing nations is already in the 10 to 50% range without nuclear war or nuclear winter. Moreover, refugees from war zones would swell local populations, increase demands on food and other supplies, and hasten the spread of disease. Political relations among surviving nations might degenerate rapidly, and local warfare could erupt, compounding the misery. Distant nations are therefore wholly justified, it seems to us, in pressing the superpowers for massive nuclear arms reductions.

The noncombatants would be faced with food shortages and starvation for years, accompanied by global pandemics, especially lethal because many human immune systems will be radiation-damaged (refs. 2.3, 12.2). Radioactive fallout, chemical toxins from burning cities, and ultraviolet radiation could be delivered in dangerous doses; and the predictability of weather and climate, upon which societies and civilizations depend, could be transformed into a prolonged and chaotic nightmare of cold, drought, and storms. According to the

SCOPE Report, noncombatant nations might ultimately suffer more casualties than combatant nations (ref. 3.11).

Consider Japan, for example. It has the strongest economy and by some standards is the most powerful nation on Earth. Imagine—we think this highly unlikely—that in a global nuclear war not a single nuclear weapon explodes on or over Japan. Nevertheless, clouds of nuclear winter smoke and radioactive fallout would be rapidly carried by the prevailing westerlies to Japan from targets in China, Mongolia, Siberia, and the Koreas. Comparatively small declines in temperature (including a single night below freezing) are sufficient to destroy the Japanese rice crop. Japan imports over 50% of its food and over 90% of its fuel. World trade would be nearly eliminated in a major nuclear war even apart from nuclear winter. If chronic nuclear winter effects last for several years, followed by severe increases in the intensity of solar ultraviolet light at the surface (from ozone layer depletion), it is not hard to see that the Japanese economy would be destroyed and most Japanese citizens might be killed. Were Japan targeted, the consequences would be even more serious. These matters are also relevant for those considering the arming of the Japanese defense forces with nuclear weapons.

Many developing nations with less stable food supplies and more fragile economies—even those at much more southerly latitudes—might be still more thoroughly destroyed. Populous nations such as Nigeria, or India, or Indonesia might collapse in a nuclear war without a single nuclear weapon falling on their soil.

The prompt and a number of the long-term consequences of nuclear war—many recognized for years—have apparently not moved the superpowers, their allies, or the nonaligned potential victim states to more than feeble action, except to accelerate the arms race. Nuclear winter, however, in which billions of noncombatants may starve to death, has helped many nations, combatant and noncombatant alike, to change policies on the issue of nuclear war (as described in the next chapter). Nuclear winter seems to have reawakened concerns about potential global apocalypse, even in distant and Southern Hemisphere nations that once thought themselves immune to—or even the potential beneficiaries of—a U.S./Soviet nuclear war.

More than 85% of all the humans on Earth live in the Northern Hemisphere. A nuclear war and nuclear winter that was wholly restricted to the Northern Hemisphere could therefore destroy most humans. If significant amounts of fine particles were carried from the Northern Hemisphere to the Southern Hemisphere (or were produced in the Southern Hemisphere), or if the enormous nuclear winter-generated hole in the ozone layer eventually moved across the Equator, or if global pandemics were sufficiently serious, then the environmental effects of nuclear war could threaten the remainder of the human species.

Until the nuclear winter theory appeared, most experts had argued that the effects of a Northern Hemisphere war would be confined to the North (discounting, of course, the possibility of nuclear targeting in the South). Fictional images of radioactive clouds spreading lethal fallout worldwide (ref. 12.3) had never been accepted by serious analysts. Nuclear winter makes it clear, however, that in a nuclear war the Earth's environment must be considered as an integrated, finely tuned biogeochemical system that can be disrupted on a global scale.

A very rough calculation by Cao Hongxing and Liu Yuhe in China suggests that the more nuclear explosions at lower Northern latitudes, the colder is the resulting global mean temperature (ref. 12.4). A major effect of nuclear winter on agriculture is indirect: an average decrease in rainfall over land in July of 50% or more for Northern midlatitudes over a wide range of optical depths, and the predicted failure for one or two growing seasons of the Asian summer monsoon rains (ref. 12.5). For large optical depths, such a failure of rainfall is found to extend to the equator.

Pittock (ref. 12.6) has written extensively on the consequences for the Southern Hemisphere of a nuclear war in the North. They are more difficult to predict, although the probability of severe effects is certainly lower than in the North. (For this reason alone, acquisition of nuclear weapons by nations in Southern midlatitudes—for example, South Africa, Australia, Argentina, Brazil, or Chile—might be especially dangerous for the human species.) Serious long-term climatic and environmental impacts due to lofted smoke and dust are likely. No three-dimensional atmospheric general circulation simulations of nuclear winter have so far included Southern Hemisphere targeting. If 1% of the world's strategic arsenals were devoted to urban targets in the Southern Hemisphere, added to the complement of Northern Hemisphere smoke that may cross the Equator, we estimate that a major Southern nuclear winter could result (ref. 12.7).

Da Silva (ref. 12.8) has proposed that a Northern Hemisphere nuclear winter, through a reduction in rainfall, could so dry the Amazon rain forest that subsequent spontaneous fires might burn a vast area; the resulting soot, he suggests, is enough to produce a secondary nuclear winter as severe as the first—but this one largely in the Southern Hemisphere. Some kind of world fire seems required to explain the global soot layer at the end of the Cretaceous epoch, 65 million years ago; it accompanies evidence for the contemporaneous impact of a 10-kilometer-diameter asteroid (or cometary nucleus) with the Earth, the presumptive cause of the extinction of the dinosaurs and most of the other species of life then on Earth (see Chapter 5, box, Impact Winter). Even in the United States, with the most advanced firefighting equipment in the world, and in the absence of nuclear war, ten thousand square kilometers of forest or more burn down each year. Nuclear war would damage forests worldwide, greatly increasing the chance of subsequent forest fires. Radiation-damaged trees are prone to infestation by insects and microbes which can turn forests into tinderboxes awaiting a spark from lightning or surviving humans. This is a danger for both the North and the South, and might lead to a significant second wave of smoke injected into the Earth's atmosphere in the year following nuclear war.

Assuming no Southern Hemisphere targeting and no "second wave," and neglecting epidemics, the cutoff of import subsidies, and increases in the ultraviolet flux, Pittock (ref. 12.6) has examined nuclear winter effects in Australia—about as far from the principal nuclear war targets as any nation on Earth. Smoke originating in the Northern Hemisphere would decrease the intensity of sunlight by some 20% for a year or more. Average temperature drops are estimated at 2 to 4°C in January (Australian summer). Rainfall would decrease across Australia by about half. When all effects, except Southern Hemisphere targeting and "second wave" soot, are included, the results might be an inability "to provide for the existing population"—i.e., massive starvation in Australia as well (ref. 12.9).

Were the severe end of the spectrum of Northern Hemisphere nuclear winters to be realized, the survival of large intact societies in the Southern Hemisphere might be the key to the eventual reemergence of a global civilization, and even to the continuance of the human species. Much more research therefore needs to be done on the longest-term (at least several years postwar) and Southern Hemisphere consequences of a nuclear conflict, including Southern Hemisphere targeting.

# CHAPTER 13

# GLOBAL POLICY IMPACT OF NUCLEAR WINTER

[We] found ourselves, by the force of events, during the last five years, in the position of a small group of citizens cognizant of a grave danger for the safety of this country as well as for the future of all the other nations, of which the rest of mankind is unaware. . . .

In the past, science has often been able to provide also new methods of protection against new weapons of aggression it made possible, but it cannot promise such efficient protection against the destructive use of nuclear power. This protection can come only from the political organization of the world.

—James Franck et al., "A Report to the Secretary of War, June 11, 1945." Reprinted in Robert Jungk, *Brighter Than A Thousand Suns* (New York: Harcourt, Brace, 1958), pp. 348–360 (The Franck Report was a secret appeal made before Hiroshima by some of the American scientists who had developed the first atomic bomb. They urged that the danger posed by the bomb be used to bring the nations of the world together to face their common peril.)

NUCLEAR WINTER HAS HELPED TO SHAKE COMPLACENCY, force agonizing reappraisals, and alter policy—not just in nuclear-armed nations but in nations without nuclear weapons, nations that were never before critical of superpower strategic doctrine, nations previously without any policy at all on nuclear war. These changed attitudes in nations without nuclear weapons have in turn cycled back to influence further the policies of the states that possess nuclear weapons.

The Delhi Declaration of 1985 by the heads of state or government of India, Sweden, Tanzania, Mexico, Argentina, and Greece refers specifically to nuclear winter as "posing unprecedented peril to all nations, even those far removed from the nuclear explosions"; decries "a small group of men and machines in cities far away who can decide our fate"; compares the peoples of the world to "a prisoner in the death cell awaiting the uncertain moment of execution"; and calls for a freeze on nuclear weapons and their delivery systems, the banning of weapons from space, and a comprehensive nuclear test ban treaty (ref. 13.1). A petition supporting this Five Continent Peace Initiative, written by one of us and signed by 95 Nobel laureates worldwide, states:

> Human technology is now able to destroy our global civilization and perhaps our species as well. The lives of all those who inhabit the Earth today, and of all generations yet to come, are in jeopardy. Nations and peoples, even those far removed from the probable nuclear war target zone, now face unprecedented devastation. The danger of nuclear war cuts across religious, economic, social, and ideological boundaries. Whatever our aspirations, prospects, and ambitions for the future, whatever our

hopes for our children and their children, all are now imperiled by the prospect of nuclear war (ref. 13.2).

It is with reference to nuclear winter that Javier Perez de Cuellar, the Secretary-General of the United Nations, implored the member states on December 12, 1984:

> As Secretary General of this Organization, with no allegiance except to the common interest, I feel the question may justifiably be put to the leading nuclear Powers: By what right do they decide the fate of all humanity? From Scandinavia to Latin America, from Europe and Africa to the Far East, the destiny of every man and woman is affected by their actions. No one can expect to escape from the catastrophic consequences of a nuclear war on the fragile structure of our planet. The responsibility assumed by the great Powers is now no longer to their populations alone; it is to every country and every people, to all of us (ref. 13.3).

In an address to the United Nations General Assembly on September 25, 1984, the New Zealand Prime Minister, David Lange, stated:

> What the scientists have already made entirely plain to all of us—plainer than ever before—is that the nuclear weapons that may have helped to maintain an uneasy peace between two great countries for more than three decades, have become a threat to the security and survival of the countries and peoples everywhere (ref. 13.4).

Nuclear winter is then explicitly described.

Mr. Lange's criticisms were equitably distributed between the United States and the Soviet Union. In the same address he queried the latter:

> In the light of the devastation that would be caused by nuclear war, many countries, of which New Zealand is one, have the greatest difficulty in understanding [the U.S.S.R.'s] current reluctance to take part in bilateral arms control negotiations with the United States.

That reluctance soon disappeared. At that same United Nations General Assembly, representatives of many other nations drew political lessons from nuclear winter (ref. 13.5).

The first use that New Zealand made of the reparations given her by France—following the incident in which French commandos sank Greenpeace's *Rainbow Warrior* in Auckland harbor, killing one of its crew (to prevent the vessel from observing France's nuclear weapons tests near Tahiti)—was to fund a study of the consequences of nuclear winter for New Zealand (ref. 13.6).

In the United Nations Special Session on Disarmament, June 1, 1988, the Prime Minister of Sweden, Ingvar Carlsson, began with a description of nuclear winter drawn from the U.N. report (ref. 3.16):

> Since the end of the Second World War, a handful of nations have acquired the capability of destroying not only one another, but all others as well. The deployment of nuclear weapons has placed humanity under a threat which is unparalleled since the beginning of history. Nuclear arms are not just a more potent category of weapons. They are unique in the sense that their use may threaten the very survival of our civilization, and of mankind itself. . . . All countries, therefore, have not only the right but also the duty . . . to prevent the ultimate disaster.

Nuclear winter drives home, Mr. Carlsson said, the need for "common security," a term coined by his predecessor, Olof Palme:

> It means that in the nuclear age, one must find security together with one's adversary. It means one cannot build a safe world on the threat of mutual annihilation. It means that one cannot reach peace by frightening other countries.

In Britain nuclear winter has influenced opposition parties that may conceivably come to power in the future. The British Labour Party's 1986 Defence Policy Statement argues for conventional rather than nuclear deterrence in Europe, partly because of nuclear winter:

> Any significant nuclear exchange would produce a 'nuclear winter' in the Northern Hemisphere. Hundreds of millions of people would die from famine, and the collapse of life-supporting conditions. In the sure knowledge of what it would do to ourselves, and our country for generations, is it reasonable to believe any longer that either we, or the Americans, would launch the nuclear weapons to halt a Soviet invasion of Europe? (ref. 9.9).

Dennis Healey, former British Minister of Defense and member of the Labour shadow cabinet, was finally convinced of the futility of the British nuclear deterrent when he understood nuclear winter (ref. 13.7).

Sometimes the policy implications of nuclear winter lead in unexpected directions. In the early 1980s and before, Greek Prime Minister Andreas Papandreou was opposed to American military installations on his country's soil—the chief reason given being the consequent likelihood that, in an East/West conflict, Soviet nuclear warheads would also find their way to Greek soil. But in 1985 Mr. Papandreou began to tone down his objections. Opposition to American bases, he said, "is weakening with the passing of time, because the nuclear winter will finish us all whether we are bombarded or not" (ref. 13.8).

In the first few years after the discovery of nuclear winter, it was too much to hope that major changes in policy and doctrine would so soon have been implemented. Nevertheless, even then, there were signs, some of them considerably more than straws in the wind. For the Soviet Union, we are told (ref. 13.9) that as early as 1984/85 the Minister of Defense and the Foreign Minister were briefed on nuclear winter by Soviet scientists (ref. 13.10). Articles on nuclear winter have appeared in *Pravda*, *Izvestia*, and many other mass circulation periodicals, and it has been discussed repeatedly on All-Union television (ref. 13.11). A lecture on nuclear winter given, pre-glasnost, by one of us at Moscow University was reported in some detail in the Soviet press (ref. 13.12). Vladimir Petrovsky, Deputy Minister of Foreign Affairs, writes:

> The global nature of that threat is abundantly clear in terms such as "nuclear winter" and "nuclear night"—describing phenomena threatening our entire planet—that have recently gained international currency. The integrity and interdependence of the world mean that security is integral and interdependent, thus making it imperative to insure that security is universal (ref. 13.13).

Several popular and technical books and reports on the subject have also been published in the U.S.S.R. (ref. 13.14). Nuclear winter and its policy implications have been discussed at many international forums by highly placed government scientists, including Yevgeny Velikhov, Vice President of the Soviet Academy of Scientists and chief scientific adviser to President Gorbachev; Velikhov explicitly states that a nuclear war in which only 100 megatons is exploded over cities could produce nuclear winter, which he says "exposes the global danger of nuclear weapons for all mankind" (ref. 13.15).

There is also some reason to think that awareness of nuclear winter has permeated to the highest levels of Soviet policy-making. For example, in a television address on the Soviet unilateral nuclear test moratorium, delivered August 18, 1986, then–General Secretary (now President) Gorbachev stated:

> The explosion of even a small part of the existing nuclear arsenal would be a catastrophe, an irreversible catastrophe, and if someone still dares to make a first nuclear strike, he will doom himself to agonizing death, not even from a retaliatory strike, but from the consequences of the explosion of his own warheads (ref. 13.16).

In many other addresses, the Soviet leader has indicated that the extinction of the human species is a possible consequence of nuclear war. For example, in an address on February 16, 1987, to an international forum in Moscow, he commented,

> For centuries, men have been seeking after immortality. It is difficult to accept that every one of us is mortal. But to tolerate the doom of all humanity, is just impossible. . . . We reject any right for leaders of a country—be it the U.S.S.R., the U.S. or another—to pass a death sentence on mankind. We are not judges, and the billions of people are not criminals to be punished (ref. 13.17).

Views held around 1980 by members of the next American Administration—including Ronald Reagan and George Bush—on the "survivability" and "winnability" of nuclear war are instructively laid out by Robert Scheer in reference 4.7. In response to the argument that the massive size of the nuclear arsenals makes strategic "parity" meaningless, soon-to-be Vice President Bush replied,

> Yes, if you believe there is no such thing as a winner in a nuclear exchange, that argument makes a little sense. I don't believe that.

That is, Mr. Bush then believed that it was possible to win a nuclear war. "We have a different regard for human life than those monsters do" was Mr. Reagan's judgment on Soviet versus American nuclear war plans. Of the nuclear policymakers in the first Reagan Administration, Scheer writes,

> As I have come to know them I have been struck by this curious gap between the bloodiness of their rhetoric and the apparent absence on their part of any ability to visualize the physical consequences of what they advocate.

It is possible that nuclear winter played a role in changing such mindsets.

After the first public announcement of the nuclear winter findings in late 1983, there was more public discussion and debate on the subject (ref. 13.18), as well as—despite its low level of funding—a much stronger research program, in the United States and the West than in the Soviet Union and the East (ref. 13.19). Many briefings at high levels were given within the U.S. Government, beginning in early 1984 (refs. 13.20, 13.21) or earlier. Some, we are told, were addressed to one-way windows, behind which were officials unwilling to have it known that they had once heard something about nuclear winter—a peculiar kind of deniability. Secretary of Defense Caspar Weinberger reluctantly issued four Congressionally mandated assessments of nuclear winter, one of which acknowledges that even a mild form of nuclear winter might kill as many people worldwide as are killed by the direct effects of nuclear war (ref. 13.22). A White House position paper declares, "If deterrence were to fail, without a shield of any kind, it could cause the death of most of our population and the destruction of our nation as we know it" (ref. 13.23). Defense Department spokesmen, including former Assistant Secretary of Defense Richard Perle, have testified before Congress, also sometimes reluctantly, that nuclear winter is a serious threat. They have often concluded, however, that the only policy implication of this threat is to strengthen existing U.S. policies, especially SDI (ref. 13.24).

Nuclear winter has been embraced by some in the U.S. defense establishment because it seems to provide an argument for wholesale conversion of the strategic arsenals to low-yield, high-accuracy weapons, some of them with ground-penetrating or burrowing capability—which, it is argued, would greatly reduce the likelihood of nuclear winter (ref. 13.25). In fact, as discussed under Option (ii) in Chapter 9, deployment of this weapons system while maintaining anything remotely like the present strategic arsenals may *increase* the chances of nuclear war and nuclear winter. At any rate, the chief attraction of this weapons system conversion is not mainly to minimize the probability of nuclear winter; it is being developed to put Soviet underground shelters and command posts at risk. There are those who believe that such weapons would permit a nuclear war–fighting capability, or that an arms race for such weapons would be economically more debilitating for the Soviet Union than for the United States. (See also box, Chapter 18.)

That President Reagan was aware of nuclear winter in his second term is made clear in the following remarks he made in an interview published in the *New York Times*, February 12, 1985:

> A great many reputable scientists are telling us that such a [nuclear] war could just end up in no victory for anyone because we would wipe out the Earth as we know it. And if you think back to a couple of natural calamities—back in the last century, in the 1800's, just natural phenomena from earthquakes, or, I mean, volcanoes—we saw the weather so changed that there was snow in July in many temperate countries. And they called it "the year in which there was no summer." Now if one volcano can do that, what are we talking about with the whole nuclear exchange, the nuclear winter that scientists have been talking about?

He also announced, in his January 16, 1984, "Ivan and Anya" speech, that "A nuclear conflict could well be mankind's last." Nevertheless, at least as of 1986, no new policy guidance on nuclear war planning had been issued to or by the Joint Chiefs of Staff because of nuclear winter, and no such changes in guidance were planned (ref. 13.26).

But beginning with NUWEP-87 (Nuclear Weapons Employment Policy, 1987), a trend away from "economic targets" can be discerned (ref. 8.19). A broad-scale review of U.S. war plans by civilian officials and experts in the Department of Defense uncovered vastly more urban/industrial targeting than was necessary for deterrence or, if the worst came, for rapid termination of the war. Preventing Soviet postwar economic recovery was accordingly de-emphasized, and, in the words of then Air Force Chief of Staff, Gen. Larry Welch, "literally thousands of industrial targets have been dropped from the SIOP" (ref. 8.10). If the prospect of nuclear winter has played some role in the adoption of more restrained targeting options by the United States, we are glad—although the lack of interest in nuclear winter by most of those responsible for NUWEP-87 and SIOP-6F suggests otherwise. In any case, cities and petroleum facilities are still targeted on such an immense scale that only a small fraction of the nuclear explosions in a general war could suffice to bring on nuclear winter. Also, changes in targeting protocols are not subject to rigorous verification, and can readily be altered again. Real or purported softening of war plans without major reductions in force levels does not constitute an adequate policy response to the prospect of nuclear winter. Nothing is known publicly about whether nuclear winter has changed Soviet war plans.

Because of a perceived conventional superiority of Warsaw Treaty Organization (WTO) over NATO forces in Europe, the West has for decades announced itself obliged to depend on nuclear weapons for extended deterrence against a Soviet conventional attack. Thus, any claim, such as that offered by nuclear winter, of global catastrophe following nuclear war has been treated by some as biased against Western policy interests. It is useful to note, though, that this cuts both ways, as Prime Minister Papandreou's remarks (above) clearly show. In the past, the Soviet Union has not been shy about warning nations allied with the U.S., and others, that accepting U.S. bases makes them vulnerable to Soviet attack in the event of war; it has proposed that, instead, such nations accept Soviet guarantees of deterrence against attack by the West: "But it can hardly matter to a state whether it is the object of a Soviet threat or the beneficiary of a Soviet guarantee if the nuclear winter is known to recognize no neutrals and will devastate all states alike.* The whole technique for attempting to change the cor-

* [It will not. As we have seen, the nuclear winter effects of a mainly Northern midlatitude war will be less severe in the Southern Hemisphere; but perhaps not so much less as to negate this argument for the South.]

## PUBLIC OPINION ON NUCLEAR WINTER

Nuclear winter has focused the attention of people all over the world on the dangers of nuclear war and in many cases motivated them to do something about it—including a young girl from Maine named Samantha Smith. Opinion polls of Soviet and American teenagers show a pervasive belief in global catastrophe following nuclear war: "Do you believe that after a worldwide nuclear war the world will be so cold and dark and radioactive (known as nuclear winter) that almost no one will survive?": U.S. students, definite or probable yes—70%; Soviet students—79%.

"Do you believe the majority of the population can survive a worldwide nuclear war if there are enough fallout shelters, food, water, and other supplies?": U.S. students, definite or probable yes—25%; Soviet students—20%.

But such beliefs are hardly restricted to teenagers. Eighty-one percent of American adults believe (or at least did so in 1985) that a few years after a nuclear war there would be no one at all left alive in the United States and the Soviet Union.

In an in-depth Colorado poll, 62% of adult respondents held that nuclear winter would follow from a 3,000-warhead nuclear exchange, and 54% believed that in a major nuclear war 3 to 5 billion people would be killed. And in a New Zealand public opinion poll, the most serious consequences for New Zealand were given as radioactive fallout, nuclear winter, and food shortages (ref. 13.33).

relation of military power (conventional as well as nuclear) and political influence in favour of the U.S.S.R. is in jeopardy" because of nuclear winter (ref. 13.27). Indeed, subsequent events have shown the emergence of a very different U.S.S.R., one apparently much less predisposed to the coercive use of nuclear weapons.

When, during the first Reagan term, nuclear winter was discovered and its policy implications first drawn, the United States, as we have said, was embracing policies of nuclear "war-fighting," "containment" (so a small nuclear war would not escalate to a "central exchange"), and even "prevailing" after global thermonuclear war. The worldwide devastation and consequent self-deterring aspect of nuclear winter was viewed in some influential American circles as an impediment to then-fashionable strategic doctrines. Although those who played the major role in the discovery of nuclear winter were six Americans and a citizen of the Netherlands (see box, "Nuclear Winter: Early History and Prehistory," Chapter 3), there were those who thought Soviet confirmation of the theory suspicious. Others worried that nuclear winter would redound to the propaganda advantage of the Soviet Union (ref. 13.28). Similar misgivings about nuclear winter's American origins have been expressed in the U.S.S.R., including in debate at the Soviet Academy of Sciences (ref. 13.29). But now that the leaders of both nations have agreed that "nuclear war cannot be won and must never be fought," any propaganda asymmetry of nuclear winter is reduced, and the national origins of those who discovered it seem scarcely relevant.

We know that it's possible to see, in the totality of the foregoing remarks, nuclear winter as a mirror in which preexisting ideological or other biases are reflected back to the beholder (cf. refs. 2.7, 2.8, 3.1). We also know there is a danger of falling into the fallacy that academic logicians call *post hoc ergo propter hoc*: If Z follows Y in time, Z must be caused by Y. But beyond that, we believe it is also possible to see significant changes in the attitudes of both the United States and the Soviet Union on a wide range of matters—from declaratory policy to force levels to targeting—affected by nuclear winter (ref. 13.30). As Nobel laureate Luis Alvarez, one of the developers of the atomic bomb, put it shortly before his death: "There is some indication that [nuclear winter] is weakening the Soviet military's long-held belief that nuclear war is survivable. . . . It has had a very salutary effect on the thinking of military planners on both sides of the world" (ref. 13.31).

The prospect of nuclear winter may, thus, have played a role in reversing long-standing American and Soviet attitudes on the prospects of surviving, or even winning, a nuclear conflict; in challenging the notion that more nuclear weapons mean greater security; and in helping to thaw the Cold War (ref. 13.32).

CHAPTER 14

# DARKNESS AT NOON

## SIX CLASSES OF NUCLEAR WINTER

Oh dark, dark, dark, amid the blaze of noon.

—John Milton, *Samson Agonistes* (1671), 80

TO CARRY OUT ANY MEANINGFUL POLICY ANALYSIS OF NUclear war, the full range of possible environmental consequences must be weighed. In this chapter, we describe six classes of nuclear winter effects, each accompanied by a different degree of darkness at noon. These classes are characterized by six different values of the absorption optical depth averaged over the Northern Hemisphere, $\tau_a$, a measure of the quantity of overhead soot remaining in the atmosphere beyond the first few days of the war. (We introduced the concept of optical depth in Chapter 7 and in reference 7.10. Again, we consider only pure absorption of sunlight by smoke particles and ignore any additional effects due to scattering of sunlight off smoke or dust.) Here we treat nuclear wars that break out in the summer half-year (roughly March to September), when the immediate, "acute" climatic consequences are more severe. Wars initiated in the winter half-year have less severe immediate consequences for the climate and for agriculture because then it's cold already. A winter war with high values of $\tau_a$, and/or with many fine particles persisting at altitude into the next growing season, will of course have serious implications as well (ref. 14.1). It now seems clear that "chronic" long-term nuclear winter effects from stratospheric smoke and dust may persist for years (ref. 3.11).

The six classes of nuclear winter we propose are:

Class I. No significant environmental effects ($\tau_a$ less than 0.1). In this case, because of low levels of attack, functioning escalation controls (if such are possible), targeting that avoids cities, and/or unusual meteorological conditions, the combined environmental effects (cold, dark, radioactivity, pyrotoxins, etc.) have impacts much smaller than the direct effects within the warring nations, and insignificant impacts in noncombatant nations (compared, say, to the disruption of the world economy). This case is clear: it is nuclear war without nuclear winter, and provides one kind of test of the robustness of any nuclear winter policy recommendations—would they make sense even in the absence of nuclear winter?

Class II. "Marginal" nuclear winter ($\tau_a$ approximately 0.5).
This represents the case of comparatively few nuclear explosions, especially near urban centers (again corresponding to the dubious proposition that a nuclear war can be "contained"), and providentially low smoke emissions combined with efficient rainout of soot (cf. Figures 2, 4). The corresponding average Northern Hemisphere temperature declines would be only a few degrees Centigrade, although precipitation could be seriously perturbed. Major agricultural disturbances might result (refs. 3.11, 3.13). Famines worse than the worst "volcanic winter" cases discussed in Chapter 7 are possible. There would be a perceptible darkening of the sky. Some semblance of a harvest might be rescued in the lower American Midwest and the southern Ukraine, unless nuclear winter–induced drought were severe. Combined with the economic dislocation from the direct consequences of the war, even a mild nuclear winter could have serious consequences for noncombatant nations. Deaths from nuclear winter–related effects worldwide, though, would probably not reach the numbers killed directly by the war.

Class III. "Nominal" nuclear winter ($\tau_a$ approximately 1).
This is near the lower end of the most plausible range of environmental effects following a full-scale nuclear war (considered here to involve some 25 to 50% of the existing strategic arsenals—exploding roughly 3,000 to 6,000 strategic nuclear weapons on each side). It carries in its wake significant cooling and darkening, drought, massive quantities of pyrotoxins generated, widespread radioactive fallout, and other atmospheric perturbations. Average land temperature drops would be about 10°C. At noon, the Sun would have about one-third its usual brightness. Months later, sunlight would return to more than its usual intensity, enhanced in the ultraviolet by depletion of the high-altitude ozone layer. Collapse of agriculture, and fam-

ine, could be widespread. Within the warring nations, these effects might generate casualties approaching those from the prompt effects of the war. Crop failures—from lowered temperatures, failure of the monsoons, and other causes—are expected in many noncombatant nations in the first growing season following the conflict. The most likely such failures would be in India, China, some African nations, and perhaps Japan (ref. 3.11). Worldwide, as many as 1 to 2 billion people could be placed in jeopardy of starvation. The Southern Hemisphere, and most coastal regions and island states, would not experience major climatic disturbances.

Class IV. "Substantial" nuclear winter ($\tau_a$ approximately 3).

This is near the upper end of the most probable range of widespread climatic impacts of a full-scale nuclear war. It carries with it extensive continental land freezing, highly disturbed rainfall patterns, widespread radioactive and chemical toxicity, and severe Northern Hemisphere ozone depletion. The light that penetrates the smoke and reaches the ground is barely enough for green plants to make a living from photosynthesis. In the first few months, the sky in daytime is deeply overcast and starless at night. Ecological impacts could be harshest in the subtropical regions of the Northern Hemisphere (ref. 3.11). The climatic disturbances could persist, in diminishing extent, for years. The greatest burdens would fall on nations in the Northern Hemisphere, particularly in North America, Europe, and Asia, but also in Northern Africa. States such as Indonesia and the Philippines could also suffer severe crop losses (ref. 3.11). In the Southern Hemisphere, subtropical zones of Africa and South America and parts of Australia could experience agriculturally significant disturbances. The radioactive fallout and pyrotoxin hazards in noncombatant countries would have extremely serious consequences, but would probably be secondary to the climatic damage. Direct nuclear targeting in the Southern Hemisphere would deepen the climatic impacts there, and is the key determinant of nuclear winter severity in the South. The increased ultraviolet flux might threaten food supplies for years after. Worldwide, several billion people would be placed in danger of starvation over several years (ref. 3.11). The global civilization itself would be at significant risk. Depending on the course of the environmental sequelae, many species could face extinction, although human extinction might be only an outside possibility.

Class V. "Severe" nuclear winter ($\tau_a$ approximately 5).

Within the range of plausible nuclear war scenarios (ref. 14.2) and smoke emission parameters, more drastic environmental disturbances could occur. Simulations with larger-than-baseline smoke emissions (and absorption optical depths) have been carried out (first in ref. 2.2) but have not recently been widely discussed. Here, deep temperature declines could occur on all major land masses in any season, even in the tropics, imperiling many species and many key ecosystems (ref. 3.11). Less than 1% of the sunlight makes it through the smoke; it is twilight at noon for months, and there is not enough light for plant photosynthesis. Severe climatic effects could persist for years. Agriculture using surviving seed stores might be reduced to medieval or even pretechnological levels of productivity. The long-term widespread environmental destruction would decisively outweigh the direct effects of nuclear war; the prognosis for the quick revival of the global civilization would be grave. These effects, added to enhanced exposure to radioactivity, pyrotoxins, and pandemics, and, later, to solar ultraviolet radiation leaking through a greatly depleted ozone shield, might imperil everyone on Earth.

Class VI. "Extreme" nuclear winter ($\tau_a$ approximately 10 or more).

This is the extreme upper limit of what is possible under preferential targeting of cities, refineries, and petroleum repositories worldwide using almost all of the strategic and tactical arsenals. For months, there is darkness at noontime—about as dark as on a clear moonlit night back before there had been a nuclear war. A Class VI nuclear winter constitutes the direst possible nuclear assault on our own species and on the rest of life on Earth (ref. 14.3).

Many scientific studies and policy analyses of nuclear winter have been restricted essentially to the "nominal" case, Class III. This is partly because of the scientific interest in exploring the transition region in which significant climatic anomalies begin to occur. But it may also be due to a deeply felt reluctance to consider a global catastrophe of such magnitude as to challenge—clearly enough for anyone to grasp—the policy and doctrine to which the United States and the Soviet Union have become habituated. Sometimes this reluctance is expressed as scientific caution.

Our analysis of the evidence suggests that Classes III and IV, the "nominal" and "substantial" nuclear winters, are the most probable outcomes of a major nuclear exchange involving urban and petroleum targets (ref. 3.14), with Classes II, V, and VI representing less likely outcomes (ref. 14.4). The fatalities from a Class III or IV nuclear winter—never mind, for a moment, more severe outcomes—are compared with those from representative catastrophes in the accompanying box. Unless the probability of Class V (or VI) can be *quantitatively* demonstrated to be not just small, but vanishingly small, risk analysis demands that we give it special attention in making decisions on policy and doctrine; i.e., the value we attach to our civilization and species is so high that even small probabilities that we are placing them in jeopardy must be taken very seriously (cf. ref. 6.1). That's the way insurance policies are written.

Clearly it would be useful for strategic planners to know just what level of nuclear winter is likely to develop from what level of nuclear war. But such a connection is beset with difficulties. Much depends on the targeting strategy: a "small" nuclear war in which hundreds of cities are burning is likely to generate a much more severe nuclear winter than a "large" one in which thousands of hardened targets are attacked, but only a few cities are burning (refs. 2.2, 14.2). There are uncertainties connected with the yield and burst height of nuclear weapons, the season (for the less severe cases, the summer half-year is assumed here), local weather conditions, the pace of the nuclear exchange, etc. In the original TTAPS study we analyzed dozens of different nuclear war scenarios and targeting plans—which, we have reason to believe, are consistent with the nuclear war plans of the United States and the Soviet Union. The key variable seems to be targeting, a spectrum ranging from a pure "countervalue" war, in which only cities are attacked, to a pure "counterforce" war, in which only hardened targets are attacked, and in which no cities or petroleum facilities burn.

# REPRESENTATIVE HUMAN CATASTROPHES

| *Cause* | *What/Where* | *When* | *Fatalities* |
|---|---|---|---|
| Nuclear power plant accident | Chernobyl, U.S.S.R. | 1986 | 100? |
| Accidental chemical explosion | Halifax Harbor, Canada | 1917 | 1,654 |
| Chemical spill and venting | Bhopal, India | 1984 | 3,500? |
| Volcanic eruption | Mt. Tambora, Indonesia | 1815 | 160,000 |
| Nuclear weapon explosion | Hiroshima, Japan | 1945 | 200,000? |
| Anomalous weather | Cyclone, Bangladesh | 1970 | 300,000 |
| Earthquake | Shaanxi, China | 1556 | 830,000? |
| Flood | Huang He River Basin, China | 1931 | 3,700,000 |
| Famine | North China | 1876–79 | 10,000,000? |
| World War I | Mainly Europe | 1914–18 | 20,000,000 |
| Plague pandemic, "Black Death" | Europe | 1347–51 | 25,000,000 |
| World War II | Worldwide | 1939–45 | 40,000,000 |

| | | | |
|---|---|---|---|
| Nuclear war | Nuclear winter, Class III or Class IV, worldwide | ? | 3,000,000,000? |

Estimated Tambora fatalities include immediate deaths (10,000), subsequent local epidemics and famine (80,000), and weather-related deaths worldwide (J. W. Wright, ed., *The Universal Almanac* [Kansas City: Andrews and McMeel, 1990]). "Black Death" fatalities worldwide are suspected to have been much greater than the European values listed here. World War I and II fatalities include both military and civilian victims (Ruth Leger Sivard, *World Military and Social Expenditures, 1987–1988* [Washington, D.C.: World Priorities, 1987]). Examples chosen are biased toward recent times, when better data are available.

Clearly a pure countervalue war and a pure counterforce war are equally unrealistic abstractions, the one because of the military need to limit the adversary's strategic retaliation, and the other because of the proximity of cities to strategic targets. A real war will be some indeterminate mix of the two cases.

In Figure 6 we give a crude measure of the connection between the number of warheads, N, detonated in a nuclear war, and the severity of the war, measured by $\tau_a$ or class number (e.g., Class III), for several target categories. In each case the severity of the climatic and other environmental effects increases as the number of detonated warheads increases. Eventually the curves flatten out as the number of warheads increases, because there are only so many targets—especially targets with large concentrations of combustible material.

The lower curve, for rural military facilities only, represents an extremely idealized nuclear war, much more stringent in its targeting than even a pure counterforce war. It assumes that only military targets far from cities, and from petroleum refineries and depots, are attacked. Because many command and control centers, primary and secondary strategic airfields, and nuclear submarine bases are near cities, this would constitute an exceptionally foolish targeting strategy—amounting to provoking a major retaliation in the course of destroying only a

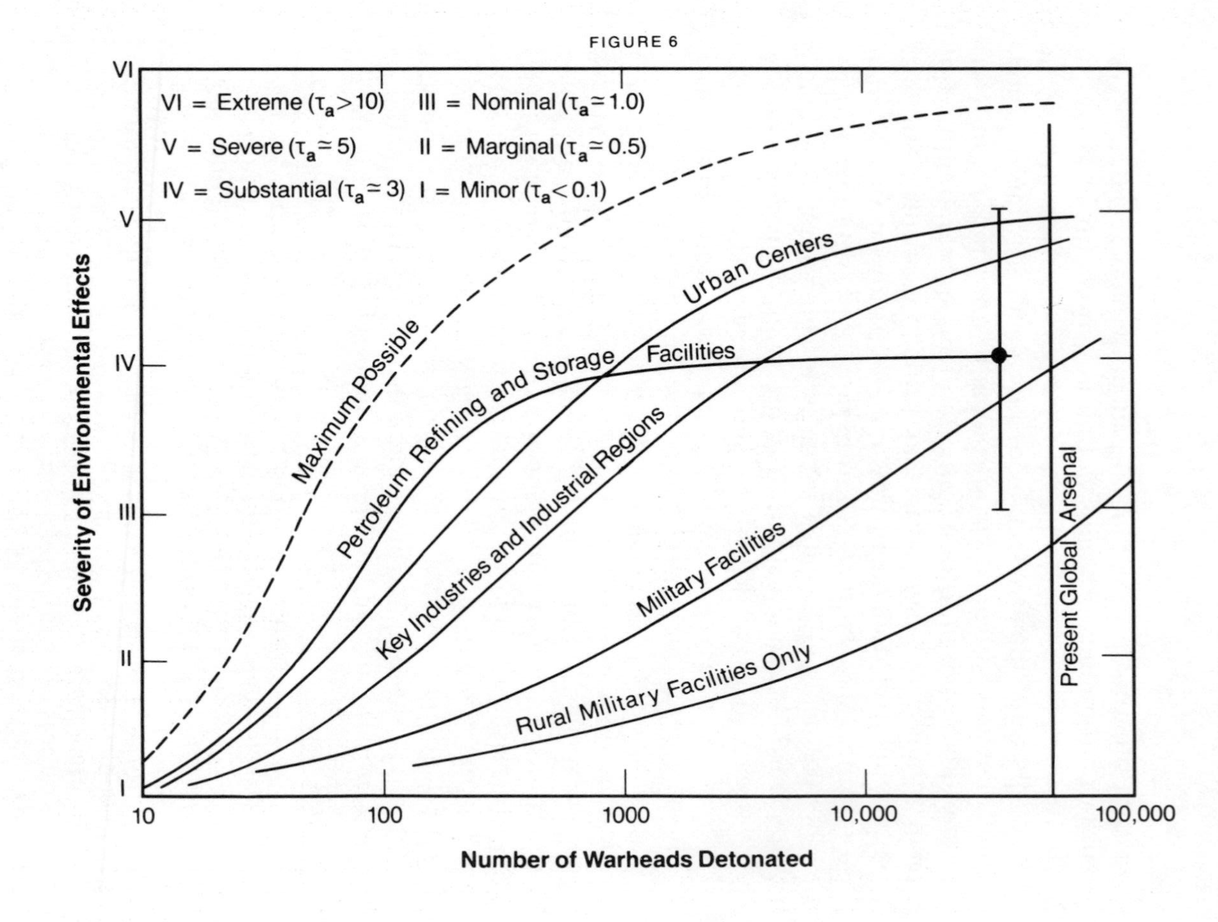
FIGURE 6
VI = Extreme ($\tau_a$>10) III = Nominal ($\tau_a \simeq 1.0$)
V = Severe ($\tau_a \simeq 5$) II = Marginal ($\tau_a \simeq 0.5$)
IV = Substantial ($\tau_a \simeq 3$) I = Minor ($\tau_a$<0.1)
Maximum Possible
Urban Centers
Petroleum Refining and Storage Facilities
Key Industries and Industrial Regions
Military Facilities
Rural Military Facilities Only
Present Global Arsenal
Severity of Environmental Effects
VI
V
IV
III
II
I
10
100
1000
10,000
100,000
Number of Warheads Detonated

The severity of nuclear winter depends chiefly on how many nuclear weapons are exploded on what targets. The more cities and petroleum facilities in flames, the worse the climatic damage is. In this schematic representation, the upper dashed curve roughly indicates the maximum likely climatic effect for any combination of targets, given present uncertainties on fuel loadings, soot emission factors, etc. A measure of increasing nuclear winter severity, Classes I to VI—ranging from insignificant to apocalyptic global consequences—is discussed in the text. The curves tend to flatten out as the number of warheads increases, because the number of targets and quantities of combustible materials are eventually nearly used up. The rural and military targets lead to substantial dust generation, but require many high-yield groundbursts (consistent with present war plans and strategic systems) to cause significant global effects. The industrial, urban, and petroleum targets are characterized by combustible materials highly concentrated at relatively few sites; this is why global nuclear winter may be generated with only a few hundred detonations or less. As discussed earlier, petroleum refineries exhibit the greatest climatic sensitivity with the fewest detonations (even for individual warhead sizes of around 1 to 10 kilotons—i.e., tactical nuclear weapons). We estimate the probable error in our estimates to be ± one nuclear winter class, as shown by the sample error bar at the right of the curve for petroleum targets. The vertical line at the right indicates the present total U.S./Soviet nuclear weapons inventory, including both strategic and tactical warheads. The British and French nuclear forces (being expanded to around 1000 and 700 strategic warheads respectively) will each be capable of causing a nuclear winter with pure urban targeting. Nuclear winter may be within the reach of China as well, with about 350 operational strategic warheads (possibly to be increased to nearly 1,000 if MIRVing the CSS-2 missile takes place as anticipated). It is clear that a few hundred warheads, if specifically aimed at fire-sensitive targets, might exceed the threshold for an environmental catastrophe unprecedented in the tenure of humans on Earth.

small fraction of the adversary's retaliatory force. Nuclear winter is here brought about because of the generation of stratospheric dust from groundbursts as well as smoke from burning vegetation; but to induce significant global effects, this requires a very large number of explosions (with yields of several hundred kilotons or more each—consistent with present and currently planned strategic systems). With such a targeting strategy, you need something like all the world's nuclear weapons, strategic and tactical, to generate a nominal nuclear winter.

The curve marked "Military Facilities" is much closer to a pure counterforce war. Only military facilities are targeted, and no urban areas are attacked "per se." However, because of the proximity or co-location of strategic targets to cities, a number of cities are nevertheless in flames. The shape of this curve has been derived assuming that great care is taken to minimize, as nearly as possible, urban targeting—consistent with the (ultimately quixotic) mission of destroying the adversary's retaliatory capability but not its civilian population.

We believe such scrupulous targeting to be unachievable in an actual nuclear war—because of inaccuracies in nuclear weapons systems that have never been tested in combat,* because of the perceived necessity to make a major commitment of nuclear forces before command and control failures; and because of the wrenching emotional stresses that would burden civilian and military leaders. In this case a nominal (Class III) nuclear winter could be generated with a little more than 10% of the world's arsenals, and the full world arsenals could produce a substantial (Class IV) nuclear winter.

Nuclear attacks on key industries and industrial regions, on urban centers, and on petroleum refining and storage facilities—all intentionally, or "per se"—are targeting strategies seldom discussed. But, as we have mentioned, they are consistent with war-fighting aims and with the goal of dominating other nations in the postwar environment. These are countervalue attacks; large-scale countervalue targeting represents the ultimate failure of escalation control. In any case, some nuclear detonations on such targets are inevitable in a large nuclear war, again because of proximity to strategic targets. These upper three solid

* U.S. ICBMs have never, not once, been test-fired from their operational silos.

curves illustrate the most unexpected and troubling implication of nuclear winter—that only a few hundred nuclear detonations, or less, seem sufficient to bring about at least a nominal nuclear winter. Only 100 small warheads devoted to petroleum refining and storage facilities would suffice. Indeed, with something like a hundred downtowns burning, or the same number of petroleum facilities, even a substantial nuclear winter seems possible.

Urban and petroleum targets are characterized by high concentrations of flammable materials in a relatively small area; this is why they have the potential to create a global nuclear winter with a modest number of detonations. As we discussed earlier, petroleum refineries exhibit the greatest climatic sensitivity for the least number of detonations (even for tactical—that is, about 1 to 10 kiloton yields—rather than strategic weapons). The British and French nuclear forces (with about 1,000 and many hundreds of strategic warheads, respectively, when their present "upgrading" and "modernization" is completed) are seen to be capable of causing a marginal and perhaps even a nominal nuclear winter with only about one-third of the warheads dedicated to urban targets in the Soviet Union and, perhaps, elsewhere. If nuclear war is waged by one side only, more total warheads would be required to produce a given effect—because the quantity of combustible materials and the resulting smoke emission tend to diminish as the more abundant smaller targets, which generally have less in them that can burn, are selected. If both sides are attacked, there are more big flammable targets.

The upper, dashed curve of Figure 6 roughly indicates the *maximum* likely climatic effect for the most dangerous combination of targeting—given adverse values of the present uncertainties on concentrations of combustible materials, soot emission factors, etc. This, and no other, is the curve of the worst case. Thus, within the uncertainties of present knowledge, it may be possible that only a few percent of the world's strategic arsenals would be able to produce a severe, Class V nuclear winter, and the apocalyptic Class VI nuclear winter may be just within reach of the full global inventory of nuclear weapons.

Also note, from the same upper dashed curve, that—with the most adverse targeting, and the unluckiest values of incompletely known physical parameters (how much soot is generated, for example, or how much of it is quickly rained out)—as few as 50 or 100 warheads detonated might cause a Class III nuclear winter.

The present total global nuclear arsenals (strategic and tactical weapons together) are shown by the vertical line to the right of the figure. If the strategic arms reduction talks (START) are fully implemented, that vertical line would move slightly, just a little more than perceptibly, to the left.

Two principal conclusions that can be drawn from this figure are: (1) everything depends on targeting, although there seems no reliable way to verify an adversary's targeting strategy; and (2) without absolutely reliable constraints on targeting or continental-scale defenses, the number of nuclear weapons in the world is 100 or even 1000 times more than what is required to produce a nominal nuclear winter. Since such constraints and defenses are unachievable, the only reliable way to prevent nuclear winter and darkness at noon is to make major and verified cuts in the global nuclear arsenals.

Massive reductions in the nuclear arsenals is thus not a favor the Americans do for the Soviets or vice versa. It is not a reward for the other side's good behavior. It is something we do for ourselves and for the human species. It is, in the most complete and literal sense, a selfish act on behalf of every nation and every person on Earth.

# CHAPTER 15

# A FURNACE FOR YOUR FOE

Be advised.
Heat not a furnace for your foe so hot
That it do singe yourself.

—Norfolk, in William Shakespeare, *King Henry VIII*, Act 1, Scene 1

WE NOW PAUSE TO TAKE ANOTHER LOOK AT DETERRENCE in the light of nuclear winter. As mentioned, it has been fashionable for some time to assert that the only purpose of nuclear weapons is to guarantee that they will never be used. "Peace Is Our Profession" has been the slogan of the Strategic Air Command—the organization responsible for all U.S. land-based missiles and all strategic bombers—since its inception.* Nevertheless, nuclear weapons also can be, and in some cases have been, employed in attempts to (a) deter conventional attack by a potential adversary, or (b) influence decisions of (or, equivalently, extort concessions from) other nations, including those without nuclear weapons (ref. 15.1). But only a small number of nuclear weapons would in principle suffice for both objectives (a) and (b), and therefore these uses of nuclear weapons could, if we wish, be maintained without threatening nuclear winter (ref. 15.2). For present purposes we consider the actual prime functions of nuclear weapons to be deterrence of both nuclear and conventional military action. Even those who advocate nuclear "war-fighting" capabilities agree that deterrence is their goal (ref. 15.3).

In terms of weapons, two distinct varieties of nuclear deterrence are currently discerned: a strategic nuclear deterrent (SND) involving long-range, generally high-yield systems and a tactical nuclear deterrent (TND) involving short- and intermediate-range low-yield systems, especially in Europe and at sea. Some believe this distinction to be illusory. Admiral Noel Gayler (ref. 15.4) argues:

* Until March 15, 1990, when it changed a little, becoming "War Is Our Profession"—to which was added the hopeful afterthought: "Peace Is Our Product."

From the weapons/forces standpoint, there is no real distinction to be drawn between TND and SND. They are on a continuum, except in literary treatments. The only real-world distinction is between nuclear and non-nuclear.

What about from a nuclear winter standpoint?

Until the INF Treaty, the U.S./NATO versions of these two deterrents were tightly coupled by the presence of the Pershing missile force in Europe. Tactical weapons represent, in the ladder of escalation, the first potential nuclear response to conventional military action: "If you can't hold 'em back, use tacs." Nominally, nuclear winter does not strengthen TND, nor does it weaken it, inasmuch as no major environmental effects would be expected from a limited tactical nuclear war, if such were possible. There are still so many tactical weapons in densely urbanized Europe, however, that an extensive tactical nuclear war in which no weapons were detonated on American or Soviet soil (were *such* a war possible) might nevertheless bring about a hemispheric nuclear winter—because a single tactical warhead can ignite the largest petroleum facility and several such warheads can set a large city burning. Moreover, any more than token strategic response to a tactical/conventional conflict could trigger nuclear winter, with debilitating consequences for both sides. A strategic first strike is clearly made less attractive by nuclear winter. However, because of possible escalation from a conventional attack to a tactical nuclear response, leading soon after to strategic war and nuclear winter, deterrence of conventional aggression is also improved by knowledge of nuclear winter. Nuclear winter lends no support to the idea of a sharp distinction between tactical and strategic deterrence.

Nuclear winter can enhance strategic deterrence in the following ways:

A. Nuclear winter increases the *uncertainty* in the outcome of a nuclear first strike, or of an exchange:

(i) Through short-term obscuration of the lower atmosphere, degrading surveillance and intelligence missions —including terminal guidance and damage assessment, necessary for successful battle management; and

(ii) Through long-term environmental effects (even for purely counterforce attacks) that may determine the eventual degree of "success" of a specific military action, or preclude anything recognizable as victory in a major nuclear conflict.

B. Nuclear winter increases the ultimate societal *costs* of employing nuclear weapons, and does so equitably by distributing the costs among the several parties to the conflict, and by assuring that the costs cannot be avoided (although some effects might be mitigated); indeed, nuclear winter is a liability in all conceivable cases, and may represent the greatest cost of the war in some, not unlikely, scenarios.

C. Nuclear winter encourages *caution* with regard to specific targeting plans (for example, where cities, or petroleum facilities, or massive barrages against mobile missiles might be involved) and positive escalation control. More generally, nuclear winter further and strongly inhibits any actions that might lead to nuclear war.

D. Thus, nuclear winter raises *doubts* about political/military concepts of "surviving" a nuclear war, much less of "prevailing" in such a conflict. For larger optical depths (Class IV nuclear winter and worse), it even calls into question the idea of national recovery from nuclear war.

E. By greatly enhancing the devastation of nuclear war, nuclear winter also lessens the perceived dependence of stable deterrence on large stockpiles (ref. 15.5). Weapons that cannot be used are ineffective agents of retaliation or coercion. Much smaller inventories can then be considered for maintaining strategic stability.

How might each class of nuclear winter affect deterrence? We are here considering the range of likely climatic responses for a given set of targeting decisions. For retaliation by the atmosphere, as for retaliation by the adversary, "the threat of retaliation does not have to be 100 per cent certain; it is sufficient that there is a good chance of it, or if there is belief that there is a good chance of it" (ref. 15.6). Without any threat of large-scale environmental consequences (Class I), nuclear winter is of course irrelevant to deterrence. In the case of marginal or nominal nuclear winter (Classes II and III), deterrence is strengthened, perhaps to the point of self-deterrence against a first strike. Moreover, given the incremental concern over even a marginal nuclear winter, attacks on many urban targets that otherwise would be destroyed might be withheld, saving millions of lives. The proximity of cities and strategic targets in turn, limits attack options, leaving the attacked nation in a much stronger position for post-war recovery. Accordingly, even the prospect of Class II nuclear winter renders the aggressor's perceived advantage in initiating nuclear war less tenable. This is increasingly so for more severe nuclear winters.

A nominal (Class III) nuclear winter additionally promises serious consequences for noncombatants. If postwar geopolitical alignments are perceived as important (ref. 15.7), then the weakening or loss of support from key noncombatants would be a serious impediment to plans for surviving or prevailing—although this desire for postwar support coexists with a desire that all other surviving nations be weak, as discussed above. There is of course grave moral ambiguity in placing at risk the lives of hundreds of millions of people in other nations in an attempt to preserve favored domestic political and economic institutions in the combatant nations. "You must die so I'll be free" is a slogan with limited appeal.

Class IV (substantial nuclear winter) would further reinforce mutual deterrence, for the reasons stated above. Clearly, if Class IV is one of the likely outcomes of a central exchange (as our present knowledge of the underlying science seems to indicate), and if this fact were widely understood, major self-deterrence would follow.

Severe nuclear winter (Class V) or worse would represent the fabled "Doomsday Machine" (ref. 16.1). No rational leader could begin to contemplate nuclear war if Class V were at all likely. Unfortunately, not all leaders are rational. The prospect of severe nuclear winter suggests that we are trusting the human future to the reliability of machines and the sobriety and sanity of military and civilian leaders into the indefinite future and for a steadily increasing number of nations. Just think of Hitler and Stalin. Madmen can take control of modern industrial states. They can even do so without gross violations of the legal norms (ref. 15.8). Over time, the probability of madmen in key political or military positions in nuclear-armed nations approaches a certainty. Severe nuclear winter implies that the present strategic arsenals and doctrines may constitute a Doomsday Machine with the timer already set and counting. What is the prudent policy response to this prospect?

# CHAPTER 16

# THE DOOMSDAY MACHINE

The "Doomsday Machine" which could exterminate us all could already be constructed. For aught we know, it *has* already been constructed.

—Bertrand Russell, *Has Man a Future*? (New York: Simon and Schuster, 1962), 69

For more than a year, ominous rumors had been privately circulating among high-level Western leaders that the Soviet Union had been at work on what was darkly hinted to be the Ultimate Weapon, a Doomsday device. Intelligence sources traced the site of the top secret Russian project to the perpetually fog-shrouded wasteland below the arctic peaks of the Zhokhov Islands.

—Opening narration of Stanley Kubrick's film *Dr. Strangelove* (Columbia Pictures, 1963; written by Stanley Kubrick, Terry Southern, and Peter George)

HERMAN KAHN OF THE RAND CORPORATION AND THE Hudson Institute—two leading think tanks that advise the U.S. military—was an influential American nuclear strategist. It was he, for example, who originated the terms "counterforce" (for destroying the adversary's retaliatory capability) and "countervalue" (for destroying the adversary's cities). His high-level classified briefings, reports, and books were instrumental in the evolution of both U.S. and Soviet nuclear policy. In his book *On Thermonuclear War*, first published in 1960, Kahn introduced the idea of a "Doomsday Machine." He believed that such a device would be "hard" to build in the 1960s, but much easier by the 1980s and '90s. It would go off, on its own, if a nuclear war were started—no matter by whom or why—and would kill "one or two" billion people. Or more. It could not be reasoned with. Once it was activated, even its builders could not alter its irrevocable purpose.

Its function was deterrence. Who would start a nuclear war knowing that Doomsday was a likely outcome? But to deter, all potential adversaries had to know about it. It would be missing the point of a Doomsday Machine to keep its existence secret.

An ideal Doomsday Machine had to satisfy the following criteria: "frightening"; "inexorable"; "automatic" (so the device "eliminates the human element, including any possibility of a loss of resolve as a result of either humanitarian consideration or threats by the enemy");* "persuasive" ("Even an idiot

* James R. Newman, in a review for *Scientific American*, described *On Thermonuclear War* as "a moral tract on mass murder: how to plan it, how to commit it, how to get away with it, how to justify it." Kahn himself commented that "It is the hallmark of the expert professional that he doesn't care where he is going as long as he proceeds

should be able to understand [its] capabilities"); and "fool-proof" (meaning that it would have a very low probability of going off *before* a nuclear war) (ref. 16.1). By these criteria, the present world nuclear arsenals *do* constitute at least a conditional Doomsday Machine—as did the arsenals of the '60s and '70s. Nuclear winters in the middle range of severity and worse will likely kill as many people as Kahn's hypothetical Doomsday Machine. Presciently, he thought that the most probable doorway to Doomsday would be "the creation of really large amounts of radioactivity or the causing of major climate changes." He did see some difficulties, though:

"The difficulties lie . . . ," Kahn wrote, in the fact that the

> Doomsday Machine is not sufficiently *controllable* . . . [A] failure kills too many people and kills them too automatically. There is no chance of human intervention, control, and final decision. And even if we give up the computer and make the Doomsday Machine reliably controllable by the decision makers, it is still not controllable enough. Neither NATO nor the United States, and possibly not even the Soviet Union, would be willing to spend billions of dollars to give a few individuals this particular kind of life and death power over the entire world.

So neither Americans nor Western Europeans "should or would be willing to design and procure a security system in which a malfunction or failure would cause the death of one or two billion people. If the choice were made explicit, the United States or NATO would seriously consider 'lower quality' systems; i.e., systems that were less deterring, but whose consequences were less catastrophic if deterrence failed."

And this caution is not limited to the general public:

> I have been surprised at the unanimity with which the notion of the unacceptability of a Doomsday Machine is

---

competently." But it is important not to regard Kahn as some anomalous moral monster because of these sentiments. His is a common, and perhaps irreducible, mode of military thinking. Clausewitz wrote of war: "This is the way in which the matter must be viewed, and it is to no purpose, it is even against one's better interest, to turn away from the consideration of the real nature of the affair because the horror of its elements excites repugnance" (ref. 16.2). The problem is not the strategists drawn to and molded by war, current or prospective, so much as the institution of war itself.

> greeted. . . . Except by some scientists and engineers who have overemphasized the single objective of maximizing the effectiveness of deterrence, the device is universally rejected. It just does not look professional to senior military officers, and it looks even worse to senior civilians. . . .
>
> The closer a weapons system is to a Doomsday Machine, the less satisfactory it becomes. . . .
>
> In addition to making clear that their weapon systems are useful—that is, that they can do something—the military must also make clear that the weapon systems are not overly powerful. . . . They must not be Doomsday Machines. . . . They must not even look like they could be a Doomsday Machine if misused, much less when used as authorized.
>
> I cannot stress [too much] how important it is that it be made clear that we are not developing and planning the use of Doomsday Machines—or even systems which, if they should be used (whether they are good deterrents or not), will destroy the defender and a large portion of the world along with the aggressor. . . .

In *Dr. Strangelove,* Stanley Kubrick puts Kahnian doctrine into the mouth of a U.S. Air Force General named Turgidson: "I'm not saying we wouldn't get our hair mussed. I am saying only ten to twenty million people killed, tops, depending on the breaks." When the death count got up to a billion, though, even Herman Kahn began to be moved by the very same humanitarian considerations that he otherwise viewed as impediments to effective deterrence. At one "gigadeath" (= 1,000 "megadeaths" = 1,000,000,000 people killed) he was willing to change his mind about strategic deterrence. Preventing the rise of Doomsday Machines was, he wrote, a "central problem of arms control—perhaps *the* central problem."

It is a bitter irony that the Doomsday Machine against which Kahn warned may have been slowly constructed because of the very policies that he advocated—and with the most dangerous possible flaw of Doomsday Machines built in: since nuclear winter had not yet been discovered, no one knew that a Doomsday Machine had been devised.

Herman Kahn died in 1983, as the early results on nuclear winter were first being discussed among nuclear strategists. He was just completing a book (ref. 16.3) that was published posthumously, with a laudatory introduction by Gen. Brent Scow-

## DR. STRANGELOVE: NUCLEAR DOOMSDAY IN POPULAR CULTURE

As a U.S. strategic bomber loaded with nuclear weapons begins an unauthorized attack on the Soviet Union, the President, in the War Room, queries the Commander of the Strategic Air Command:

> PRESIDENT: General Turgidson: When you instituted the human reliability tests, you assured me there was no possibility of such a thing ever occurring.
>
> TURGIDSON: [*Offended*] Well, I—er—don't think it's quite fair to condemn a whole program because of a single slip-up, sir.

But since a small nuclear attack is going to happen anyway, Turgidson proposes taking advantage of the "opportunity" by converting it into a massive first strike. He has not reckoned with the fact that this will set off the Soviet Doomsday Machine, because the Russians haven't yet gotten around to telling anybody about it.

—From Stanley Kubrick's film *Dr. Strangelove*, written by Stanley Kubrick, Terry Southern, and Peter George (Columbia Pictures, 1963).

croft. The editors go so far as to add in a footnote, "If the nuclear winter theory turns out to be correct, it would have implications for some of the scenarios discussed in this book." This seems excessively laconic. As the remarks quoted above indicate, the only consistent Kahnian policy response to the discovery that we may inadvertently have built a Doomsday Machine is to destroy it. Even a conspiracy of madmen at the helms of the various nuclear-armed nation-states, however unlikely that might be, should not be able to visit nuclear winter on the human species.

The only way to achieve such a degree of safety is to reduce the world nuclear stockpiles to a level at which nuclear winter cannot occur. This is what it means to dismantle the Doomsday Machine. Massive, worldwide destruction of nuclear weapons is therefore essential for national security—the very icon at whose altar the worldwide buildup of nuclear weapons was laid. As Kahn emphasized, there are compelling political and strategic reasons to make sure that not a hint of a Doomsday Machine sits in the nuclear arsenals: "They must not even look like they could be a Doomsday Machine if misused, much less when used as authorized."

Massive nuclear arms reduction is a serious undertaking. It enters uncharted waters. It is not without its dangers. But "against a great evil," John Stuart Mill said, "a small remedy does not produce a small result. It produces no result at all." If we find ourselves anywhere near constructing a Doomsday Machine, we must stop and dismantle it, Kahn stressed, even if in so doing we decrease the reliability and credibility of deterrence. Fortunately, as we argue below, properly configured strategic arsenals at comparatively low levels can increase, not decrease, crisis stability and national security.

CHAPTER 17

# IS INFINITY ENOUGH?

## MINIMUM SUFFICIENT DETERRENCE (MSD)

A war, therefore, which might cause the destruction of both parties at once . . . would permit the conclusion of a perpetual peace only upon the vast burial-ground of the human species.

—Immanuel Kant, *Perpetual Peace* (1795), I, 6

MAJOR CUTS IN THE WORLD NUCLEAR ARSENALS ARE IN the interest of everyone on earth. They give no special advantage to the Americans or the Russians or anybody else. They are not driven by politics or ideology or national allegiance—only by our interest in staying alive. If we like, we can accept the posture of nuclear deterrence. We can even consider it a good idea to threaten some other country's largest cities. But don't we want to make sure that no conceivable circumstance, no computer malfunction, no mad leader, no intelligence or communications failure, *nothing* could destroy our global civilization and endanger our species? This isn't an argument about who's right in the quarrels between the big powers; it isn't even a question of nationalism versus world order. It's only a matter of everyone—including the billions of people with no part in whatever dispute or mistake or misunderstanding that would lead to nuclear war—having a right to a future.

In the early 1980s, one of us asked a leading scientific adviser to the Soviet government why the U.S.S.R. felt compelled to have as many nuclear weapons and delivery systems as the U.S., when a much smaller number would suffice for deterrence. He pulled out a notebook and wrote down his answer: "$\infty = \infty$," infinity equals infinity.

It is now widely recognized (ref. 17.1) that existing nuclear forces far exceed what is actually required for deterrence, a situation often described as "overkill." Nevertheless, whatever anxieties you may have (or, more important, whatever anxieties your adversary may have) about the reliability and effectiveness of your arsenal are calmed if you have manyfold redundancy—effectively an infinite arsenal. Reducing the arsenals, therefore, necessarily raises anxieties about the stability of deterrence in a crisis. Would both sides be convinced of the inescapability of devastation if retaliation relied upon many fewer weapons?

The comparatively small British, French, and Chinese arsenals—each with a few hundred warheads—represent a *de facto* demonstration, in the real world, that small strategic arsenals *can* constitute adequate deterrence, or at least that national leaders believe they can. The one clear example we have of the United States and the Soviet Union stepping back from the brink of nuclear war is the Cuban Missile Crisis of 1962. Each side was deterred by the nuclear forces of the other. But the Soviet nuclear arsenal then comprised only about 300 warheads. "We were deterred, plain and simple," concluded Defense Secretary Robert McNamara. "The lessons of the missile crisis are simple. . . . It takes very, very few [weapons to deter]."

The prospect of nuclear winter even at its worst does not by itself remove the danger of nuclear war, because the world we live in is replete with accident, unlucky coincidence, incompetence, and madness. Nuclear winter does, though, tend to undermine the current doctrine of deterrence. With nearly 60,000 nuclear weapons in the world, how could the threat of a massive first strike or of massive retaliation be believable, if it inexorably puts the attacker in mortal peril? Nuclear winter suggests that deterrence is credible only with much smaller weapons inventories. If a threat to use the current massive arsenals is empty, or perceived to be empty—because of self-deterrence—and if, despite the danger, a small nuclear war is able or even likely to escalate to a central exchange (ref. 9.3), then the American and Soviet nuclear forces are stripped of much of their vaunted deterrent function. Their willingness to engage in even a "limited" nuclear war is in doubt. Shouldn't partisans of deterrence, then, become advocates of minimum sufficiency?

These facts and implications motivate us to ask how force structures could be modified to maintain or strengthen deterrence while reducing the risk of global climatic catastrophe (ref. 17.2). It has been argued that "minimum sufficiency" (see box) in the nuclear arsenals combined with flexible targeting is morally a far superior posture to the present predicament (ref. 17.3). President Dwight Eisenhower went so far as to argue that "if we cut back on armaments to where only a retaliatory force is left, war becomes completely futile" (ref. 17.4).

The potential value of *unilateral* steps can be debated, with nearly persuasive arguments on both sides. In a dizzying succession of unilateral steps in the late 1980s, the Soviets announced that they would make massive withdrawals of troops and armor from Europe (they are well on their way to implementation); remove all overseas military bases by the year 2000; dismantle the Krasnoyarsk radar because it is in violation of the Anti-Ballistic Missile (ABM) Treaty; cut tank production in half; slow their modernization of nuclear weapons; and no longer militarily intervene in other nations. The U.S.S.R. has not tested anti-satellite weapons since August 1983; it ended its invasion of Afghanistan; for eighteen months it halted all nuclear weapons tests, hoping (in vain) for a responsive U.S. moratorium; it has made major steps toward democracy (with a few lurches backward), and permitted, in neighboring countries, still more major steps that severely undermine the coherence and military effectiveness of the Warsaw Treaty Organization. The Soviets have—with no perceptible harm so far—embraced unilateral arms control measures. They cannot continue indefinitely, however, without a response in kind by the United States. They have their conservatives, hard-liners, militarists, nationalists, paranoids, and strategists who remember Munich and Hitlerian blitzkrieg, just as the United States does. But it is unnecessary for the U.S. to consider unilateral steps. There are now uniquely promising opportunities for joint U.S./Soviet *bilateral* steps, eventually to be expanded to include the other nuclear-armed and nuclear-capable nations. This is beginning to happen, but with a scope and pace still incommensurate to the seriousness and urgency of the problem.

The prospect of nuclear winter puts at least all Northern Hemisphere nations in a comparably vulnerable state and therefore on notice. Every nation on Earth has an urgent self-interest in making sure that no sizable nuclear war—one involving the detonation of a few hundred warheads or more—ever occurs. The notion that a "small" nuclear war could be kept "contained," that "escalation controls" could keep it from swiftly evolving into a global nuclear conflict, is little more than a pious hope (ref. 9.3). It would be foolish and deadly as a cornerstone of national security. The danger is so grave and so

# WHAT TO CALL IT?

What shall we call a strategic arsenal small enough to avoid nuclear winter, but large enough to provide a real measure of deterrence? This is closely related to a much older question: What is the minimum strategic force that can provide adequate deterrence? This was called "Minimum Deterrence" by naval officers and civilian scientists attempting to justify the introduction of the Polaris nuclear submarine in the late 1950s (cf. ref. 17.14). Their advocacy was tinged with interservice rivalry: It was argued, on grounds that still have much merit, that a small, invulnerable submarine retaliatory force was by itself a wholly adequate deterrent (cf. ref. 9.14), and that it made the much more vulnerable Air Force bombers and fixed-site missiles superfluous, or even dangerous. But "because the word 'Minimum' carried a connotation of gambling with the nation's security for budgetary reasons, it was changed to 'Finite' (which had the connotation of wanting enough and no more and also suggested that the opponents wanted an infinite or at least an unreasonable amount)" (ref. 5.3, footnote, p. 14). There is the problem, though, that Finite Deterrence is clearly a misnomer; all deterrence is finite, even none at all. Infinity is unattainable.

In seeking an alternative, we have offered the term "Canonical Deterrent Force" (CDF) [R. P. Turco and C. Sagan, "Policy Implications of Nuclear Winter," *Ambio 18* (7), 1989, 372–376]—which we intended in the mathematical sense of basic or standard, but which, we are reminded, also suggests an agreed-upon, although arbitrary, canon of belief; the ambiguity then lies in the fact that there are dissenters who hold to other faiths. Even more important, the phrase does not convey the idea behind the canon. Similar problems, at least in English, apply to the term "reasonable sufficiency," the historically significant advocacy by Mikhail Gorbachev of, in his words, "preserving the overall balance, but at the lowest possible levels" (cf. Raymond L. Garthoff, "New Thinking in Soviet Military Doctrine," *Washington Quarterly*, Summer 1988,

131–158): People differ about what is reasonable; Gen. A. I. Gribkov, then Chief of Staff of the WTO Combined Command, defined reasonable sufficiency as essentially whatever forces the U.S. had (Gribkov interview, "Doctrine for Assuring the Peace," *Krasnaya Zvezda* [*Red Star*], September 25, 1987; Garthoff, *ibid.*). Other analysts, Soviet and American, have proposed "defensive deterrence," or "fundamental deterrence" (Stephen Shenfield, "Minimum Nuclear Deterrence: The Debate Among Soviet Civilian Analysts," Center for Foreign Policy Development, Brown University, November 1989).

We propose instead to return to the phrase "minimum deterrence," but to insert the key qualifying adjective: "minimum sufficient deterrence," or MSD. Equivalently, we will use "minimum sufficiency." We wish to lay as much stress on "sufficient" as on "minimum." At least then the debate is focused on what is the minimum arsenal sufficient for deterrence (and for avoiding nuclear winter).

open-ended that the only sure guarantee is to destroy almost all the nuclear weapons on Earth—multilaterally, reliably, verifiably. All this is possible. There seem to be no serious technical impediments. The chief problems are political. The nations with the largest nuclear arsenals have the greatest need, as well as the greatest moral obligation, to take foolproof measures guaranteeing that nuclear winter can never occur. If they wish, they can still preserve their reliance on strategic deterrence. And since the transition from tactical to strategic war is likely to be continuous and inevitable, they might, if they are so inclined, maintain extended deterrence to safeguard their "vital interests" with strategic weapons—very few strategic weapons.

When, in the remainder of this book, we explore ways to reach minimum sufficiency, we do not pretend that these are anything like roadmaps. They are rough sketches only, intended to stimulate and encourage better artists and draftsmen. Let's look at the details a little.

Clearly there are safe and unsafe, stabilizing and destabilizing, ways to reduce the nuclear arsenals. There are steps that

## MARTIAN OVERKILL

Soon after [Robert S.] McNamara took over as Secretary of Defense [1961], one of his more cynical Assistant Secretaries "explained" the position to me in this way. "Don't you see?" he asked. "First we need enough Minutemen [silo-launched missiles] to be sure that we destroy all those Russian cities. Then we need Polaris [submarine-launched] missiles to follow in order to tear up the foundations to a depth of ten feet. . . . Then, when all Russia is silent, and when no air defenses are left, we want waves of aircraft to drop enough bombs to tear the whole place up down to a depth of forty feet to prevent the Martians recolonizing the country. And to hell with the fallout." It was not long before he retired from his post in the Pentagon.

—Lord Solly Zuckerman, *Nuclear Illusion and Reality* (New York: Viking Penguin, 1983). Zuckerman was chief scientific adviser to the British Ministry of Defence.

seem prudent in the short term, but which are roadblocks to long-term progress. As in the arms race itself, there are political impediments to change—including the common-sense prenuclear-age perception that reducing a nation's arsenal makes it weaker. Nevertheless, some steps seem clear. For example, it would be safest to destroy the most destabilizing weapon systems first—especially "MIRVed" missiles, those with many warheads.

We propose several approaches to be taken (together, in appropriate proportions); they are discussed in more detail later:

(1) Multilateral destruction of the most vulnerable and destabilizing strategic systems (e.g., silo-based MIRVed missiles) and comprehensive deMIRVing of what is left—back to one nuclear warhead per missile;

(2) Phasing out of short-flight-time intermediate range delivery systems in Europe and Asia, as has already been partially accomplished under the terms of the INF (Intermediate-Range Nuclear Forces) Treaty;

(3) Gradual elimination of all forward-based tactical nu-

clear forces, including short-range missiles, that are in danger of being overrun in a conventional invasion (and therefore vulnerable to the "Use 'em or lose 'em" temptation);

(4) Reduction, balancing, and substantial pull-back and demobilization of NATO and WTO (Warsaw Treaty Organization = Warsaw Pact) conventional forces in Europe (this requires larger WTO than NATO force reductions, as are now underway —see below);

(5) Abandonment of the Strategic Defense Initiative (SDI) as currently constituted, and any comparable Soviet programs, and maintenance of the military role in space exclusively for communications, weather forecasting, surveillance, launch warning, and treaty compliance missions (ref. 17.5);

(6) Reduction, and the earliest possible phasing out, of all nuclear weapons testing as the minimum sufficiency regime begins to be established; and

(7) Strategic force levels reduced to around 100–300 warheads each for the U.S. and U.S.S.R., with substantial reductions for other nations.

While these proposals are broad, they have now all been offered at the highest levels in discussions and negotiations between the United States and the Soviet Union; they have all been discussed (and to varying degrees endorsed) by specialists; and they are all, in our opinion, technically and politically achievable. Justifications for all of these measures (except for the numerical values in [7]) have been made and widely discussed without explicit reference to nuclear winter. We believe that the spectrum of possible climatic outcomes of nuclear war greatly increases the cogency and urgency of taking these steps, but that they make sense even without nuclear winter.

Clearly there are important issues to be resolved about pace and coordination, in order that no party has a perceived or real, even if temporary, strategic advantage that might lead to miscalculation. At all times in the divestment process, force structures—both nuclear and conventional—must be configured consistently with the military goal of crisis stability, and appropriately balanced between (and among) the potential antagonists. It is also essential to establish machinery for verification, including, as needed, intrusive on-site verification. Soviet acceptance of the Natural Resources Defense Council's seismic network (including American technicians at the Soviet test site near Semipalatinsk) and subsequent official observation of underground tests on each other's soil; bilateral agreement on intrusive inspection by each side of the production facilities and deployed weaponry of the other, as agreed in the INF treaty; inspection of secret weaponry of the other side by the Chairman of the Joint Chiefs of Staff and the Chief of the Soviet General Staff; and bilateral implementation of demand inspection of military maneuvers in Europe under the Stockholm Accords (ref. 17.6), all suggest that verification no longer constitutes anything like the obstacle to a comprehensive nuclear settlement that it once did.

We will examine force structures beginning in the next chapter. But first we take a closer look at the size of a minimum sufficient deterrent (MSD) force.

An MSD is usually defined as the smallest effectively invulnerable force capable of delivering a devastating retaliatory blow following an unimpeded first strike—to which we add the criterion of being incapable of inducing a nuclear winter. From the beginning, some have argued that no larger force was needed for deterrence. In early 1945, before Hiroshima, Leo Szilard—one of the prime movers of the Manhattan Project—drafted a message for President Franklin Roosevelt, who did not live to see it. It read in part,

> The existence of atomic bombs means the end of the strong position of the United States in this respect. From now on the destructive power which can be accumulated by other countries as well as the United States can easily reach the level at which all the cities of the "enemy" can be destroyed in one single sudden attack. . . . Outproducing the "enemy" might therefore not necessarily increase our strength greatly (ref. 17.7).

The father of U.S. nuclear strategy, Bernard Brodie, commented:

> Superiority in numbers of bombs is not in itself a guarantee of strategic superiority in atomic bomb warfare. . . . If 2,000 bombs in the hands of either party is enough to destroy entirely the economy of the other, the fact that one side has 6,000 and the other 2,000 will be of relatively small significance (ref. 17.8).

Minimum sufficiency was also recognized as early as 1945 by many of the Los Alamos scientists who had built the first atomic bombs (ref. 17.9), and in 1946 by former Vice President Henry Wallace:

> So far as winning a war is concerned, having more bombs —even many more bombs—than the other fellow is no longer a decisive advantage. If another nation had enough bombs to eliminate all of our principal cities and our heavy industry, it wouldn't help us very much if we had 10 times as many bombs as we needed to do the same to them (ref. 17.10).

George F. Kennan, then head of the State Department Policy Planning Staff, wrote to Secretary of State Dean Acheson on January 20, 1950,

> We may regard [nuclear weapons] as something superfluous to our basic military posture—as something which we are compelled to hold against the possibility that they might be used by our opponents. In this case, of course, we [should] take care not to build up a reliance upon them in our military planning. Since they then represent only a burdensome expenditure of funds and effort, we [should] hold only the minimum required for the deterrent-retaliatory purpose (ref. 17.11).

Kennan also quoted a speech by Soviet Ambassador Andrei Vyshinsky on November 10, 1949, before the U.N. General Assembly, in which Soviet nuclear weapons procurement was said to be only for minimum deterrence.

"What you want is *enough*," explained President Dwight Eisenhower to the fire-snorting Commander of the Strategic Air Command, Gen. Curtis LeMay. "A deterrent has no added power once it has become completely adequate" (ref. 17.12). And in the last interview before he died, Soviet President and long-time Foreign Minister Andrei Gromyko argued for minimum sufficiency even if the United States maintained much larger arsenals. This should have been Soviet policy from the beginning, he said, but no political or military leader, and no scientist, had the courage to suggest so radical a step. But "today we have the right to be more clever and bolder" (ref. 17.13).

Historically, however, expert opinion has varied widely on what constitutes an MSD force. In 1959, Adm. Arleigh Burke, the Chief of Naval Operations, argued:

> In making our retaliatory force secure from enemy attack, we do not need great numbers of missiles and bombers. Whether the U.S.S.R. has one-half as many or several times as many missiles as the United States is really academic as long as we have the assured capability to destroy Russia and as long as the Soviets know it and are really convinced of it (ref. 17.14).

He urged an invulnerable submarine missile force with a few hundred warheads. In 1961 the President's arms control and science advisers advocated 300 single-warhead Minuteman ICBMs; the Secretary of Defense, 700 ± 100; the Secretary of the Air Force, 1,450; the Air Force Chief of Staff, 2,950; and the Commander of the Strategic Air Command, 10,000 (ref. 17.15).*

The detonation of 10 to 100 modern thermonuclear warheads on any nation would (with urban/industrial targeting) produce unacceptable economic damage; a similar number of explosions could have a dominant influence on the outcome of a major military operation (with targeting of ground and sea forces). Using technologies that minimize vulnerability (see below), the *maximum* number of warheads on each side needed to guarantee unacceptable retaliatory damage is certainly less than 1,000; that is, at least a factor of ten below the number of warheads in the present strategic arsenals (ref. 17.17). If the MSD warheads and delivery systems have high survivability and reliability, only about 100 might be required on either side (ref. 17.18).

Such small MSD forces have been considered before in the West, and not only in the Kennedy-Johnson years (ref. 9.14): In his second day in office, President Jimmy Carter proposed to

* Perhaps he would have asked for ∞ had it been possible. A few thousand more than 10,000 strategic weapons is where the United States—and the Soviet Union—finally got to. Plus about 15,000 tactical weapons each, most of them more powerful than the Hiroshima and Nagasaki bombs. The early advocates of minimum sufficiency lost this debate. Hardly any of the scientists on the Manhattan Project imagined there would one day be between 10,000 and 100,000 nuclear weapons in the world (ref. 17.16).

the Soviets deep mutual cuts in the strategic arsenals (which was promptly rejected—foolishly, Soviet spokesmen now say). And in his first meeting with the Joint Chiefs of Staff (ref. 17.19) Carter asked for a study of an MSD comprising a few hundred weapons (which was also promptly rejected—but the public record seems to hold no current JCS reconsideration of the wisdom of this judgment). The current British strategic doctrine is based on the "Moscow criterion," the belief that the ability to obliterate Moscow reliably constitutes, by itself, adequate deterrence (ref. 17.20); it is consistent with McGeorge Bundy's "existential deterrence" (ref. 9.14; see box). Minimum deterrence is "subcutaneously embedded" (ref. 17.21) in the Scowcroft Presidential Commission recommendations and in the government report *Discriminate Deterrence* (ref. 17.22). And, at least in words, it has been endorsed by NATO: "As Europe changes, we must profoundly alter the way we think about defense. . . . We seek the lowest and most stable level of nuclear forces needed to secure the prevention of war." (ref. 17.23).

Nevertheless, some experts believe that force reductions below many thousands of strategic warheads would be risky for maintaining extended deterrence, or even for strategic deterrence (e.g., ref. 9.8). But worldwide arsenals of thousands of strategic warheads could generate, with high likelihood, at least a marginal nuclear winter (Figure 6); a global arsenal of a few hundred such weapons would be much less likely to generate serious climatic consequences, even if there were abundant countervalue targeting. [Such small inventories would also make it much less likely that in internal ethnic conflict—e.g., in the Soviet Union or China—nuclear weapons would be captured by regional nationalists (ref. 17.24).] We propose setting retaliatory forces at levels that nearly eliminate the possibility of nuclear winter for all future national policies and targeting doctrines. The specific nuclear winter objective in the transition to minimum sufficiency is *to generate a stable deterrence in which the probability of nuclear war is extremely small, but where, if the worst happens, there would be no significant global climatic consequences.* Nuclear winter not only provides additional arguments for minimum sufficiency, but also helps to establish what requisite MSD force levels are.

The ratified Intermediate-Range Nuclear Force (INF) agree-

## THINKING ABOUT THE UNTHINKABLE

There is an enormous gulf between what political leaders really think about nuclear weapons and what is assumed in complex calculations of relative "advantage" in simulated strategic war. Think tank analysts can set levels of "acceptable" damage well up in the hundreds of millions of lives. . . . They are in an unreal world. In the real world of real political leaders —whether here or in the Soviet Union—a decision that would bring even one hydrogen bomb on one city of one's own country would be recognized in advance as a catastrophic blunder; ten bombs in ten cities would be a disaster beyond history; and a hundred bombs on one hundred cities are unthinkable.

—McGeorge Bundy, National Security Adviser to Presidents Kennedy and Johnson, "To Cap the Volcano," *Foreign Affairs*, October 1969, 9–10. See also Bundy, *Danger and Survival: Choices About the Bomb in the First Fifty Years* (New York: Random House, 1988).

ment and the proposed (rep. 17.25) START (Strategic Arms Reduction Talks) agreements are both important first steps towards an MSD regime. However, INF divestments constitute only some 3% of the global nuclear arsenals, and the fissionable materials from these weapons are not being destroyed, but rather are being reprocessed into new nuclear weapons. There are some 10,000 nuclear weapons left in Europe, *outside* of the Soviet Union. Even the proposed START agreement—involving the dismantling of perhaps a third of the U.S./Soviet strategic arsenals (ref. 17.26) and perhaps even the destruction of the warheads—while representing an unprecedented nuclear arms reduction, is only a beginning: It would leave the nuclear superpowers with enough strategic weapons to destroy five times over every city on the planet with a population in excess of 100,000 people.

There will be an inclination, especially if we get to implement START, to congratulate ourselves, to think we have enough momentum now to coast into a world safe from nuclear weapons, to turn our attention to other matters. But START would leave the nuclear superpowers with dozens of times the number of strategic weapons needed to initiate nuclear winter. START is essential, but it is only—as the acronym suggests—a first step.

# CHAPTER 18

# WHAT KINDS OF WEAPONS?

## STRATEGIC FORCE STRUCTURES

Wait for the day when the sky will pour down blinding smoke, enveloping all men: a dreadful scourge. Then they will say: "Lord, lift up this scourge from us. We are now believers."

—*The Koran*, 44:1, N. J. Dawood, trans. (New York: Viking Penguin, 1974)

IN STRATEGIC WEAPONS SYSTEMS, TECHNOLOGY OFTEN drives doctrine rather than the other way around. When it becomes possible to devise new means of destroying or intimidating a potential enemy, strategists can readily be found to offer compelling reasons why the new weapons are urgently needed—no matter how lethal or effective the existing weapons may be. But the adversary nation also has such scientists and strategists. So, frequently, the nation making the innovation has only a momentary military advantage, until the other side catches up; then, at great cost, they both find themselves back where they started from, with neither possessing anything like a decisive military advantage—except that the world has now become, because of the new weapons system, a much more dangerous place. Moreover, each new weapons system on one side cries out to be targeted by additional weapons on the other—leading to a runaway arms race.

Weapons procurement and the evolution of force structures are prominently driven by short-term goals. The longer-term implications are often left for the rest of us to worry about. But in a prudent nation, the arsenals would reflect the national purpose for which they are intended, and that includes both short-term and long-term goals. Here is one typical statement of that purpose:

> The fundamental objective for the American strategic forces during the McNamara period and since has been to deter a nuclear attack against the United States and its allies by maintaining nuclear forces that could ride out even a full surprise attack by the Soviet Union and still be able to retaliate overwhelmingly (ref. 18.1).

In pursuing this objective and its Soviet equivalent, the U.S. and the U.S.S.R. have accumulated nearly 60,000 nuclear weapons and in the process put our civilization at risk. Ironically, this goal can be achieved with very small forces—if they are of the right kind.

We have stressed that as few as 100 city centers or 100 petroleum facilities in flames may be enough to bring about nuclear winter. By no means would all nuclear weapons necessarily be used if a war were to occur in a minimum sufficiency regime, and not all those that were used would be exploded over cities. If we took seriously the wish to make a world so safe that not even a conspiracy of madmen in control of the nuclear-armed nation-states could generate even a nominal nuclear winter, then the total number of nuclear weapons in the world would have to be very small—perhaps less than 100.

But the number of 100 burning cities or (especially) petroleum facilities is obviously inexact. There are many uncertainties. Yield is one, in that a high-yield airburst will burn more of a city than a low-yield airburst. Not all strategic targets will be cities or petroleum facilities. Not all warheads will hit their targets, because of imperfect accuracy, technological failures, or terminal defenses. Some weapons might be withheld for purposes of future negotiation or coercion. There are vagaries of weather. We cannot know beforehand the season of the war. These and other uncertainties starkly outline why any estimate of where the nuclear winter "threshold" is must be imprecise.

If we nevertheless must estimate, even if roughly, where that threshold is—and the stakes are so high that it is essential to have some idea of the answer—we must also balance our wish that if we err, it be on the side of safety, against our wish to make the force large enough to satisfy the demands of deterrence. We have tried to specify that threshold with great caution (Chapter 7). Indeed, we may have been too cautious in that smaller optical depths might still cause major disruptions in agriculture (Class II nuclear winter). Our best effort to achieve such a balance leads us to envision a minimum sufficient deterrence regime in which the United States and the Soviet Union have about 100 nuclear warheads each, with the rest of the world having, at most, an equal number. But we cannot be sure. We have much more confidence that 1,000 warheads on each side is too many to prevent nuclear winter; and that 10 on each side, at least in the view of many, is too few for deterrence (ref. 18.2). A world with no nuclear weapons—where strategic deterrence is perfect, extended deterrence nil, and the probability of nuclear winter zero—is a very different matter, to be addressed later.

We will shortly discuss the possible nature of MSD forces—at first from a purely technical standpoint, i.e., ignoring the economic, psychological, and political consequences of a continuing arms race, even if that arms race were devoted to new weapons intended to ensure the most stable minimum deterrent attainable. The types of invulnerable strategic forces we have in mind could be deployed in one or more of several modes. Single-warhead delivery systems are envisioned in each case, as these are the most stabilizing: In such a regime, one offensive missile can in principle destroy *at most* one opposing warhead. By contrast, in the current strategic regime, with 10-warhead missiles, a single offensive missile can in principle destroy up to 100 opposing warheads (10 × 10: ten warheads launched by a single missile, each warhead destroying ten warheads in each unlaunched adversary missile. Allowing for targeting errors, 50 is a more likely number than 100). This is why the deployment of MIRVs (Multiple Independently Targeted Reentry Vehicles—each reentry vehicle capable of carrying its own nuclear weapon) is fraught with danger in a time of high tension. Attack the other side's warheads while they're still on the ground, MIRVing counsels, before they destroy your cities. MIRVing argues for first strike. MIRVing skews strategists away from arms reductions and towards arms races. MIRVing is a mistake.

There was a complex array of imperatives that led to the American introduction of MIRVed missiles: misplaced concerns that the Soviets might be developing an effective defense against ballistic missiles which needed to be overwhelmed by many more warheads; domestic politics, in which a Defense Secretary intent on resisting pressures for the development of new strategic rockets had to find a sop to throw to the hawks; the utility of MIRVing as a "bargaining chip" to bring the Soviets to the table for arms control negotiations, or—something rather different—MIRVs as a major step in the endless struggle to target every potential war-fighting asset of the Soviet Union; advancement and promotion built into the military and the defense industry; a way of saving the taxpayer's hard-earned dollar, at least in the short run; the excitement of groundbreaking technological developments; and love of country (ref. 18.3).

The fact that MIRVing destabilizes the strategic relationship by providing powerful incentives for first strike seems not to have been seriously contemplated by any of the multitude of decision-makers involved—although since deployment there have been plentiful enough misgivings expressed, and wistful longings (for example, by former Secretary of State Henry Kissinger) for the good old days when each missile had only one warhead. Nor do we know of any reluctance on the part of the Soviets, who, after the American lead, quickly advanced the research and development program that led to their own MIRVed missiles. Even today, almost every new ballistic missile system in the process of being deployed by the United States and the Soviet Union (e.g., MX and SS-24) is heavily MIRVed (ref. 18.4). The history of MIRVing is as good a lesson as any of the opportunistic, self-propulsive tendency—blind to long-term consequences—of the nuclear arms race.

Merely cutting the arsenals on both sides, while increasing the MIRVing of the remaining missiles, could *de*crease crisis stability; and, although the contention is controversial, some analysts argue that the START treaty as discussed at the very end of the 1980s had just such an undesirable property (ref. 18.5) and would increase the incentives for first strike. This possibility stimulates debate: Is the decrease in stability of such a START treaty acceptable because of the forward momentum it provides for further arms control? Or is any treaty that decreases stability worse than no treaty at all? How do we balance short-term and long-term objectives? But here, especially with the enormous residual arsenals even after implementing START, nuclear winter comes to the rescue—by making such first strikes suicidal. Nuclear winter works to decrease crisis instability. For long-term strategic stability, though, a major first step must be to deMIRV the arsenals, and this should be the focus of START II.

A stable MSD force might comprise:

(i) Single-warhead ICBMs in silos.

(ii) Mobile single-warhead land-based missiles. The planned U.S. Midgetman is a paradigm of such a system, combining mobility with substantial hardness—immunity to blast—to ensure survivability. (The Soviet SS-25 approaches the paradigm, but with more limited hardness.) Deployments on road-mobile launchers, railroad cars, or among multiple bunkers are also possible. However, especially in democracies, the idea of launch-ready nuclear-armed missiles roaming the highways and railroad tracks has been known to provoke consternation, protest, and even political action.

(iii) Deep underground basing mode. This would place the retaliatory missiles in subterranean bunkers much deeper than any current silos and hardened against any possible nuclear attack. Tunnels and shafts to the surface would be excavated when needed to launch the missiles. There are, however, problems with this technology, still in its early stages of development (ref. 18.6).

(iv) Small submarines, each with one or perhaps a few unMIRVed ballistic missiles (ref. 18.7). They could patrol not far from the home shores, hiding in deep water off the continental shelf. They would be compact and quiet, and even less detectable than modern strategic submarines. There are no known or foreseeable technologies that would make strategic submarines visible in the ocean depths. But even if they became visible, they would have roughly the same protection because of their mobility as do road-mobile or rail-mobile systems on land. And, unlike today, no rogue submarine crew could by itself initiate nuclear winter.

With force structures of this sort on both sides, the nuclear deterrent would be invulnerable to a first strike or preemption by opposing forces. Strategic stability would be robust. In such an MSD regime, the missile inventories would be roughly balanced. Then, only one warhead is available against each opposing warhead. But according to the mathematics of weapons survivability, the imperfect accuracy of missiles and the hardness of targets imply that roughly two (or more) warheads would be needed (on average) to guarantee the destruction of a single opposing missile. Neither side would have enough warheads to muster a successful first strike.

Imagine a world in which no nuclear weapon can hide from

potential attack, and where the arsenals on both sides are static, heavily MIRVed, and equal. Then, you might be able to launch a first strike so that every two of your warheads would destroy, say, ten of your adversary's warheads. You might therefore decide that there's some advantage in a crisis to launching a massive first strike. But if, in contrast, arsenals are static, unMIRVed, and equal, you must now expend two warheads to destroy every one of your adversary's warheads. If, finally, the arsenals are mobile (or hidden), unMIRVed, and equal, you must now expend more than two—perhaps many more—warheads to destroy one of your adversary's. If you launch a massive first strike, you may use up all your warheads while leaving the other side with half or more of its arsenal—to retaliate against your cities. It is easy in this circumstance to recognize beforehand that a first strike is stupid as well as criminal. The deMIRVed force structure has deterred you.

Moreover, it is highly improbable that a technological breakthrough would allow any great advantage to any one side. Even with perfect surveillance and accuracy, one missile could destroy only one warhead. Diversification of weapons systems, and—just possibly—the implementation of strictly limited terminal defenses (see below), would reduce vulnerability still further.

Communications, command, and control ($C^3$) of MSD forces could be much simpler and more robust than for most current strategic or tactical forces. Mobile missiles would have direct radio communications links hardened against the electronics-frying electromagnetic pulse (EMP) emanating from the airburst of a nuclear weapon. Communications with submarines patrolling a small region of the oceans adjacent to a given nation are also simpler and more reliable than worldwide communications with far-ranging strategic submarines. So-called decapitation strikes against the national military and civilian leadership are less probable when $C^3$ is more robust. Any inadvertent or unauthorized missile launch would represent less than a global catastrophe.

For these reasons, minimum sufficiency seems a much safer (as well as cheaper) response to the dangers of nuclear winter than the leading feasible military alternative—the conversion of the world strategic arsenals to nuclear weapons of very high accuracy and very low explosive yield. (See box. For a brief

# CAN NUCLEAR WINTER MAKE NUCLEAR WAR MORE LIKELY?

The objection is sometimes raised—more often than we would have guessed—that nuclear winter makes a "small" nuclear war acceptable by suggesting that the worst can be avoided as long as climatic effects are small. We have difficulty following this argument, because (a) if the nations were deterred from nuclear war before the discovery of nuclear winter, they would (other things being equal) be even more strongly deterred today; and (b) the likelihood that a "small" war would be contained and not escalate to a central strategic exchange and nuclear winter is much too high to risk (ref. 9.3).

A more coherent variant of the argument posits that nuclear winter will propel the arsenals to a very different force structure, and *then* a large-scale nuclear war can be "safely" fought. E.g.,

> . . . if nuclear winter effects are assumed real and significant in their political implications, they will simply push the United States and the Soviet Union in the direction of increased accuracy and lower-yield weapons even faster than they are already moving that way. The end result of this may well be to make nuclear war appear much more wageable and controllable than it now appears to be, and deceptively similar to past wars. [Michael F. Altfeld and Stephen J. Cimbala, "Targeting for Nuclear Winter: A Speculative Essay," *Parameters: Journal of the U.S. Army War College 15* (3), 1985, 8-15.]

But if such "usable" nuclear weapons systems coexisted with the "unusable" (because of nuclear winter) weapons systems of current vintage, an extremely unstable strategic balance would ensue (cf. ref. 2.3 and Chapter 9). Compare, then, the dangers of this scheme—favored by some in the military—for force restructuring in response to nuclear winter with the dangers of a minimum sufficient deterrent. If the existing arsenals must be almost entirely dismantled before the low-yield, high-accuracy systems can be introduced—why not just stop then? That's minimum sufficiency. You're there.

discussion of strategic defense, a far less feasible alternative, see below.)

An MSD mix of forces has been proposed by others—for example, by Feiveson, Ullman and von Hippel (ref. 18.8), who suggested 2,000 warheads on each side, distributed with 1,000 on long-range bombers, 500 on ballistic missile submarines, and 500 on ICBMs and cruise missiles. They proposed, as we do, eliminating all tactical and battlefield nuclear weapons. For many years the prolific American physicist and weapons designer Richard Garwin has urged an MSD of about 1,000 warheads on each side. He suggests (ref. 18.9) a U.S. force composed of "400 warheads in the form of single-warhead small ICBMs in soft (vulnerable) silos; 400 warheads divided among 50 small submarines; and 200 warheads carried on 100 aircraft as air-launched cruise missiles, two per aircraft." We indicate below our reservations about bombers and cruise missiles. Garwin stresses that each side should be permitted to select the force structure which, in its opinion, promotes greatest stability (ref. 18.9). These earlier proposals for an MSD regime have been made chiefly out of concern about the prompt effects of a nuclear war with tens of thousands of warheads, a concern that nuclear winter greatly exacerbates—leading us to advocate still smaller arsenals, with a different strategic mix. A promising and productive set of proposals on what constitutes minimum sufficiency and how to get there has been emanating from the Soviet Union (see box).

Such a wholesale reconfiguration of the present strategic forces of the United States, the Soviet Union and, necessarily, the lesser nuclear powers has many pitfalls. For one thing, if procurement practices follow the usual pattern, MSD could be very expensive. The full development worldwide of this technology could cost hundreds of billions of dollars. It is perfectly possible to imagine a strategic force a fraction of the present size which costs as much as the present force (see, e.g., Alexander H. Flax, in ref. 9.7). The ingenuity of the designers and manufacturers of nuclear weapons systems is fully equal to this

# THE MINIMUM SUFFICIENCY DEBATE IN THE U.S.S.R.

Minimum sufficiency seems to have been contemplated at high levels in the U.S.S.R. at least since the time of Premier Nikita Khrushchev (e.g., Alexander Yanov, "An Avoidable 20-Year Race," *New York Times*, October 10, 1984). But it has been considered in earnest only since the middle-1980s. President Gorbachev asked the physicist and space scientist R. Z. Sagdeev to lead a study to determine the size of an MSD consistent with Soviet national security. The Sagdeev report set such a force level at less than 5% of the present Soviet strategic arsenal (Lee Dye, "Soviets Seek Spark to Fire Space Goals," *Los Angeles Times*, July 26, 1988, and Roald Sagdeev, private communication)—or roughly 500 weapons.

A very simple such MSD force has been proposed by Soviet scientists (ref. 8.21 and A. A. Kokoshin, "A Soviet View on Radical Weapons Cuts," *Bulletin of the Atomic Scientists*, March 1988, 14–17): 500–600 warheads on each side, all in "light, single-warhead ICBMs, with some portion of these being mobile." They express reservations about ballistic missile submarines because of possible command and control problems in a crisis. With the Soviet single-warhead road-mobile SS-25 ICBM operational since 1985, such an MSD force configuration might reduce the need for the U.S.S.R. (but not for the U.S.) to develop new MSD hardware. Perhaps this has something to do with Soviet misgivings about submarines under minimum sufficiency. But stationing submarines not far off the continental shelf, as we advocate, greatly improves the reliability of communications; and the decreased vulnerability of an underwater deterrent tends to offset any residual imperfections in command-and-control. The MSD regime is envisioned as an intermediate step, of unspecified duration, between something like START (30 to 75% cuts in strategic weapons) and a later total abolition of nuclear weapons.

This provocative proposal has sparked an important debate in the Soviet Union involving groups of scientists, foreign service officials, military officers, and others. (See, e.g., Stephen Shenfield, "Minimum Nuclear Deterrence: The Debate Among Soviet Civilian Analysts," Center for Foreign Policy Development, Brown University, November 1989). The details but not the desirability of minimum sufficiency are the focus of the discussion. All alternative MSD regimes proposed involve arsenals of a few hundred weapons on each side; the debate is mainly about which weapons systems maximize stability and how much emphasis to give to second-tier nuclear powers. A group of "young diplomats" believe that Soviet mobile land-based missiles may not be able to evade future U.S. reconnaissance satellites, precision-guided missiles, and bombers—which, if true, is destabilizing, because in a crisis it provides inducements for Soviet preemptive attack, or at least launch on warning of an attack underway; the diplomats opt instead for a submarine MSD force.

Another area of debate concerns how far the Soviet Union might go in unilateral arms reduction steps (assuming a recalcitrant United States unwilling to make deep cuts). The well-known political commentator and former party official Alexander Bovin wonders if "we have not hypnotized ourselves with a cult of parity" ("Readers Ask, *Izvestiya*'s Political Observer Replies," *Izvestiya*, April 16, 1987). At least some of the concerns about unilateral reductions inferred by Shenfield—e.g., vulnerability to a U.S. first strike—are eased by nuclear winter.

The whole tenor of these debates on mimimum sufficiency seems at least as vigorous and innovative as those underway in the United States. This is true despite the fact that "Soviet science traditionally has been and still remains rather rigid and conservative, reacting only slowly to new critical problems. This rigidity has characterized the attitude of a considerable part of the Soviet scientific community toward international security problems." (Roald Sagdeev, "Glasnost and the New Journal," *Science and Global Security 1* (1, 2), 1989, v–vi.) Presumably we can expect much more new thinking as Soviet society continues to derigidify. Mightn't it be a good idea to bring representatives of the various U.S. and Soviet points of view together—in an informal workshop setting—to see if mutual progress can be made in working out their common nuclear destiny?

challenge. Of course, if it would stabilize the nuclear arms race and prevent nuclear war, it would be money well spent. Moreover, the cost *could* be modest compared with the strategic systems it replaces. For example, roughly $20 billion buys us 10 midget submarines with a few unMIRVed missiles each, plus retrofitting 100 Minuteman ICBMs so they are deMIRVed. This price tag is comparable to the *annual* costs of personnel, facilities, and infrastructure for the present U.S. nuclear arsenal. It would be about 1% as expensive as the proposed continental-scale Star Wars system, and 10% as expensive as planned strategic and tactical force "modernizations"—all of which the MSD regime renders unnecessary. Costs could also be reduced by joint U.S./Soviet development of the new technology, although this seems impractical on a variety of grounds.

However, the main problem with the development of new weapons systems to achieve MSD is not cost but the fact that it keeps the arms race going. It sustains the motivation on both sides to develop new nuclear weapons and delivery systems. Instead of encouraging the phasing out of weapons production facilities, it maintains the very military/industrial/government weapons laboratory establishments that have driven the nuclear arms race in the first place. It diverts significant fiscal and intellectual resources needed for repairing the ailing U.S. and Soviet civilian economies. And it raises difficult issues of verification—for example, to determine that permitted and not prohibited weapons are being developed or manufactured in a given facility. The fewer new weapons systems that are introduced, the easier it will be to verify treaty compliance. This was one of the original justifications for the nuclear freeze proposals of the early 1980s. All this brings us to what we believe is an important issue to be addressed in the search for the optimal approach to MSD, namely: Is there a way to deMIRV without dismantling all existing delivery systems and replacing them with a completely new array of delivery systems?

• • •

The U.S. arsenal still includes 450 Minuteman II launchers, each armed with a single warhead, each with the maximum range of all U.S. land-based ballistic missiles (ref. 8.1); the Soviet Union is said to have about 100 mobile single-warhead SS-25s, about 60 single-warhead SS-13s (Mod 2), and a number of single-warhead SS-11s (ref. 18.11). In addition, 540 Minuteman IIIs and 138 SS-17s are triply MIRVed. Thus we can see the bare possibility, instead of constructing entirely new arsenals of land-based strategic weaponry, of maintaining and refurbishing existing single-warhead intercontinental ballistic missiles and deMIRVing some that are lightly MIRVed. The remainder of the arsenal would be destroyed. Retrofitting the MX or SS-18 for a minimum sufficiency regime may have to be excluded precisely because of their large payloads or throw-weights, which might allow clandestine reMIRVing.* Because the inventories on both sides are already so extensive and varied, there would be ample opportunity to mix ICBM weapons systems and still stay under the limits of a nuclear winter–informed Minimum Sufficiency. It will always be the case that further improvements in accuracy are possible, and that new capabilities can be developed, such as burrowing to attack deeply buried command posts, leadership shelters, and weapons systems. Such capabilities would be foresworn in an MSD regime. But if the point is not only to achieve an MSD but also to stop the nuclear arms race, some of the multitude of existing strategic missiles, each maintained at or reduced to a single warhead, may be found adequate.

The disparity between reality and the ideal of a single warhead for each strategic delivery system is greatest for existing ballistic missile submarines. They are equipped with vertically mounted tubes into which fit the MIRVed ballistic missiles. The U.S. Poseidon has 16 missile tubes; Poseidons equipped with tenfold MIRVed C-3 missiles therefore have $16 \times 10 = 160$ nuclear warheads. Those with C-4 missiles carry only $16 \times 8 = 128$ warheads. The Trident submarine with C-4 missiles carries $16 \times 12 = 192$ warheads. The Trident D-4 may carry $16 \times 14 = 224$ highly accurate warheads. (Not all launchers may

* Although inspection protocols to prevent reMIRVing have been suggested (e.g., by Robert Mozley, in von Hippel and Sagdeev, ref. 19.20).

boost warheads; some might contain, e.g., backup communications or navigation satellites.) The U.S. arsenal currently includes 36 such submarines. On the Soviet side the number of warheads per submarine ranges from 16 for the Yankee I, with 16 single-warhead SS-N-6 missiles, to the Typhoon-class submarines, each with perhaps 180 warheads (ref. 18.11). In all, the Soviet ballistic missile submarine fleet has about 60 submarines, but with a much smaller fraction of them at sea at any given moment than is the practice for the U.S. submarine fleet.*

The notion of taking a weapons system which is able to destroy 150 or 200 "targets" and reducing its capability to only a few "targets" seems on purely military grounds to be a significant waste of resources (ref. 18.12). But the real question, we claim, is this: To achieve an MSD regime, is it more cost-effective to build a new fleet of minisubs or to use the existing ballistic missile submarine fleet? Because such submarines are intrinsically far less vulnerable to attack than are ground-based missiles, it would still be stabilizing to permit a few warheads per submarine (not too many, so port visits do not offer an irresistible temptation for crisis preemption), rather than one only—provided there are foolproof ways to make most of the missile launch tubes inoperable, and to make sure they stay that way. Proposed methods include filling and sealing some number of the launch tubes or cutting a segment of the launch section out of the hull; to preserve the invulnerability of such submarines, constraints may also have to be imposed on peacetime antisubmarine warfare practices (ref. 18.8).

Strategic bombers are designed to carry very large nuclear weapons payloads—up to 28 warheads in a mix of gravity bombs, short-range missiles, and long-range air-launched cruise missiles. The conversion of a B-1B or B-2 bomber to carry a single warhead, or a few, seems wildly cost-inefficient, and such bombers have nothing like the invulnerability of ballistic missile submarines. In addition, because strategic bombers can take off from many major air bases (500 estimated

* Pause just a moment to think about this again—a single submarine able to destroy something like 200 cities, each city full of men, women, and children. Right now, this minute, the oceans of the Earth are prowled by dozens of such submarines, each controlled by a few naval officers with Zeus-like powers—life and death over tens of millions of people, or more.

for the Soviet Union [ref. 8.4]) and can be diverted to numerous secondary airfields, the existence of strategic air forces on one side provides a justification for large nuclear strike forces on the other. Further, many of the airfields suitable for strategic bombers are in or near cities, making this weapons system particularly dangerous in the context of nuclear winter (ref. 18.13). An MSD regime may therefore be inconsistent with the bomber arm of the strategic triad (ground-based missiles, submarine-based missiles, aircraft). In contrast, ballistic missile submarines on station are much less vulnerable to attack, now and into the foreseeable future, and no attack against submarines at sea—so far as we can see—can make any contribution whatsoever to nuclear winter: Even a barrage of nuclear explosions can't make soot out of seawater. (In-port ballistic missile submarines—the normal U.S. stay at sea is about 60 days—put at risk a few coastal cities.) For this combination of reasons, we believe that serious consideration should be given in an MSD regime to finally phasing out the strategic nuclear bomber and its bombs and cruise missiles.

It has been argued at high levels since the Eisenhower Administration that bombers are, by many criteria, much less effective than ICBMs as strategic weapons delivery systems —although in the United States dogged efforts carried to extraordinary lengths have been undertaken to justify, develop, and deploy new strategic bombers, despite the wishes of at least two Presidents (ref. 17.19). Among the justifications proffered (see box) is the fact that, compared with missiles, bombers fly very slowly; the B-2, in fact, travels about as fast as a subsonic commercial airliner. This is said to give time to think twice about nuclear war in an international crisis. The presence of human crews aboard bombers, but not aboard missiles, is also cited as an advantage by those understandably chary about entrusting the global civilization to the judgment of robots. But the silo and submarine launch crews are no less human, and in an MSD regime—because there is no strategic advantage to a first strike—there is plenty of time to consider any crisis carefully before reaching for your missiles. The only compelling justification for manned strategic bombers would be as a hedge against the possibility that the missiles—most of the missiles, all of the missiles—might not work. If this is a real possibility, it speaks to systematic waste, incompetence, and criminal

# AN APOLOGIA FOR THE STRATEGIC BOMBER

[Gen.] Thomas D. White, the Air Force chief of staff . . . presented the Air Force's case for the B-70 [the bomber that became the B-1B]: The nation could not rely wholly on missiles, none of which had ever been fired in combat. Missiles could not be recalled, as airplanes could. Bombers could lift off and remain airborne while awaiting orders, thus giving the president a range of options in a crisis. Bombers would complicate the enemy's problem, forcing it to defend against several different kinds of attack. Finally, bombers as a demonstration of military might had a powerful psychological effect on friend and foe alike. . . .

[Then President Dwight] Eisenhower said, "We are talking about bows and arrows at the time of gunpowder when we speak of bombers in the missile age."

"There is a question," he [White] implored, "of what is to be the future of the Air Force and of flying. This shift [to missiles] has a great impingement on morale. There is no follow-on aircraft to the fighter and no new opportunity for Air Force personnel."

At the moment that General White lamented the decline of the Air Force, the service possessed 1,895 bombers, including 243 brand-new B-52s with several hundred more scheduled to be built. The Air Force had control of all three land-based ICBM systems. . . .

On January 11, 1960, General White served notice that he would fight the president on the B-70. . . .

Eisenhower told Republican leaders, "All those fellows in the Pentagon think they have some responsibility I can't see . . . I hate to use the word, but this business is damn near treason."

—Nick Kotz, *Wild Blue Yonder: Money, Politics, and the B-1 Bomber* (New York: Pantheon Books, 1988), 34–36.

cover-up on a scale so staggering that we believe we can safely reject it. (Tests of missiles on ballistic trajectories and the roughly 90% success rate in the launch of interplanetary space vehicles by both superpowers support this judgment.) If so, the justifications for the strategic bomber seem very thin.

Cruise missiles are air-breathing (not rocket-propelled) terrain-hugging vehicles that can be launched from the ground, from aircraft aloft, and from ships (including submarines) at sea. Because they are dual-capable—able to accommodate both conventional and nuclear warheads—they pose particular, although not insuperable, verification problems, and were one reason that progress on a START treaty was not announced at the Moscow Summit of 1988 or the Washington Summit of 1990 (ref. 9.6). We advocate abolition of nuclear-armed cruise missiles, using whatever verification measures may be needed. Conventionally armed cruise missiles—such as the French *Exocet*, which skims the water at an altitude of about 3 meters, and which was used successfully by Argentina to sink the British destroyer *Sheffield* in the Falklands/Malvinas war (ref. 18.14)—pose an additional danger, since in a time of crisis they could be rearmed with nuclear warheads. Approaches to this problem include inspection of all cruise missiles, tagging, sealing weapons inventory controls, and wholesale bans. The former astronaut Sally Ride and her colleagues have outlined a method for elimination of all nuclear-armed sea-launched cruise missiles, and argue that covert circumvention of a properly designed cruise missile treaty might be very difficult to achieve (ref. 18.15).

In a nuclear winter context the yield or explosive energy of a nuclear weapon is important: A 1- to 10-kiloton airburst can burn a large borough; the medium-sized cities of Hiroshima and Nagasaki were torched by 10- to 20-kiloton warheads. A 1-megaton airburst is quite enough to ignite a major metropolis (see frontispiece). This suggests the possible importance of placing limits on yields, were it possible. Modern nuclear weapons pack an enormous yield into a very small volume; large increments in yield are achievable with surprisingly small increments in weight. For counterforce targeting, a highly accurate missile does not require a high-yield warhead to destroy its target. This is one reason for the trend over the last decade or two (especially in the U.S. arsenal) toward lower yields per warhead (a trend now being reversed). But the greater the hardness of a missile silo, the larger is the yield (or the accuracy or burrowing ability) required to destroy the target. Limiting yields, if we wish to pursue it seriously, may necessitate inspection of all deployed warheads. But the dangers that verification is intended to avoid are so extraordinary that verification measures almost certainly will also have to be extraordinary.

In summary, a technically optimum approach to configuring a minimum sufficient deterrent force is to halt the growth of the existing nuclear arsenals, phase out systems in place, and phase in, instead, a range of strategic MSD systems, each with one warhead. This solution, although costly, should be much less expensive than current arms "modernization" programs, and, if properly organized, might also address the central problem of ending the nuclear arms race—by defining an asymptotic final configuration of forces and force levels consistent with strategic deterrence. Serious consideration must also be given to de-MIRVing (down to single warheads) appropriate components of the existing strategic forces, including land-based and submarine-launched ballistic missiles. The phasing out of strategic bombers, which are particularly ineffective in an MSD regime, would make sense in either case (ref. 18.16). To the extent that existing launchers are retrofitted, perhaps transitionally, the resulting mix of forces might, in range and accuracy, be somewhat less than optimally destructive, but in a stable MSD regime this is hardly a significant disability. In either case, great prudence and mutual (eventually multilateral) restraint would be needed to prevent the transition to an MSD regime from becoming a new kind of arms race (ref. 18.17).

# CHAPTER 19

# HOW DO WE GET TO MINIMUM SUFFICIENCY?

## SOME MILESTONES

*Yol bolsun* (May there be a way).

—a traditional farewell among Turki-speaking nomads in the desolate and roadless wastelands of Central Asia

IMAGINE THAT, WITH SOME FITS AND STARTS, CURRENTLY discernible global trends continue. Imagine that the United States and the Soviet Union agree to go down the path toward minimum sufficiency. Perhaps they are not committed to go all the way just yet. Perhaps they want to try out small, safe steps to test the waters. At first, probably, the strategic arsenals will be cut by one-third or one-half, as discussed in the START talks, and conventional forces in Europe will be further reduced, at least partly unilaterally. Verification protocols will be tested. Data on public attitudes to diminished arms will be gathered. Some incremental fraction of the military budget will become available to the civilian economy and to other urgent needs of the society. Some military scientists will be cycled back to industry and commerce. Some attempts will be made—by exaggerating a potential threat, perhaps, or by contriving an incident—to reverse the trend. If, nevertheless, arms reductions seem to be proceeding without significant hitch, further cuts will be implemented. Other nations will join. The process might begin to accelerate.

And yet, divestment must be done with great care—so that no nation, even temporarily, accumulates a real or perceived advantage. DeMIRVing schedules, for example, will have to be closely synchronized. Intrusive inspection and verification are essential. Divestitures of strategic arms must be accompanied by destruction of tactical weapons and reductions of conventional forces, especially in Europe. Toward the end of the process, an important new issue regarding terminal strategic defense arises. These matters—all connected with the middle stages of arms reductions down to minimum sufficiency—are the subject of this chapter.

Verification of an MSD regime poses special difficulties connected with the small size of the forces (see box). For example,

## MINIMUM DETERRENCE IN 1787

The American debate on minimum sufficiency began at the beginning. During the Constitutional Convention of 1787, in the discussion of

> . . . that dangerous institution, the standing army, [Elbridge] Gerry of Massachusetts actually proposed that the Constitution limit the size of the army to two or three thousand men. . . . Washington killed this idea by muttering audibly from the chair that they should next make it unconstitutional for any enemy to attack with a larger force. [Hugh Brogan, *The Pelican History of the United States of America* (London: Penguin Books, 1986), 205.]

The force and clarity of Washington's *reductio ad absurdum* is evident. This is why bilateral and eventually global agreements and stringent verification protocols are needed for MSD.

if there are nominally only 100 missiles and warheads on each side, then cheating might appear to be a real danger. What if an additional 100 missiles, say, could be hidden somewhere in the vast territory of the other side? The key issue, however, is not whether the nations could cheat, but whether they could cheat on a scale large enough to affect the strategic balance.

Each side—through comprehensive satellite observation of industrial plants and deployments, and other means—has today a very good idea of the other's weapons inventories. When the INF Treaty was implemented, an opportunity presented itself to compare the actual inventories with what had been derived from intelligence data. The concordance was excellent for the larger, longer-range, more readily detectable SS-20s. (The Soviet list of shorter-range SS-23s and ground-launched cruise missiles offered up for destruction was somewhat larger than the U.S. had figured [ref. 19.1].) The testimony of many intelligence officials is that each side knows the other's inventories to surprising accuracy.

Under current treaty, at times previously agreed, the covers are taken off the missile tubes on fleet ballistic missile submarines so the reconnaissance satellites of the other side can take a look. Under Article V of the SALT I Interim Agreement, deliberate concealment measures that impede satellite reconnaissance are prohibited. When an "environmental tent" (actually a prefabricated shelter) was erected over Minuteman silos undergoing "modernization" the Soviets soon complained (to a joint commission established to arbitrate such claims) that this was a violation of the SALT I Treaty, and the "tents" were modified. After U.S. complaints about a similar Soviet "pattern of concealment," Soviet actions changed and the charge was not repeated. These incidents occurred not in the atmosphere of heady goodwill after Gorbachev's accession to power, but in the gray Brezhnevian interval between 1974 and the first Reagan term of office. When both sides agree that cooperation in arms control is in their best interest, extraordinary precision in verification is possible. And if some small fraction of the wealth and talent that is currently devoted to making new weapons systems were redirected to safely getting rid of the old ones, still more thorough and reliable verification would be possible.

Since verification technology depends critically on observations from Earth orbit, serious arms reductions require rigorous prohibitions against antisatellite weapons, and against any strategic defense system—whatever its ostensible function—that can be used to shoot down or destroy surveillance satellites. In this, as in many other respects, Star Wars (SDI) is inimical to the establishment of a minimum sufficient deterrent (ref. 19.2).

To protect against the possibility of a clandestine strategic force, there are carefully designed verification procedures that could be mutually agreed upon. As mentioned earlier, many key verification procedures in the INF Treaty, SALT I and II, and the 1986 Stockholm Protocol are being implemented; verification is now recognized as plausible and achievable.

We suggest the following measures:

(i) Rigorous and unambiguous protocols to inspect the dismantling and destruction of warheads (ref. 19.3) and delivery systems. This might occur at or near the site of deployment, or at facilities jointly staffed by all nations party to the treaty. One suggestion is that the weapons-grade fissionable uranium and plutonium removed from the warheads be used to generate electricity in existing commercial fission power plants (ref. 19.4)—the ultimate in beating swords into plowshares.

(ii) Unhindered space-based surveillance of deployed forces to determine weapons characteristics, numbers, movement, testing, and so on—an intensification of the current routine monitoring of strategic forces. This requires the cooperation of each side with the reconnaissance protocols of the other.

(iii) Continuous on-site monitoring of all places at which fissionable material is known to be mined or refined, and where weapons are assembled, stored, or deployed (ref. 19.5). Were the U.S. and U.S.S.R. deeply committed to massive arms reductions they would be developing, and sharing, new monitoring technologies—which would be of further use when nuclear arms reductions extend to other nations.

(iv) Guaranteeing widespread public understanding on each side of the illegality, unpatriotic stupidity, and danger of a sequestered strategic force, and establishing reprisal-free means for citizens to report questionable facilities in their own nations.

(v) U.S. general-purpose mobile inspection teams stationed in the U.S.S.R., and Soviet teams in the U.S.A., with rapid access to any sites deemed suspicious. Some Americans, enjoying more advanced technologies in many fields, are reluctant to allow such access (ref. 19.6). They fear military and industrial espionage. Exclusions, in areas that could not be central to weapons development, could doubtless be negotiated. But having made the point for decades that Soviet objections to verification were the obstacle to arms control, it is unseemly for Americans to gulp when the Soviets come round to their point of view.

(vi) Scrupulous adherence to the present ban on encrypting data radioed back during strategic missile tests.

(vii) Use of intelligence information from other sources. Could either side be absolutely sure there was not a mole in, say, their Politburo Defense Committee or National Security Council? Could the "benefits" of cheating—

# CAN CITIZENS MAKE IT HARD TO CHEAT AS THE ARSENALS GET SMALLER?

Leo Szilard, the first person in human history to understand that a nuclear chain reaction is possible, later made heroic efforts to understand how massive cuts in the strategic arsenals might come about. In 1961 he proposed a way by which citizens might help guarantee that their own governments abide by arms reduction treaties:

> When the agreement was signed and published, the Chairman of the Council of Ministers would address the Russian people and, above all, the Russian engineers and scientists, over radio and television and through the newspapers. He would explain why the Russian government had entered into this agreement and why it wished to keep it in force. He would make it clear that any secret violation of the agreement would endanger the agreement and that the Russian government would not condone any such violation. If such violations did occur, as they well might, they would have to be regarded as the work of overzealous subordinates whose comprehension of Russia's true interests was rather limited. In these circumstances, it would be the patriotic duty of Russian citizens in general, and Russian scientists and engineers in particular, to report any secret violations of the agreement to an agent of the International Control Commission. In addition to having the satisfaction of fulfilling a patriotic duty, the informant would receive an award of $1 million from the Russian government. A recipient of such an award who wished to enjoy his wealth by living a life of leisure and luxury abroad would be permitted to leave Russia with his family.
>
> The Russian scientists pointed out that by repeating the same thesis over and over again, as they well knew how to do, the Russian government could create an atmosphere which would virtually guarantee that Russian scientists and engineers would come forward to report secret violations.

> The Russians further proposed that agents of the International Control Commission maintain establishments in all Russian cities, and in the larger cities several establishments. An informant could simply walk into such an establishment with his whole family and make a deposition. [Szilard, "The Disarmament Agreement of 1988," *The Voice of the Dolphins* (New York: Simon and Schuster, 1961), 57–58.]

Of course, identical arrangements, spelled out by treaty, would be made in the United States and other nuclear-armed nations.

As much as we admire Leo Szilard's inventiveness, it occurs to us that this might be an arrangement ripe for subversion—e.g., by intentional false alarms. Somewhat different arrangements might be more workable. Any successful arrangement could work only in conjunction with a range of other verification methods, as we have outlined.

given the secure retaliatory force on the other side—balance the possible consequences of being found out?

Even if verification were not foolproof, a properly designed arms reduction protocol would guarantee enormous difficulties for whichever side sets out to achieve a significant clandestine advantage. Even major cheating—say *doubling* the forces on one side (cf. ref. 18.8)—could not seriously threaten the retaliatory capability of a well-designed MSD (due to the nature of the mathematics of weapons survivability), even if the other side knew precisely where every missile was. Depending on the MSD force configuration, special verification problems might have to be solved—e.g., for nuclear-armed sea-launched cruise missiles, or for guaranteeing that deMIRVed boosters are not clandestinely *re*MIRVed. But the solutions to these problems do not seem to be beyond reach. We see verification as an essential, achievable, and stabilizing element of a minimum sufficiency regime.

There is a necessary tension between MSD forces that are invulnerable, mobile, or hidden—as required for strategic stability—and forces that are subject to verification by such means as we describe here. But this is nothing new: Submarines that spend most of their days at abyssal depths, hiding from reconnaissance satellites, at agreed-upon times surface and open their tubes to inspection from space in accord with treaty protocols. An important criterion for the design of MSD forces is establishing an optimum balance between these conflicting criteria.

An extended system of strategic defenses to protect the citizens of a vast nation from ballistic missiles, as originally envisioned in the Star Wars or SDI program, would be highly destabilizing (ref. 19.7) as well as ruinously expensive—even if it were possible, which it is not. An effective strategic defense against a general attack of many thousands of warheads and perhaps a million decoys is widely regarded as an illusion (ref. 17.5). And, as already noted, what a Star Wars system *can* do, readily enough, is shoot down satellites—and this capability is inconsistent with the essential role of treaty verification by reconnaissance satellites in a regime of minimum sufficiency.

One still fashionable justification for strategic defense is that, while it may not be able to protect the civilian population of the United States, "it undermines the confidence that an attack will succeed" (ref. 19.8). But the same remark is true, in a different way, of nuclear winter. Therefore, to the extent that generating strategic uncertainty is a valid justification for strategic defense, nuclear winter reduces the need. We also note published concerns about at least two separate mechanisms by which SDI might bring on its own variety of nuclear winter (ref. 19.9).

Limited *terminal* defenses (protecting hardened missile silos, or the traverse area of mobile missiles) to strengthen the survivability of an MSD might conceivably be stabilizing and desirable—provided the additional expense could be justified, and the defensive system could not be used as an effective antisatellite weapon (which seems to imply severe constraints on the range and acceleration of the missiles). For example, the single-site 100 interceptors already permitted by the Antiballistic Missile (ABM) Treaty might enhance MSD survivability and thus deterrence. Such defenses would be non-nuclear and require no space-based weaponry. Unlike Star Wars, terminal defense over limited areas—a missile silo, say, or part of a coastline—is relatively inexpensive and well within our technological capability. The *Galosh* system of 100 rocket interceptors around Moscow is an example; but much more capable, high-acceleration, very-short-range systems are possible. The offensive missiles would have to be verifiably stripped of any penetration aids. Note that the forces we are describing are strictly consistent with the ABM treaty: For 100 interceptors to be useful, we do not want very many more than 100 land-based launchers. On the other hand, development and continuing refinement of any but very short-range area-defense systems poses the destabilizing threat of breakout: the fear that one side will develop effective regional strategic defense in the weapons-sparse MSD regime and in effect disarm the other side. For this reason, the *cons* of MSD terminal strategic defense may outweigh the *pros*. In either case, the present ABM treaty should be preserved.

> Nor was I ever able to see any military reality in what is now referred to as theater or tactical warfare. . . . The men in the nuclear laboratories of both sides have succeeded in creating a world with an irrational foundation, on which a new set of political realities has in turn had to be built. They have become the alchemists of our times, working in secret ways which cannot be divulged, casting spells which embrace us all.

These words were spoken by Lord Solly Zuckerman, former chief science adviser to the British Defence Ministry (ref. 19.10). Such sentiments are still considered irresponsible or (literally) unspeakable in many of the corridors of military power.

Despite the profound recent political and social changes in Eastern Europe, there are over 10,000 tactical warheads and nuclear "munitions" in Western and Eastern Europe, apart from the U.S.S.R. Tactical nuclear weapons are intended to counter regional aggression. They are currently the most likely potential triggers for a global nuclear war. Today's U.S. tactical weapons are deployed to: (a) bolster the Western conventional forces in Europe; (b) act as a tripwire (politely called "linkage" or "coupling"), reassuring European NATO allies of a U.S. commitment to defend against a Soviet attack (ref. 19.11); (c) conduct regional warfare in the European and Asian theaters;

**1** Two one-megaton-yield nuclear warheads explode over the New York metropolitan area. Dark, sooty fires are set substantial distances away.

[Painting by Chesley Bonestell. Courtesy, Frederick C. Durant III, Bonestell Space Art.]

**2** A low-yield nuclear weapon explodes over the Nevada desert in the "Climax" test, June 4, 1953. This weapon had a yield roughly four times that of the bomb that destroyed Hiroshima. [Courtesy, U.S. Department of Defense.]

**3** The black, sooty smoke from the burning of a small modern building. A great deal of such smoke, at high altitude, prevents ordinary sunlight from reaching the ground and warming the Earth, but permits infrared thermal radiation to escape to space and cool the Earth—optimum properties for generating nuclear winter.
[Courtesy, B. Williamson, Lawrence Berkeley Laboratory.]

**4** Hiroshima, after its destruction on August 6, 1945, by a roughly twelve-kiloton nuclear weapon dropped by the United States. This explosion and the bombing of Nagasaki two days later put a prompt end to World War II. In this first nuclear "war," all available weapons in the stockpile (both of them) were used. In a modern nuclear war, destruction would be much greater and the

killing much more widespread—in part because there would very likely be nowhere from which to receive help and few places to which to escape; and in part because of nuclear winter, whose effects would only begin to be felt days after the war was over.
[U.S. Strategic Bombing Survey photo; courtesy, Popperfoto, London.]

**5** Successive pictures of the same strip of landscape and sky on Mars, as taken by *Viking 1* in its extended mission in the early 1980s. A "sol" is a Martian day, a little longer than an Earth day. The days around sol 1742 witnessed a dust storm and a lowering of surface temperature. Such correlations, connected with the mechanism of nuclear winter, amount to a few °C temperature drop for a typical heavy dust storm.

[Image courtesy NASA and The Planetary Society.]

**6** The Sinai Peninsula at the boundary of Israel and Egypt, just after a nuclear war begins. Cities and petroleum facilities are in flames. The resulting black, sooty clouds are blown in a common direction by the prevailing winds. At other targets, larger smoke plumes are carried to higher altitudes. The smoke merging from many such plumes is predicted to generate nuclear winter.

[Painting copyright © 1983 by Jon Lomberg. Courtesy of the artist.]

**7** A nuclear war begins. We are looking down on the North Pole. The boundary between day and night cuts across the Arctic ice cap. Each bright flash of light represents a fireball from a nuclear explosion. Only about 100 of the many thousands of explosions in such a war are being detonated at the moment depicted. North America, mainly in daylight, is at bottom. The Eurasian landmass, in darkness, is at left. Europe is at top right. Enormous clouds of smoke can be seen, carried eastward by the prevailing westerlies. From this distant view it is difficult to determine who started the war.

[Painting copyright © 1983 by Jon Lomberg. Courtesy of the artist.]

**8** The same view of the Earth about a week later. Again the Arctic ice cap is below us, and again the ribbon of twilight cuts across it. Smoke from the fires set by thousands of nuclear explosions has now risen to high altitude, spread, and merged, covering much of the Northern Hemisphere. Light levels and temperatures have plummeted. In some places the smoke is still patchy; it continues to spread north toward the Pole, and south toward and across the Equator. Even now there are a few fresh nuclear explosions—perhaps ordered by the commanders of ballistic missile submarines, unaware that the war is "over."

[Painting copyright © 1983 by Jon Lomberg. Courtesy of the artist.]

**9** Ten days after the war. Another view of the world after nuclear war—this one from an observing station high above the Equator. The smoke has covered most of the Northern Hemisphere. Outriders of smoke reach as far south as Patagonia. But Central and South America have, so far, escaped the worst of the smoke. Note, at northern midlatitudes, the patches of very-high-altitude, brighter smoke: This is dust from high-yield groundbursts that works to cool the Earth through a different mechanism—by reflecting more sunlight back to space, before it reaches the dark smoke.

[Painting copyright © 1983 by Jon Lomberg. Courtesy of the artist.]

**10** Equatorial view of the Western Hemisphere some weeks after a nuclear war. The main smoke pall, although still patchy, has reached Argentina and Chile. This and the last three figures depict one of many nuclear war scenarios. In many models the main smoke pall would by this time be much less patchy, although not as much smoke would have reached so far into the Southern Hemisphere (unless nuclear weapons were exploded there).

[Painting copyright © 1983 by Jon Lomberg. Courtesy of the artist.]

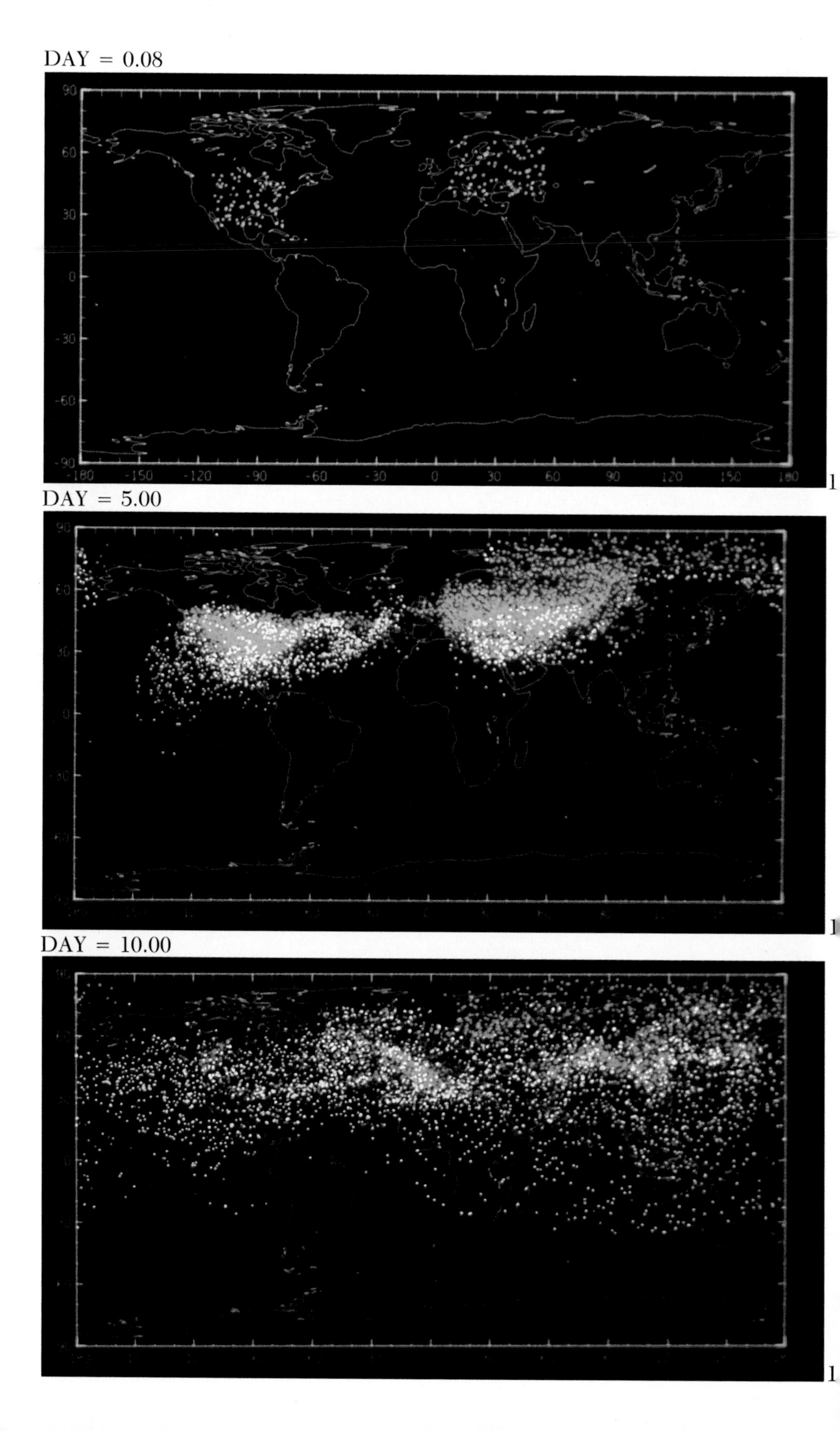
DAY = 0.08
90
60
30
0
-30
-60
-90
-180
-150
-120
-90
-60
-30
0
30
60
90
120
150
180
DAY = 5.00
DAY = 10.00

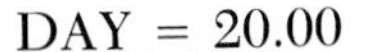


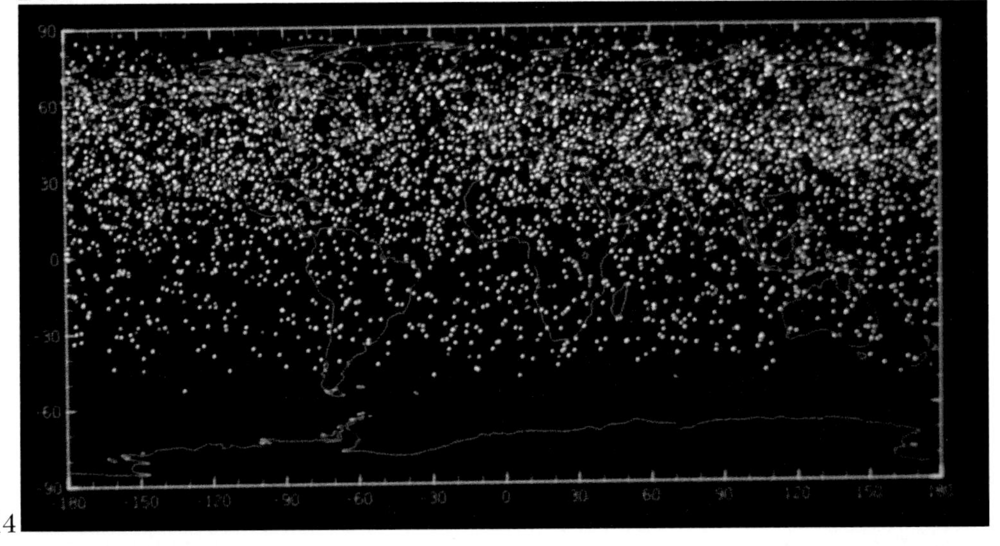


14

DAY = 150.00

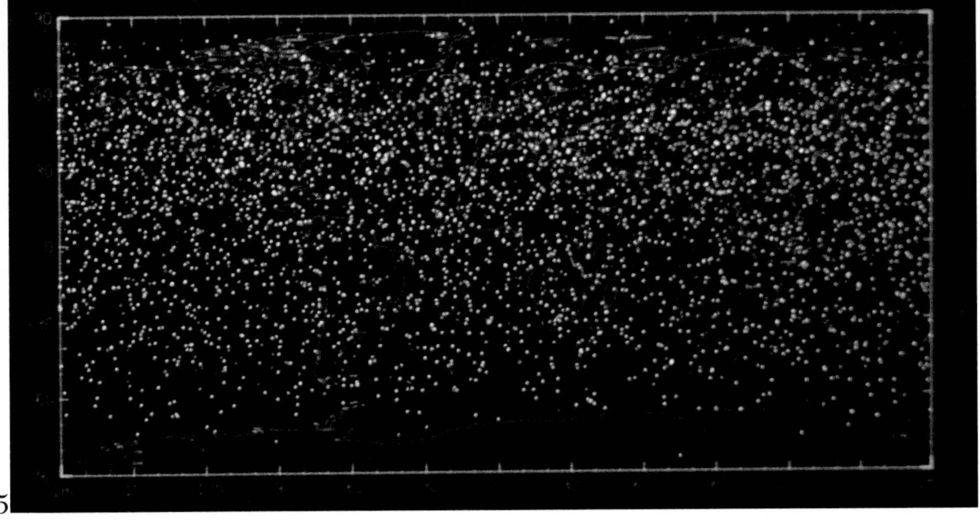

15

**11** Smoke produced in a hypothetical nuclear war, with fires and smoke emission arbitrarily confined to the continental U.S., Europe (excluding the U.K.), and European U.S.S.R. Each colored dot represents a certain fixed mass of smoke. All the dots together show the extent of smoke injection two hours after the war. It is assumed that the war takes less than an hour; i.e., that all the weapons are detonated essentially simultaneously. Blue dots indicate smoke below 10 kilometers altitude (6 miles). Yellow dots indicate smoke, including stratospheric smoke, above 10 kilometers altitude. The accumulating smoke, still mainly in the lower atmosphere, will when spread over a hemisphere be enough to generate a Class III ("nominal") nuclear winter. The subsequent movement of this smoke, both vertically and horizontally, as calculated by a state-of-the-art three-dimensional general circulation model, is shown in Plates 12–15.

**12** Spread of nuclear war-generated smoke of Plate 11, five days after the war is over. For computational convenience each dot, representing a blob of smoke driven by solar heating and the circulation of the atmosphere, is allowed to move, but, in order that we can follow the net effect, is not allowed to spread and become a larger but more diffuse cloud. As before, yellow dots indicate smoke at more than 10 kilometers altitude, mainly in the stratosphere. By five days, the smoke covers nearly all Northern midlatitudes, and a Class III nuclear winter is well under way.

**13** Spread of smoke ten days after the nuclear war assumed in Plate 11. Much of the Northern Hemisphere is now covered, and outriders of smoke have arrived in the Southern Hemisphere. Significant temperature declines and obscuration of the Sun are widespread.

**14** Spread of smoke nearly three weeks after the nuclear war assumed in Plate 11. Smoke has now reached most parts of the globe. Semitropical, tropical, and Southern Hemisphere latitudes are covered by stratospheric smoke (represented by yellow dots), although the density of smoke (represented by the abundance of dots) is less as we go south. There is less lower atmospheric smoke (blue dots) than in previous figures because of rainout and fallout.

**15** Worldwide distribution of (almost entirely stratospheric) smoke five months after the nuclear war of Plate 11. Depending on targeting, more widely distributed and/or larger nuclear wars could generate more smoke and more severe cold and dark than in this nominal nuclear winter.

[Diagrams Plates 11–15 courtesy Robert C. Malone and Gary A. Glatzmaier, Los Alamos National Laboratory.]

**17** Average temperature changes about a week into a Class III ("nominal") nuclear winter produced by a July war between the superpowers. Additional cooling by dust is ignored. The largest temperature declines shown are over 25°C (or 45°F). These are average temperatures over about a week; maximum daily temperature drops would be larger. Note nuclear winter effects beginning to propagate into East Asia and sub-Saharan Africa. Temperature declines are already serious enough to wipe out grain production in the Soviet Ukraine and in the American and Canadian Midwest.

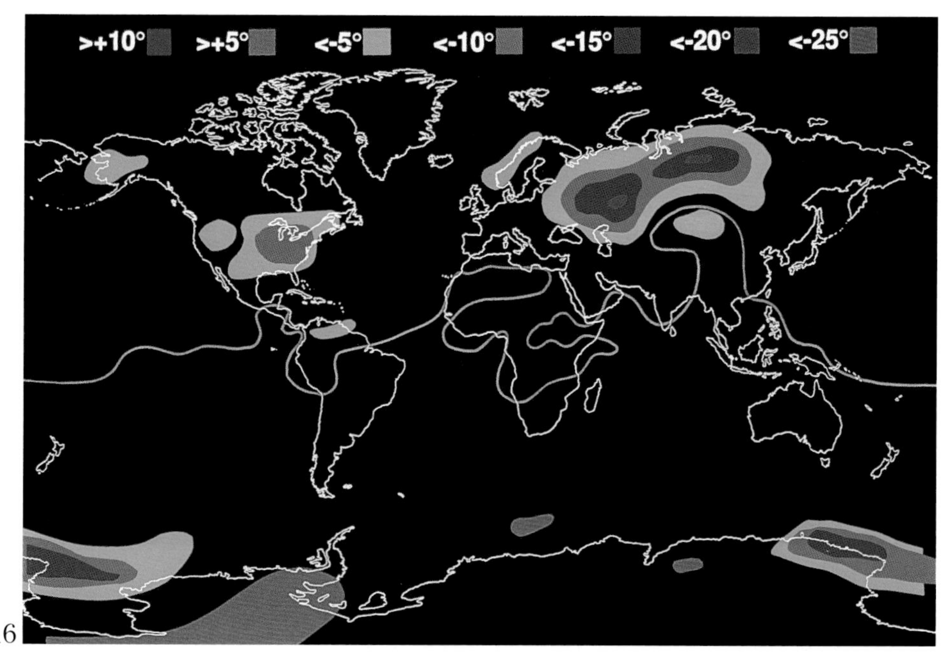


**16** Average nuclear winter temperature changes around a week (average of days 5 through 10) after a July nuclear war. Enough smoke is assumed to be generated by nuclear explosions to produce a Class II ("marginal") nuclear winter. Additional cooling by dust is ignored. Temperature changes (in degrees Centigrade or Celsius) are color-coded at top: "<" means "less than" and ">" means "more than." The largest temperature declines shown (more than a 20°C or 36°F drop) occur in the American Midwest and Soviet Central Asia. These are rough weekly averages; maximum *daily* temperature drops would be larger (generally a further cooling of 10°C or more: see Figure 1). Note the sudden chill in Venezuela.

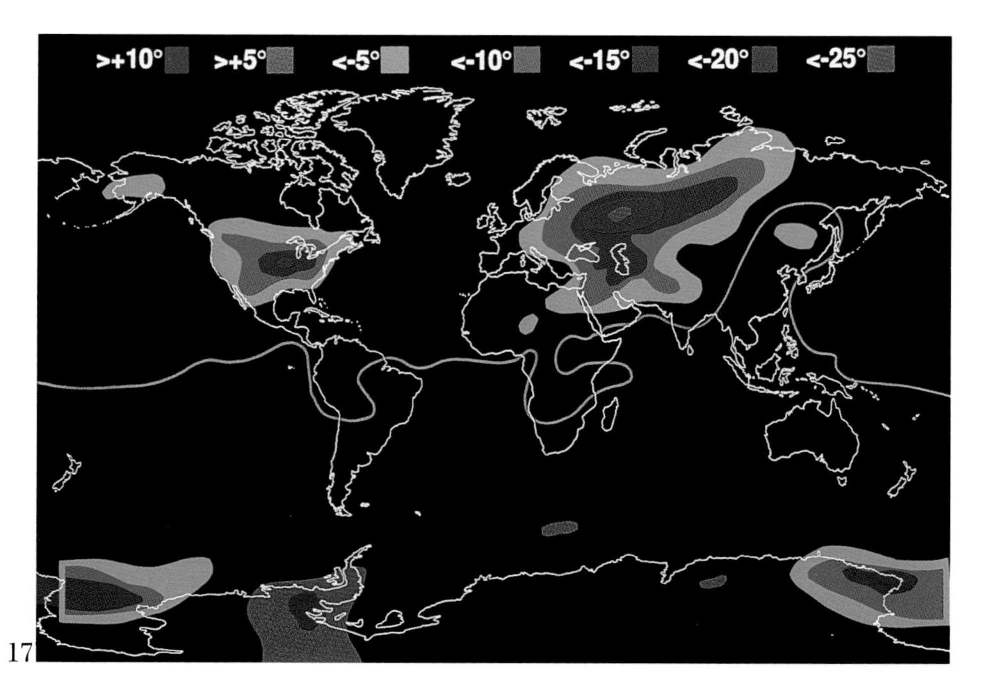

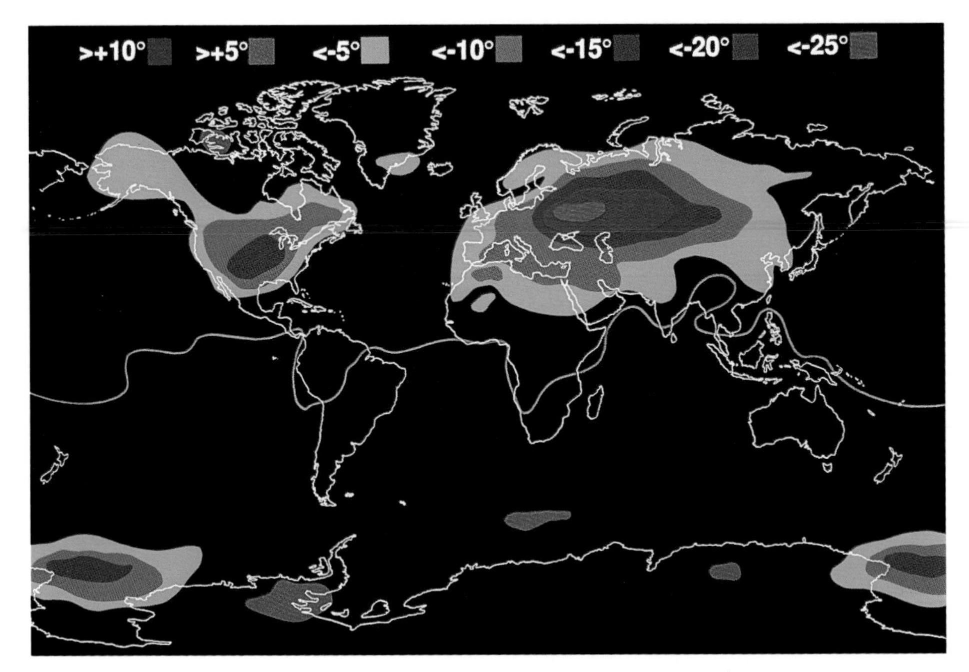


**18** Average temperature changes about a week into a Class IV ("substantial") nuclear winter, produced by a July war between the superpowers. The smoke emissions in this case are greater than in the previous two cases, but are just as likely to occur for the same weapons detonated at the same targets. Additional cooling by dust is ignored. Major temperature drops have gripped most of North America and the Eurasian landmass. Temperature declines of over 25°C (45°F) prevail over an area that extends roughly from Kiev to Moscow to the Urals. These are rough weekly averages; maximum daily temperature drops would be larger (generally a further cooling of 10°C or more). Extensive areas, in both the Eastern and Western Hemispheres, would be—at least intermittently—below freezing in midsummer. No comparable studies have ever been performed for the more severe Class V or Class VI nuclear winters.

[Data for Plates 16–18 courtesy Robert C. Malone and Gary A. Glatzmaier, Los Alamos National Laboratory. Map prepared for this book by Bell Production Services, Toronto.]

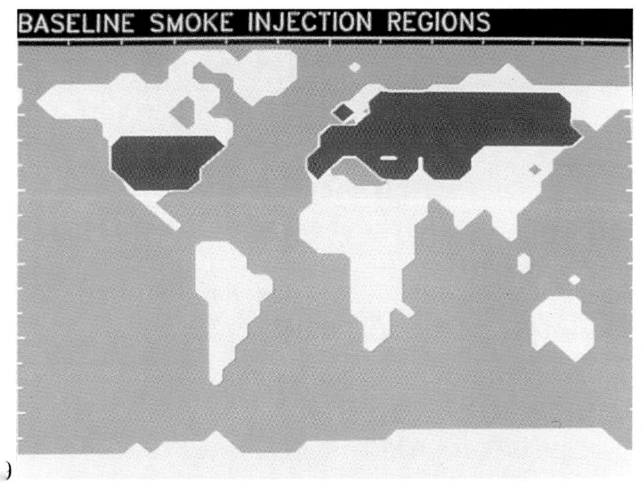


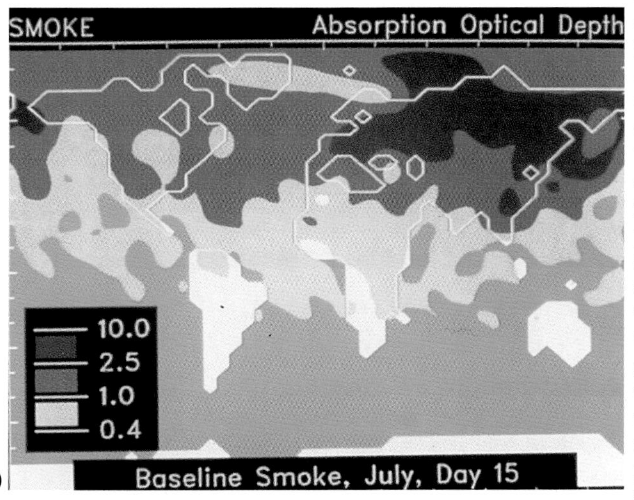


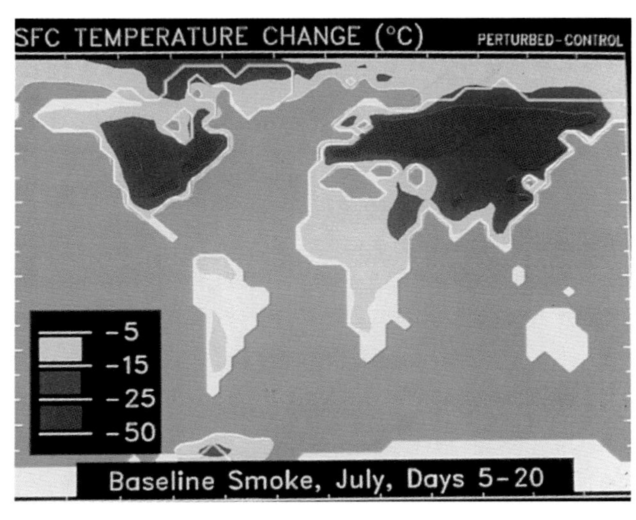


**19–21** Nuclear winter three-dimensional general circulation models, circa 1985, performed at the National Center for Atmospheric Research (NCAR) in Boulder, Colorado. Plate 19 shows the assumed smoke emission zones in the Northern Hemisphere, which are restricted to NATO and WTO states. Plate 20 shows—for day 15 after a July war—the distribution of smoke in terms of absorption optical depth, $\tau_a$, for a "baseline" nuclear exchange. Large portions of the Soviet Union show $\tau_a$ between 2.5 and 10, and tendrils of smoke with $\tau_a$ between 0.4 and 1 reach to South Africa. Plate 21 shows, averaged over days 5 through 20, the resulting temperature changes—between a 15° and a 50°C drop in the target zone. More modern three-dimensional general circulation models show declines in the same zone about half as large for comparable optical depths (cf. Los Alamos National Laboratory estimates in Plates 16–18). All recent models, including those at NCAR, are in rough agreement. The largest remaining uncertainty is the choice of $\tau_a$: How much smoke will be distributed where for a given nuclear war?

[Figures courtesy NCAR and the Center on the Long-Term Consequences of Nuclear War, Natural Resources Defense Council, Washington, D.C.]

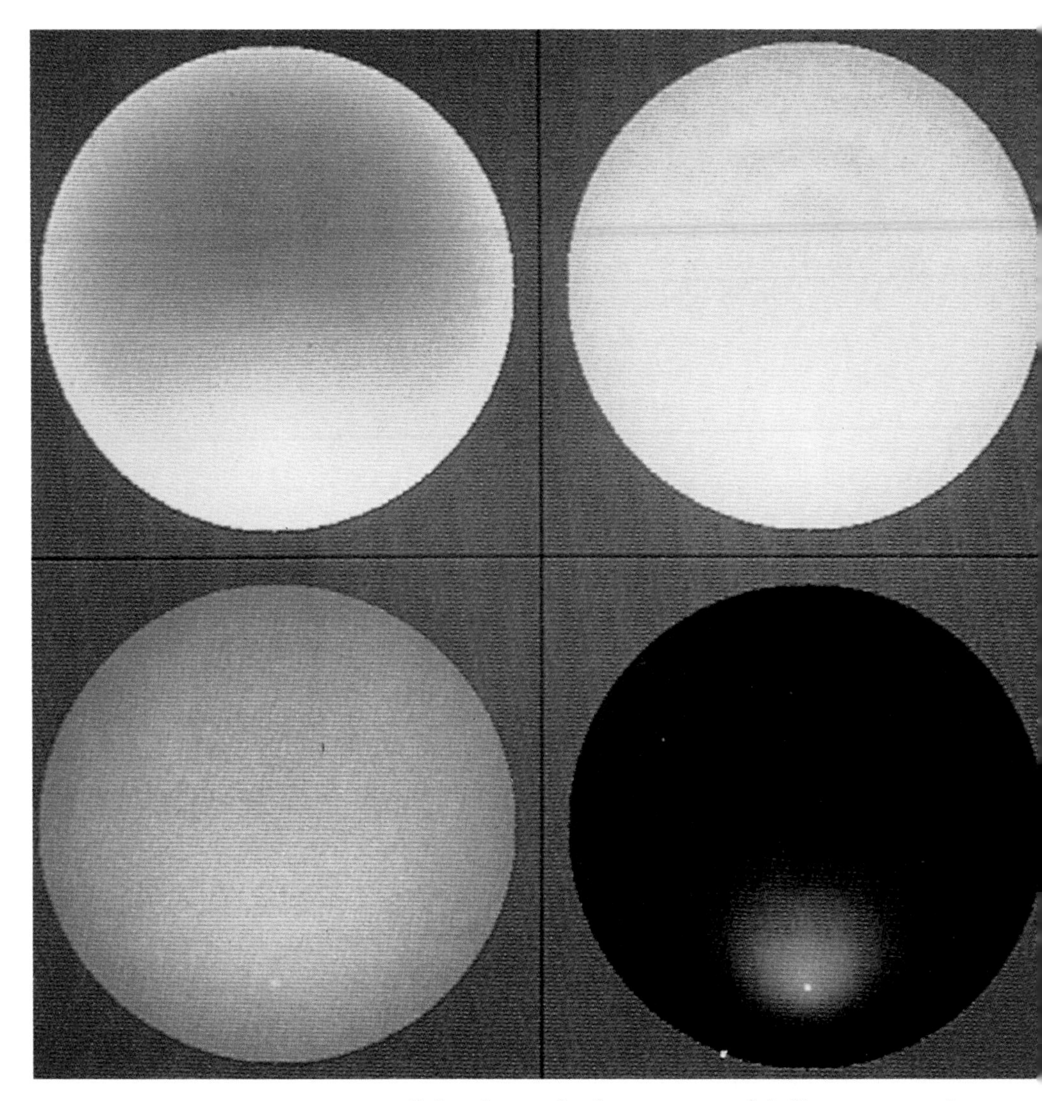

**22** Representations of the sky with obscuration of different optical depths, as if photographed by an all-sky camera. The zenith is at the center of each circle, the horizon at the periphery. *Upper left:* Normal sky. *Upper right:* Cloudy sky. *Lower left:* Sky with twenty times the normal concentration of aerosols. *Lower right:* The sky in nuclear winter, TTAPS baseline case.

[Courtesy, T. Nakajima and H. Ogawa, Upper Atmosphere Research Laboratory, Tohoku University, Sendai, Japan. *Bulletin,* Tohoku University Computing Center, *19* (1986), 78–83.]

and (d) carry out naval operations. Intermediate-range nuclear delivery systems (e.g., FB-111 aircraft) in Europe and Asia complement the tactical nuclear forces, allowing operations that extend to rear echelons, military bases, and depots.

Under an MSD agreement, the entire concept of limited nuclear warfare using tactical weapons would be repudiated, and essentially only strategic nuclear deterrence would be left. This raises important questions of balance in conventional forces and of how strategic weapons might be used to dissuade or counter conventional aggression.

We suggest: (1) strictly verified withdrawal and destruction of all intermediate-range nuclear delivery systems (already accomplished for intermediate-range ballistic missiles under the INF Treaty); (2) strictly verified withdrawal and destruction of all tactical nuclear weapons; (3) balancing of the conventional forces in Western Europe* by reduction and restructuring of WTO and NATO forces; and (4) establishment of limited use declaratory policy for the MSD weapons in their role of deterring large-scale conventional military aggression.

The removal of all forward-based tactical nuclear weapons is already under consideration by the superpowers and has several advantages: It would account for the withdrawal of half to two-thirds of all the nuclear weapons on Earth; it would eliminate the weapons most likely to be used in a conflict and that are under the least rigorous command and control; it would destroy the weapons most likely to be overrun in a conventional attack—and the weapons, therefore, most vulnerable to the "use 'em or lose 'em" dilemma of field commanders; and it would set the stage for (or accompany) major conventional force reconstruction and strategic nuclear force reductions.

Before the 1980s, U.S. tactical nuclear weapons in Europe could, if used early or preemptively, effectively counter any Soviet conventional superiority. But in the 1980s both NATO conventional forces and the Soviet tactical nuclear arsenal were built up. As a result, U.S. tactical nuclear weapons became primarily directed against Soviet formations *before* they cross the border (either before or after war has begun). The NATO ver-

---

* More than half the planet's advanced conventional weaponry is currently concentrated in the European Central Front (ref. 19.12), despite the decay of WTO's military capability.

sion of this doctrine was called "Airland Battle 2000." It was bound to be destabilizing, because it is virtually indistinguishable, in Soviet eyes, from preparations for a tactical first strike. Thus both sides acquired a tactical nuclear weapons offensive capability, a dangerous development. On the other hand, the revolution in Eastern Europe makes a tactical nuclear attack by either nation increasingly implausible. The case that tactical nuclear weapons are essential to deter a conventional attack is substantially weakened. But there are no guarantees as long as the weapons are there. Policies change. Accidents happen.

A transitional solution is to verifiably establish a nuclear weapons-free corridor in Central Europe. It might be about 100 kilometers wide, and filled only with light infantry of the two blocs, or by a U.N. peacekeeping force, or maybe just by civilians going about their daily business. It would provide much longer warning times and much diminished incentives for the early use of tactical nuclear weapons. Such proposals, including some with considerably wider corridors, have already been offered by WTO (ref. 19.13). Denuclearization should be much more stable when it applies to entire nations (e.g., the two Germanys plus Czechoslovakia) than to thin buffer zones.

In the United States, the National Commission on Integrated Long-Term Strategy urges that "arms control policy should give increasing emphasis to conventional reductions" (ref. 17.22). Former U.S. arms control Ambassador Jonathan Dean has outlined a ten-year plan in which six selected armaments categories in WTO and NATO forces would be reduced to 50% of present NATO holdings, and ground and air personnel would also be reduced to half. Such a force structure is expected to save some $75 billion per annum for the United States, and a comparable amount for the other NATO nations (ref. 19.14).

WTO nations have asked for the removal of Soviet troops, and the U.S.S.R. has acceded and begun to comply. On December 7, 1988, at the United Nations, President Gorbachev announced the planned unilateral withdrawal and demobilization of some 500,000 Soviet troops (including 90,000 officers), 10,000 tanks, 8,500 artillery pieces, and 800 tactical aircraft by 1991. This declaration presents a historic opportunity for massive mutual conventional force reductions in Europe. The recent transition to democracy in Eastern Europe eliminates the perceived need for Soviet forces to police doctrinal adherence and political docility in WTO nations. The historic opening of the border between the two Germanys in November 1989 and German unification greatly reduces the likelihood of WTO or NATO conventional aggression in Central Europe. The Pentagon now estimates that the risk of conventional war in Europe is the lowest it has been since 1945. A National Intelligence Estimate judges that Soviet force reductions, the increasing openness of Eastern European society, and U.S. intelligence improvements have lengthened the warning time of a major WTO attack on NATO to several months (ref. 19.15). CIA and Defense Department assessments conclude that such an attack is now highly unlikely and, if it occurred, could be repelled by NATO conventional forces (ref. 19.15). A massive surprise conventional attack from East to West in Europe—whatever its likelihood may once have been—is now highly improbable, on grounds both of opportunity and of motivation. The argument for extended deterrence is eroding. "We are no longer adversaries," says NATO (ref. 17.23). The Cold War is over.

In his 1990 State of the Union address, President Bush called for the two nations to reduce their troop levels in Central Europe to 195,000 men each (ref. 19.16). The proposal means cutting U.S. forces by 60,000 men and women and Soviet forces by 370,000; in addition, 30,000 American troops in Southern Europe were to be exempt from the troop cuts. After gulping once on this latter proviso, the Soviet Union agreed. Other serious and substantive U.S. responses to Soviet proposals and actions on conventional force reductions are urgently needed.

The balancing of conventional forces in Europe—including infantry, armor, tactical and intermediate range aircraft, naval airpower, and cruise missiles—is a specialized problem that we will not comment on here, except to note that it is almost certainly achievable (ref. 19.17). Any small residual imbalances would be irrelevant in an MSD regime—in part because of the inherent and historical advantages of conventional defense over conventional offense, which requires of a successful aggressor numerical superiority, according to common military wisdom, of a factor of 1.5 to 3 over the defender. Moreover, with the force structures outlined here, the overall deterrence of the combined nuclear and conventional forces would be synergistic and reinforcing.

The restructuring of conventional forces in Europe would also be important in maintaining the strong alliance between the U.S. and Western Europe. The proposed removal of U.S. nuclear weapons from Europe has occasionally been criticized as an abandonment of the defense of Western Europe. But with the deployment of the minimum deterrent forces outlined here, the net military security of Europe would be increased: Nuclear deterrence would be enhanced, conventional forces balanced, crisis stability greatly increased, and nuclear winter put virtually out of reach. Tactical nuclear weapons and extended deterrence would be rendered unnecessary; indeed, that is almost the case today. Given the reservations that many nations have about U.S. nuclear policies, deep reductions in nuclear forces accompanying the establishment of an MSD regime would tend to strengthen some American political positions worldwide.

In suggesting the establishment of MSD forces that surely represent a major departure from current strategic postures, we must, reluctantly, discuss appropriate policies on targeting and use—recognizing that foolproof verification and enforcement of such protocols cannot be assured by technology alone. Potential targets in an MSD regime include principal military installations; conventional force formations; primary $C^3$I ("I" for intelligence) and leadership centers; key industrial complexes; and major urban/economic centers. Some selection among these categories would be required, but the number of weapons in the arsenal would be too few to bring about nuclear winter. In one sense, the MSD policy of retaliation is roughly equivalent to the policy of mutually assured destruction (MAD), in that some urban targets would be included at least implicitly in the target set. Surely, no assurances that cities are omitted would be credible. If only because of the many fewer warheads available in the world in an MSD regime, the *number* of cities threatened would be much less than is the case with the present force structures. But a greater *fraction* of the deliverable warheads might be targeted on cities in an MSD regime, in order to maximize deterrence with limited forces. That is what the present small British, French, and Chinese forces do. City targeting remains a ghastly and hateful necessity, but it is intrinsic to the very nature of nuclear weapons. This is one reason that, despite all the technical, political, and psychological impediments, many long for a world that has abolished the things altogether (Chapter 22).

Because of the structure of the MSD forces, weapons would be under strict, continuous, positive control, and after an initiation of hostilities could—should circumstances warrant—be withheld without fear of destruction while diplomatic negotiations were pursued. This makes for a much less hair-trigger nuclear arsenal, and therefore for a much less perilous world.

U.S. (and British and French) nuclear forces are sometimes still viewed as a counterpoise to what has (often and erroneously) been described as "overwhelming" Soviet superiority in conventional forces in Europe. However, there is increasing evidence that NATO conventional forces were for many years fully adequate to turn back or significantly delay a WTO surprise attack or long-preparation attack on Western Europe (see box and ref. 19.18)—even before the strain on WTO from the revolutions in Eastern Europe and unilateral Soviet troop withdrawals (ref. 19.19).

The Soviets have proposed (ref. 19.13) a long-term program of asymmetrical reductions in conventional forces in Europe—perhaps even extending from the Atlantic to the Urals—whereby WTO personnel and armor are withdrawn and demobilized in much larger numbers than are NATO forces, and a large central zone of Europe is cleared of forces; the proposal also includes reduction of NATO and WTO forces by a million men and women in five years. The process has, in a way, begun with the agreements in principle in February 1990 for the two nations to remove 430,000 troops from Central Europe, with a USSR/USA withdrawal ratio of more than 6 to 1. (Cf. ref. 19.16.) This does not prevent war, but it makes surprise attack much less likely to occur or to pay any dividends. Under these circumstances the perceived need for battlefield or tactical nuclear weapons becomes less urgent, and a pledge of no-first-use of nuclear weapons entails, to a much greater degree than today, no forfeiture of a militarily significant option. This, in turn, slows horizontal proliferation of nuclear weapons. The breaching of the Berlin Wall and the raising of the Iron Curtain help make such steps stabilizing and practical.

. . .

# "MUTUAL PRECLUSION": THE JOINT CHIEFS OF STAFF ON CONVENTIONAL WAR IN EUROPE

Overall, the four . . . key intangibles—personnel quality, alliances, technology, and industrial mobilization capabilities—. . . favor the United States and its allies and would help NATO overcome wartime quantitative force advantages of the WP [Warsaw Pact = Warsaw Treaty Organization = WTO]. These factors complement NATO's formidable and improving force capability and raise serious questions about the WP's ability to prevail over NATO in a conventional conflict.

. . . The assessments indicate that US and allied ground, air, and naval forces should be capable of mounting a strong defense that might frustrate the Soviet Union and its allies.

. . . The primary deficiency in the theater results from the significant force ratio advantage currently enjoyed by WP forces. Additionally, currently programmed NATO force structure, modernization, and sustainability improvements would likely be offset by WP improvements even if recent Soviet arms reduction proposals are implemented.

. . . Although these deficiencies might preclude NATO from achieving its stated objectives, NATO forces could also preclude the WP from achieving its objectives.

—From the Joint Chiefs of Staff, *1989 Joint Military Net Assessment* (Washington, D.C.: Department of Defense, 1989), 5–8 through 5–14 *passim*. This assessment was published before new estimates (ref. 19.15) on the unlikelihood of a European war were issued, before the effective destruction of the Berlin Wall, and before the advances of democracy in Eastern Europe.

The foregoing discussion samples some of the issues that we believe may be central in finding the path to an MSD regime. They are presented here partly to give a flavor of what such a process might be like, and partly to stimulate or provoke the reader into foreseeing additional problems and/or devising novel solutions. Of course, we do not pretend our discussion is complete, or even that it includes most of the important questions. But we do know that there are large numbers of smart people all over the world longing for an opportunity to make a difference, to help unspring the trap, to make a safer world. We wish to encourage them.

Private organizations can make significant contributions. A landmark is the Cooperative Research Project on Arms Reductions by the Federation of American Scientists and the Committee of Soviet Scientists Against Nuclear War (ref. 19.20): American and Soviet scientists working together, addressing the details of the path to minimum sufficiency—which fork in the road to take, what bumps to avoid. The American half of the effort was financially guaranteed by one person. Individual scientists and scholars can make important contributions. There are now Peace Studies programs and departments in many universities, where such topics are researched and taught. The U.S. government is advertising for scholarly help in arms control and disarmament (ref. 19.21), and the Soviet Academy of Sciences has for the first time allocated funds to support civilian scientists who wish to change careers toward arms control and disarmament (ref. 19.22).

It's hard to think of more worthy work. We hope that many established scholars, but particularly large numbers of young people, will consider devoting a part of their lives to finding the path.

In fact, there are many paths. Each will have its own benefits and its own burden of risks. The problem before us is to map the terrain, chart the possible paths, make sure some are less risky than our present course, and then find the safest and swiftest route to minimum sufficiency.

# CHAPTER 20

# SKETCH OF A NEAR-TERM STRATEGIC PATH FOR THE UNITED STATES

For though management and persuasion are always the easiest and the safest instruments of governments, as force and violence are the worst and the most dangerous, yet such, it seems, is the natural insolence of man that he almost always disdains to use the good instrument, except when he cannot or dare not use the bad one.

—Adam Smith, *The Wealth of Nations* (1776), Part V, Chapter 1

ENORMOUS BUDGETARY DEFICITS ARE PLAGUING THE U.S. economy; unprecedented new civilian expenditures are needed to preserve the national well-being; there is a widely recognized need to reduce military spending. The United States is now in an advantageous position to help solve these problems while at the same time enhancing global stability (ref. 20.1) and taking the early steps toward minimum sufficiency. Among the very first steps toward a full MSD regime as discussed in preceding chapters would be these: Delay the mass production and deployment of new and expensive strategic nuclear systems—as long as the Soviets, who have strong and similar motivations, do the same—and pursue negotiations to achieve massive and equitable reductions in the strategic arsenals that render such new weapons systems superfluous. The existing massive strategic retaliatory forces are robust, and even with deep cuts can remain so for decades.

At the same time, Soviet willingness to negotiate could be accommodated in a broad plan for mutual and substantial arms reductions. Based on our previous discussion, we suggest that the U.S. proposal—*in exchange for comparable Soviet reductions* consistent with the differences in force structures—should contain the following elements:

(a) Major and joint NATO/WTO conventional force reductions in Europe beyond current tentative and preliminary steps. The negotiations could move much faster in an environment of strategic force reductions and increasing military stability. Nuclear and conventional force reductions go hand in hand, among other reasons because a significant reduction and demobilization of WTO conventional forces weakens the NATO argument that extended deterrence is necessary.

(b) Terminate the MX ("Peacekeeper") missile program, with

# "BACK TO BEING CIVILIZED"

Every U.S. administration since the early 1950s—even the most bellicose—has announced itself in favor of reducing the global strategic arsenals. The same President who declared the U.S.S.R. "the focus of all evil in the modern world" could also muse, just after the nuclear winter findings were first publicly discussed:

> If both of us would say, "We have heard the scientists talk about how the world itself could be destroyed. As long as we maintain things so that neither side is able to start a war with the other, why don't we reduce our arsenals?" And if we start down that road of reducing, for heaven's sake, why don't we rid the world of these weapons? Why do we keep them? . . . Let's get back to being civilized. [President Ronald Reagan, interview, *Time*, January 2, 1984, 37.]

The conversion of these sentiments into policy was soon evident:

> We are vigorously pursuing the negotiation of equitable and verifiable agreements leading to radical reductions in the existing nuclear arsenals of the U.S. and Soviet Union. [Department of Defense, Office of the Assistant Secretary of Defense—Public Affairs, "Department of Defense Views on Major Nuclear Issues and Proposals to Freeze Nuclear Forces at Current Levels," August 1985; reprinted in Donna Uthus Gregory, ed., *The Nuclear Predicament: A Sourcebook* (New York: St. Martin's Press, 1986).]

These attitudes were partially a response to the then politically powerful grass-roots movement to freeze the world nuclear arsenals at their then-current levels and force structures. The Administration in its public statements professed the higher moral ground of arms reductions rather than freeze. The arms cuts were often depicted as occurring under the umbrella of an impermeable continental-scale SDI defense (which we now know to be a forlorn hope). But as the Freeze and Star Wars movements have faded, the "radical reductions" have remained and become increasingly central to U.S. strategic policy. (A useful history of the Freeze movement is Douglas C. Waller, *Congress and the Nuclear Freeze: An Inside Look at the Politics of a Mass Movement* [Amherst: University of Massachusetts Press, 1987].)

its ten independently targeted warheads per nosecone. Highly-MIRVed weapons are destabilizing and dangerous.

(c) Retire the B-1B strategic bomber fleet over a period of ten years in concert with overall cuts in strategic systems. In modern times, the bomber wing represents the weakest and most vulnerable element of the U.S. strategic triad, and bombers have special liabilities connected with nuclear winter (Chapter 18).

(d) Terminate the B-2 Stealth bomber program. The role of slow penetrating bombers in the future strategic balance is very doubtful (ref. 17.19), and the enormous expense of these aircraft will ultimately be countered by comparatively modest investments in air defense.

(e) Terminate the D-5 Trident missile submarine program. The C-3 or C-4 missiles—and, indeed, the Poseidon boats—provide adequate retaliatory, although not first-strike, capability.

(f) Place on hold all cruise missile production. These weapons substantially complicate arms reduction negotiations.

(g) Restrain the SDI program to essential research—far short of development or deployment—an action that would open the door more widely to strategic negotiations while guarding against the possibility of technological breakout.

(h) Install "command-destruct" systems (like those on all rockets launched from Cape Kennedy) on all cruise and ballistic missiles, so they can be destroyed by coded radio signal in the case of accidental or unauthorized launch.

(i) End all tests of nuclear weapons. Further nuclear testing is not essential for weapons reliability or crisis stability (ref. 20.2). The U.S.S.R. has repeatedly stated it would join a U.S. test moratorium.

(j) Negotiate a comprehensive nuclear test ban treaty with the Soviets. Such an action would provide the United States with a major and readily implemented early accomplishment.

## TAKING YES FOR AN ANSWER

> Despite public avowals that nuclear war is unwinnable and that strategic arms arsenals must be dismantled, the Soviets continue deploying new generations of intercontinental ballistic missiles, submarine-launched ballistic missiles, and manned strategic bombers. [U.S. Department of Defense, *Soviet Military Power: 1989* (Washington, D.C.: U.S. Government Printing Office, 1989), 42.]

This is perfectly true. But it is equally true for the United States.

> It would be possible for the Soviets to sign a START treaty, as currently envisioned, and essentially retain the capability to conduct strikes against critical nuclear, other military, and political and economic targets. [*Ibid.*, 47.]

This is also true. And equally so for the United States.

Each side now retains such huge numbers of nuclear weapons that substantial cuts can be made without significantly diminishing the prospects for mutual annihilation, should a nuclear war break out. For that, much deeper cuts are needed. But it hardly follows from this that current levels of nuclear forces are desirable or prudent. For decades each side has accused the other of insincerity when it proposed major arms reductions. It is now time to take yes for an answer.

Highly reliable verification is now possible with a combination of on-site seismic and other monitoring and satellite surveillance.

(k) Redirect the personnel and money long focused on making weapons of mass destruction toward urgently needed civilian programs. Shut down the environmentally dangerous production reactors for fissionable material at Savannah River, South Carolina, and Hanford, Washington, and the Rocky Flats, Colorado, facility that produces the plutonium triggers for thermonuclear weapons. Redirect the national laboratories at Livermore, Los Alamos, and Sandia to develop new, environmentally sound global energy options. Dangerous and pervasive mismanagement of weapons reactors and fissionable material has now been acknowledged. The operation of such reactors is unnecessary as strategic nuclear systems are destroyed and fissionable and fusionable materials are recycled as needed from old warheads (ref. 20.3). A small residual nuclear weapons capability would have to be retained—a regime of minimum sufficient deterrence works only so long as the deterrence is credible. At least for some time, as a hedge against treaty noncompliance by some other nation in an MSD regime, the expert capability for building nuclear weapons must, unfortunately, be retained.

(l) Vigorously pursue further major bilateral strategic nuclear force reductions in negotiations with the Soviets.

(m) Develop a plan to convert the arsenals at the earliest possible date to a robust minimum sufficient deterrence nuclear force based on new or retrofitted systems with strict verification protocols. Our broad suggestions were outlined in Chapters 18 and 19. The ingenuity of the nation's scientists and engineers can be gainfully employed to find the optimum safe, effective, and affordable path from nuclear glut to strategic sufficiency.

(n) Implement and strengthen nuclear nonproliferation treaties, and ban the production of weapons-grade plutonium in civilian power reactors. A United States finally in compliance with Article VI of the Nuclear Nonproliferation Treaty (ref. 20.4), a powerful nation clearly serious about reversing the global nuclear arms race, could have moral and political leverage unknown in decades.

With respect to all these proposals, from (a) to (n), similar remarks apply to the Soviet Union (ref. 20.5).

Of course, there are many other positive actions that could be taken, including jointly staffed nuclear crisis centers and joint advances in news coverage, education, science, trade, space exploration, tourism, and steps to deal with such global problems as global warming, ozonosphere depletion, and the AIDS pandemic (ref. 20.6). However, short-term military and diplomatic steps such as those proposed above—if carefully coordinated and offered as a package that defines a future course for the world's armed forces—would gain for the United

States unparalleled credibility and international standing, and for the world, unprecedented strategic stability and a low probability, even in the worst case, of nuclear winter. We believe all of these bilateral steps are practical; most of them save vast sums at a time of looming fiscal need and financial crisis; and all of them bolster deterrence and stability. They *increase* the national security of both nations. In the long term, the strength and prestige of the United States and its allies (and the U.S.S.R. and its allies) will mainly derive from the economic health and well-being of its people—which are now threatened by a crushing burden of nonproductive military costs.*

* The moment NATO/WTO major conventional force reductions were under way; the Iraqi invasion of Kuwait and the resulting Gulf crisis applied a brake to U.S. arms reductions [Susan F. Rasky, "New Deployments in Gulf Slow Drive for Deep Cuts in Military Budget," *New York Times,* August 12, 1990, 10; Alan Riding, "Allies Reminded of Need for U.S. Shield," *ibid.;* Richard L. Berke, "Peace Dividend: Casualty in the Gulf?" *New York Times,* August 30, 1990]. These events underscored the utility of a multinational peacekeeping force to counter regional aggression, and provided a timely reminder of the dangers of continuing horizontal proliferation of nuclear weapons.

# CHAPTER 21

# OTHER NUCLEAR STATES

If the enemy is an ass and a fool and a prating coxcomb, is it meet, think you, that we should also, look you, be an ass and a fool and a prating coxcomb?

—Fluellen, in William Shakespeare, *The Life of King Henry the Fifth,* Act IV, Scene 1

In order to be sheltered against every act of hostility, it is not sufficient that none is committed; one neighbor must guarantee to another his personal security.

—Immanuel Kant, *Perpetual Peace* (1795), II, Introduction

THE GEARS OF THE WORLD NUCLEAR ARMS RACE MESH. The train of causality binds up the planet in a common peril. The United States invented nuclear weapons because it feared that Nazi Germany would build them first. The Soviet Union developed nuclear weapons to offset the American advantage. Britain and France generated their arsenals to deter the Soviet Union. China's nuclear weaponry was a response to American nuclear forces and later was sustained as a counterpoise to Soviet nuclear forces. India needs nuclear weapons because of nuclear-armed China, which invaded India in living memory (ref. 21.1). Pakistan needs nuclear weapons if India is going to have any. Israel needs nuclear weapons because of the threat of an Islamic nuclear weapon (sometimes called "The Sword of Allah"). Israel, like every nation mentioned, also wants a nuclear arsenal to provide extended deterrence. Israeli nuclear weapons prompt other Moslem nations besides Pakistan to pursue their own nuclear capability (ref. 21.2). There is evidence that Israel has for years supplied nuclear weapons technology to South Africa. The prospect of South African nuclear weapons—if development is not halted—will doubtless provide a cogent reason for nations of Sub-Saharan Africa—especially Nigeria—to acquire nuclear weapons (ref. 21.3). And with the United States and the Soviet Union plagued by vulnerable or enfeebled economies, their world status propped up by vast arsenals, is it clear (ref. 21.4) that Germany (especially a reunited Germany) and Japan will forever deny themselves nuclear weapons?

One of the many dilemmas of nuclear weaponry is this: Nuclear weapons might be acquired by one nation to discourage their use by another, but once in hand they beckon, they call to their possessors. There is a nearly irresistible temptation in a serious but nonnuclear crisis to use them for coercion and intimidation. Since this inclination is more carefree when confronting nations that lack their own nuclear weapons, after a while the deprived nations catch on and the possession of nuclear weapons becomes contagious. Such "horizontal proliferation" increases the chances of superpower conflict, global war, and nuclear winter. The only way for the superpowers to maintain MSD forces without continuing to provoke a worldwide nuclear arms race is never again to use nuclear weapons to coerce. Declarations to this effect would be nice, but they become credible only when mutual trust among nations reaches much higher levels than have been the norm in recent decades. MSD cannot safely be achieved unless closely coupled with conventional force reductions and major advances in international amity. Astonishingly, these conditions are beginning to be satisfied. The very fact of the superpowers entering a path they acknowledge might lead to an MSD regime would be a clear sign of new standards in international relations. Such standards can also become contagious.

The danger caused by horizontal proliferation, especially of strategic weapons, is greater in an MSD regime, and such proliferation could raise the force levels of what is perceived to be a minimum sufficient deterrent. On the other hand, the stakes are so high, and the resources available to the superpowers so great, that adequate incentives could doubtless be found even for the most recalcitrant state. Among the nations of greatest current concern are Israel, Iraq, South Africa, India, Pakistan, North Korea, Brazil, and Argentina.

By the beginning of the 1990s, nations known to have medium- and/or long-range ballistic missiles are the United States, the Soviet Union, Britain, France, China, Israel, Saudi Arabia, and India. The Israeli home-grown entry is called the *Jericho II*, range 1,500 kilometers. The Saudi Arabian entry is the Chinese CSS-2, range 2,700 kilometers. The Indian missile, with a range of up to 2,400 kilometers and a payload of about 1 ton, is called the *Agni*. Nations with medium-range missiles under development are said to include Argentina (*Condor II*, range 900 km), Brazil (*SS-1000*, 1,200 km), Egypt (*Badr-2000*, 900 km), Iraq (*Al-Abbas*, 900 km), and Taiwan (*Skyhorse*, 960 km) (ref. 21.5). Many other nations are developing shorter-

# MAO ZEDONG AND BLOWING UP THE EARTH

Rearing out of the Earth,
A portent across the sky—
. . . You that have drained your cupful
Of this world's radiant spring,
You blow up wild . . .
Plunging the universe into a convulsion of cold.

—"Kunlun," October 1935; in *Reverberations: A New Translation of the Complete Poems of Mao Tse-tung*, translated and with notes by Nancy T. Lin (Hong Kong: Joint Publishing Company, 1980), 37.

Nations without nuclear weapons which find themselves in competition with nuclear-armed nations, have a natural inclination to minimize the seriousness of nuclear war—at least for foreign (and domestic civilian) consumption. In 1946 Stalin's public view was,

> I do not consider the atomic bomb such a serious force as several political groups incline to think it. Atomic bombs are intended to frighten people with weak nerves, but they cannot decide the outcome of a war.

But as the Soviet Union acquired such weapons, this view became more muted. By the late 1950s, with Stalin now dead and Soviet inventories growing, nearly apocalyptic predictions on the consequences of nuclear war began to emanate from Moscow—but mainly about the consequences for the West. The U.S.S.R. would survive nuclear war, it was said—because of Socialist discipline, perhaps, or the large area of the country.

A 1950 Chinese view:

> The atom bomb is one of the modern weapons which possess the greatest destructive power. . . . Except for causing effects of destruction bigger than those produced by ordinary bombs, however, such a weapon

> can produce no other effects. The final decisive force to destroy the enemy's fighting power is still not the atom bomb but strong and vast troops. . . . To countries with a fighting will and with vast territories such as the Soviet Union and China, the atom bomb's usefulness is even smaller.

Less than a decade later, the Soviet leadership was horrified by such views, even though they had been first advanced by Stalin. Russia had now tested nuclear weapons. They knew—even though only in part—what the things could do.

In his efforts to minimize the "paper tiger" of American nuclear weapons, Chinese Communist Party Chairman Mao Zedong sometimes took a stance of Olympian detachment:

> Even if the U.S. atom bombs were so powerful that, when dropped on China, they would make a hole right through the Earth, or even blow it up, that would hardly mean anything to the universe as a whole—though it might be a major event for the solar system.

And—who knows?—even for China.

In the same discussion, as in many others, Mao commented: "The United States cannot annihilate the Chinese nation with its small stock of atom bombs." The Chinese Army newspaper in 1967 editorialized that "atom bombs, guided missiles, and hydrogen bombs, all in all, are nothing much to speak of."

In such remarks—whose practical consequence was to drive the United States to build a still bigger arsenal—Chairman Mao was mistaken. A related opinion often attributed to Mao goes (in demographically updated terms): "If the United States or the Soviet Union lose 250 million people in a nuclear war, they are destroyed forever. If China loses 250 million people, she is still the most populous nation on Earth." But nuclear winter does not work by killing a fixed number of people in each nation. Those nations that *have* more people will *lose* more people. Situated in the same northern midlatitudes as the United States and the Soviet Union, Chinese losses in a superpower nuclear war, even in the absurdly optimistic case that not a single nuclear weapon would fall on Chinese territory, would enormously depopulate China. Fortunately, official Chinese understanding of the consequences of nuclear war, now including nuclear winter, has greatly advanced since Mao's time (ref. 21.12).

More generally, systematic efforts need to be made to guarantee that the present and potential leaders of all nuclear-armed nations really understand the consequences of nuclear war. It is said that upon Nikita Khrushchev's accession to power in the Soviet Union, Defense Minister Marshal Georgi K. Zhukov insisted that Khrushchev personally witness a nuclear weapons explosion (now prohibited, above-ground, by treaty). But it is not necessary to see an explosion to appreciate its effects. Former Secretary of State Dean Rusk describes the learning experience, for the same period, from the American side:

> One thing happened early in the Kennedy Administration that ought to happen with every new leader of a nuclear power. President Kennedy assembled a group of about six senior colleagues—McNamara was one of them, I was one of them—and we spent an afternoon going through an examination of the total effects of a nuclear war, both direct and indirect . . . We had the help of an expert staff that spent all its time studying this thing. And they went through it with charts and things like that, and it was quite an experience. And when we got through with it, President Kennedy asked me to come back to the Oval Office with him, and as we got to the door he looked at me with a strange little look on his face, and he said, "And we call ourselves the human race."
>
> Now, I think there ought to be a committee of scientists drawn from the United States, Great Britain, France, the Soviet Union, and China who would be available to spend at least a day with every new leader of any of these countries to be sure that that leader understands what people are talking about when they talk about nuclear war.

[James G. Blight and David A. Welch, *On the Brink: Americans and Soviets Reexamine the Cuban Missile Crisis* (New York: Hill and Wang, 1989), 183.]

range missiles. Many of these rockets can carry nuclear warheads.

Preventing proliferation takes export controls. It would be powerfully aided by a worldwide comprehensive nuclear test ban treaty (it's hard to be sure your new bomb works if you've never exploded even one). [Many nations favor such a treaty (refs. 13.1, 21.6)] But what it fundamentally requires is arms reductions from the top—those nations that first established the arms race gear-train, that *drive* the gear-train, have the responsibility, and increasingly the incentive, to stop the engine and demolish the links.

Eventually all nuclear-armed and nuclear-capable nations will have to cooperate in resolving the nuclear dilemma. This includes the extension and strengthening of the Nuclear Nonproliferation Treaty [assuming U.S. and Soviet compliance with its Article VI (ref. 20.4)] and closer international supervision of nuclear reactors and waste products, from which weapons-grade fissionable material may be refined.

If we are to approach—much less get below—nuclear winter threshold arsenals, and if the U.S. and U.S.S.R. feel compelled to have many times more nuclear weapons than any lesser nuclear power, then the second-tier nuclear states have a vital role to play. They cannot bring about major U.S./Soviet arms divestitures, but they can probably prevent such divestitures from happening. We have previously discussed (refs. 9.9 through 9.13) the prospects that the three largest nuclear powers after the United States and the Soviet Union would reduce their nuclear forces after a significant demonstration of the willingness of the superpowers to do so (for example, after the implementation of START). Expressions of willingness have been made by the Chinese and French Governments, and by the Labour Party in Britain.

Perhaps the United States and the Soviet Union—as the powers dismantling by far the largest number of nuclear weapons —would, at least initially, insist on retaining the lion's share of the world nuclear arsenals. Perhaps each would argue for an inventory nearly equal to the sum of the weapons in the arsenals potentially arrayed against it. But this requirement may be softened. A group of Soviet analysts associated with the Foreign Ministry offers for discussion an MSD regime in which the second-tier powers each have half as many nuclear weapons as the U.S. or U.S.S.R. (ref. 21.7)—e.g., the following inventories: U.S., 600; U.S.S.R., 600; Britain, 300; France, 300; China, 300. If, say, Britain and France continue to be aligned with the United States, this would amount to a 2:1 disparity, which in an MSD regime might still be acceptable. This is of course not an official Soviet position, but it suggests a promising flexibility. (If China were included, the disparity would become 2.5:1, and this might provide an argument for reducing the non-superpower arsenals a little more: say, U.S. = U.S.S.R. = 600; Britain = France = China = 200. Avoiding nuclear winter drives us to smaller arsenals than in these Soviet proposals, perhaps U.S. = U.S.S.R. = 100; Britain = France = China = 50 or less.) It seems unlikely that the United States and the Soviet Union, together with many other nations, could not find convincing arguments and incentives for reluctant nations to make appropriate divestments of their nuclear weapons.

Israel may present a more difficult problem. Estimates, since the important revelations by Mordechai Vanunu, of the size of the Israeli nuclear arsenal—so far, apparently, entirely gravity bombs—range from 40 to 200 weapons (ref. 21.8). But the development of the *Jericho II* intermediate-range missile and the launching of an artificial satellite by Israel are intended to convey the prospect of Israeli nuclear weapons deliverable by rocket. (Even the name of this weapons system disquietingly conveys the threat of cities destroyed and their inhabitants slaughtered; ref. 21.9.) Israel's long-standing official response to inquiries has been this precise and invariable formulation: "Israel will not be the first nation to introduce nuclear weapons into the Middle East" (ref. 21.10). Since it is now clear that Israel *does* have nuclear weapons (ref. 21.11) and *has* introduced them into the Middle East, how shall we understand this repeated assurance? Might Israel have noted American and Soviet nuclear weapons in their Mediterranean and/or Persian Gulf fleets and used this as a justification for its own nuclear arsenal? If so, withdrawal of American and Soviet nuclear weapons from the Middle East—which would be a natural consequence of an MSD regime—might assist Israel in reducing its arsenals. More likely, nuclear weapons—and increasingly their intermediate-range delivery systems—were introduced by Israel as a counterpoise to an enormous conventional superiority, at least on paper, in men and armor of its Moslem neighbors (and the "first nation to introduce" formulation may have been the least mendacious answer available consistent with the perceived claims of national security).

Thus, the Israeli position is a kind of microcosm of what used to be called the NATO conventional force "dilemma"—i.e., at least tactical nuclear weapons purportedly needed to counter conventional superiority. If so, major cuts in the Israeli nuclear forces may require some combination of asymmetrical force reductions (but reductions on both sides), reinforced conventional defenses, greater cooperation in finding a just resolution of the issue of Palestinian nationhood, and international efforts to eliminate the causes of terrorism in the Middle East. But it will mainly require almost as dramatic an improvement in Israeli-Moslem relations as is currently in progress between East and West. Both improvements were considered equally unlikely not so long ago. Major progress by the United States and the Soviet Union toward an end to the nuclear arms race and the establishment of an MSD regime should serve as an example of what changes are possible in the relations of contending nation-states. The United States and the Soviet Union could provide by their actions a further role model for other nations. In addition, if substantial progress toward MSD is made, the principal powers will have redoubled motivation to address the problems of the Middle East. In a longer-term perspective, movement of the world energy economy away from fossil fuels—in part because of concern about greenhouse warming—is likely to generate profound changes in the Middle East.

In these last three chapters, we have briefly described a new strategic regime that might be established in the next decade. It is designed to provide much more stable nuclear deterrence than we enjoy today and comes close to eliminating any chance of a cataclysmic nuclear winter. The changes in force structure we envision and a very crude representation of their timing are summarized in Figure 7: the strategic launchers are de-MIRVed; tactical weapons are eliminated; the yield of strategic weapons goes down; the funds spent for strategic defense are

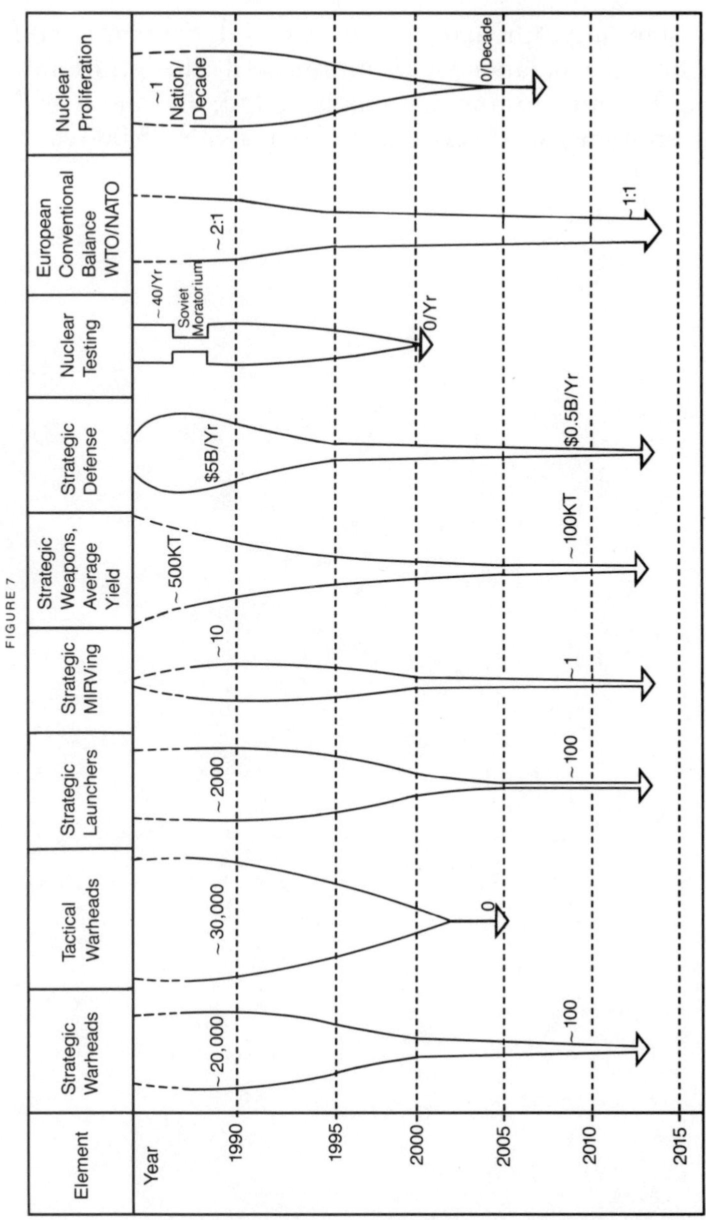


Schematic indication of how various components of the existing U.S. and Soviet nuclear force structures and related activities might be reduced so that a minimum sufficient deterrence (MSD) regime is approached around the turn of the millennium, January 1, 2001. Dashed lines near top for each component indicate previous trends. "~" means approximately; KT means kiloton; B means billion.

cut substantially; nuclear tests are banned; conventional forces in Europe are balanced and diminished; the proliferation of nuclear weapons to new nations is halted; and the world arsenal of strategic warheads goes from some 25,000 to a few hundred or less.

# CHAPTER 22

# ABOLITION

Great perils have this beauty, that they bring to light the fraternity of strangers.

—Victor Hugo, *Les Misérables*, Part IV (1862)

Faith, like a jackal, feeds among the tombs, and even from these dead doubts she gathers her most vital hope.

—Herman Melville, *Moby Dick* (1851), Chapter 7

THE WORLD IS INFESTED WITH NUCLEAR WEAPONS AND haunted by the prospect of their use. Is there any realistic chance, perhaps in a somewhat more distant time, of their complete abolition? Could we wipe the things off the face of the planet? Abolition would certainly reduce the danger of nuclear winter to the vanishing point (ref. 22.1). But abolition has its own dangers.

As arsenals become very small, smaller than the levels of minimum sufficiency, the leverage provided by a few nuclear weapons becomes correspondingly greater; and the danger grows that some nation has sequestered a handful of nuclear weapons and will use them for intimidation or for an act of ideological or religious or ethnic fanaticism (ref. 22.2). The fear of such leverage is a formidable obstacle to abolition, as was first articulated, so far as we know, in 1946 by Frederick S. Dunn (ref. 22.3):

> In a world made bombless by treaty, the first to violate the treaty would gain an enormous advantage. Under such conditions the opportunities for world dominance would be breathtaking! Hence we come to the paradox that the further the nations go by international agreement in the direction of eliminating bombs and installations, the stronger becomes the temptation to evade the agreement. The feeling of security which one imagines would come from a world without bombs would seem to be a fleeting one.

An abolition regime would require sustained prior experience in treaty verification with intrusive inspection, and enthusiastic public support for the preceding MSD regime. Perhaps it would draw sustenance from the obvious economic benefits in a world no longer enslaved by the insatiable demands of a conventional/nuclear arms race. It would require trust—based not on naive or pious hope, but on words and deeds extended over at least a generation or two (ref. 22.4). Our planet still seems far from such a state, although lately encouraging signs of progress have been in evidence.

There are many who believe that advances in conventional weaponry have made large-scale non-nuclear warfare far more deadly than, say, World War II; that nuclear weapons have a special utility in discouraging direct military confrontation among nuclear-armed nations (ref. 22.5); and that their abolition may therefore be a fatal mistake. The heightened horrors of a modern conventional superpower war would provide their own measure of dissuasion, but nothing approaching nuclear deterrence. Some go so far as to conclude that the presence of nuclear weapons and, especially, the prospect of nuclear winter have already made major wars obsolete (ref. 22.6)—a condition that abolition would reverse.

In contrast, another prominent school of thought holds that retention by the United States and the Soviet Union of any strategic weapons at all is not only morally reprehensible but also a fatal invitation to other nations to build up their nuclear forces so they'll be able to play with the big boys—an ambition far more difficult to fulfill in the present era of superpower nuclear weapons glut (ref. 22.7). In the long run, minimum deterrence leads, it is said, to a world in which the superpowers have many fewer nuclear weapons and everybody else has more (although—a point relevant to nuclear winter—with global arsenals still very much less than today's). Vertical proliferation is replaced by horizontal proliferation. Only a world without nuclear weapons can, it is argued, be safe.

Even if the world moves swiftly toward minimum sufficiency, we do not think that a reduction of the current global nuclear inventories to zero by the year 2000 (as proposed on January 15, 1986, by M. S. Gorbachev and endorsed by other governments and the British Labour Party) or some comparably near date is likely. There are simply too many nuclear weapons in the world and they have been embedded too deeply into the way governments and leaders think. Changing thinking on this scale will take more than a few years. But it may just be possible to reduce the arsenals to something approaching minimum sufficiency near the beginning of the new millennium (ref. 22.8), as was suggested by one of us some years ago (ref. 2.3; Figure 7).

If we are ever to abolish nuclear weapons, along the way we will have to pass through the force levels corresponding to minimum sufficiency. MSD is a way station on the path to abolition —from which we can safely survey the landscape and the remaining obstacles. Getting to MSD will itself clarify and smooth out the terrain. Then we can realistically judge the variety of proposals, ranging from verified abolition of national arsenals while instituting a small transnational MSD force, to "weaponless deterrence"—full global abolition with deterrence provided through the capability to rearm quickly should circumstances warrant (ref. 22.9). As massive arms reductions are implemented, new approaches to abolition are likely to emerge.

In the traditional relationship between scientists and political leaders, suggestions about new ways of killing large numbers of people are often embraced and rewarded, while appeals for restraint after the weapons have been brought forth are rejected or ignored. Then, the scientists are told, they are naive and out of their depth; they are trespassing beyond their competence. The invention of weapons is a scientific matter, the politicians and bureaucrats say, but the *use* of weapons is a political matter. So the 1939 advice of Albert Einstein and Leo Szilard to President Franklin Roosevelt was accepted and nuclear weapons were devised and tested; but the warnings by these very same scientists in 1945 and later—and their accurate forecasts of a runaway nuclear arms race with the Soviet Union —were rejected by American politicians with annoyance or contempt. So also Andrei Sakharov's genius in devising the Soviet thermonuclear weapon was appreciated, even revered; but his insights into the dangers of his creation were unheeded or scorned by Soviet politicians for thirty years.

It's a little like—we recognize that the analogy is imperfect —an impetuous teenager given a high-performance sports car who cannot be made to read the manual, sit still for a discussion of safety precautions, or even pass the driving test. He doesn't want to know the dangers. He just wants to feel the wind in his face, hear the engine roaring, and impress his peers. Soon he wants an even later model car—if one is available. But sometimes adolescent risk-taking and an occasional brush with death can help bring us to more mature attitudes, feelings, and thought processes. Perhaps nations, like people, can grow up.

In the draft address which turned out to be his last written words (we have already quoted from it in Chapter 5), Albert Einstein professed his belief that, should the quarrel among the nuclear-armed nation states degenerate into war, "mankind is doomed." He then made clear why he thought change would be so difficult:

> Despite this knowledge [of the consequences of nuclear war], statesmen in responsible positions on both sides continue to employ the well-known technique of seeking to intimidate and demoralize the opponent by marshaling superior military strength. They do so even though such a policy entails the risk of war and doom. Not one statesman in a position of responsibility has dared to pursue the only course that holds out any promise of peace, the course of supranational security, since for a statesman to follow such a course would be tantamount to political suicide. Political passions, once they have been fanned into flame, exact their victims (ref. 22.10).

But such a course may no longer be tantamount to political suicide. Global economic interdependence, the unexpected opening up of the Soviet Union and Eastern Europe to the rest of the world, the emerging European Union, the improving effectiveness and acceptability of the United Nations and the World Court, and the growing political success of environmental agendas—as well as the linking of the world through telephone, television, facsimile machines, and computer networking—are all working in the same direction. Suddenly there has emerged a potent set of positive and negative incentives—carrots and sticks—drawing and driving the nation-states together (ref. 22.11).

Because it threatens the largest number of people and because its dangers have global venue, the prospect of nuclear winter is not least among these influences. Never before in human history has there been such a degree of shared vulnerability. Every nation now has an urgent stake in the activities of its fellow nations. This is most true—because here the danger is greatest—on the question of nuclear weapons. Nuclear winter has alerted us to our common peril and our mutual dependence. It reaffirms an ancient truth: When we kill our brother, we kill ourselves.

We have entered a most promising time—not just because the walls are tumbling down, not just because money and scientific talent long devoted to the military will now become available for urgent civilian concerns, but also because we finally are becoming aware of our unsuspected—indeed, awesome—powers over the environment that sustains us. Like the assault on the protective ozone layer and global greenhouse warming, nuclear winter is a looming planetwide catastrophe that is within our power to avert. It teaches us the need for foresight and wisdom as we haltingly negotiate our way out of technological adolescence.

From the halls of high Olympus, where strange dooms are stored for humans, there is reason to hope that, in our time also, there is a way out—a path where no man thought.

# NOTES AND REFERENCES

1.1:

Vaclav Smil, *Energy, Food, Environment: Realities, Myths, Options* (Oxford: Oxford University Press, 1987). But while properly calling attention to failed scientific prophecy, Smil rejected the warning that chlorofluorocarbons (CFCs) imperil the ozone layer at just the moment that entirely compelling evidence was becoming available (see our Chapter 4).

2.1:

E.g., Dr. Harold A. Knapp, a U.S. Atomic Energy Commission official who investigated the deaths from radioactive fallout of thousands of sheep in Utah and Nevada following a 1953 aboveground U.S. nuclear explosion. In 1963 Knapp concluded that radioactive strontium-90 in fallout could dangerously work its way up the food chain—from grass to cows to babies—and be concentrated in the bones of children following bomb tests, to say nothing of nuclear war. The A.E.C. promptly suppressed his report, and Knapp resigned. ("Harold A. Knapp, Nuclear Test Expert, Dies at 65," by Glenn Fowler, *New York Times*, November 11, 1989, 33.) There are many similar cases, in which the continuance of the nuclear arms race was considered more important than the health of the citizens the bombs were supposedly protecting, in the recent history of the Soviet Union, Britain, France, China, and Israel—as well as the United States. Cf. "U.S. Sees a Danger in 1940s Radiation in the Northwest: First Such Concession" by Keith Schneider, *New York Times*, July 12, 1990, A1.

2.2:

R. P. Turco, O. B. Toon, T. P. Ackerman, J. B. Pollack, and C. Sagan, "Nuclear Winter: Global Consequences of Multiple Nuclear Explosions," *Science* 222, 1983, 1283–1297. TTAPS is an acronym constructed from the surnames of the authors.

2.3:

The first assessment of the policy implications of nuclear winter (Carl Sagan, "Nuclear War and Climatic Catastrophe: Some Policy Implications," *Foreign Affairs*, Winter, 1983/84, 257–292) reached the following conclusions:

(a) A major strategic attack, even if there were no retaliation from the adversary nation, may generate catastrophic climatic consequences, at least hemisphere-wide; if these potential consequences are understood beforehand, they work to deter such an attack.

(b) Sub-threshold war—sizable nuclear attacks intended to discourage nuclear retaliation by the adversary, by injecting a precisely determined amount of fine particles into the Earth's atmosphere, so that the climatic response is just short of nuclear winter—is a fool's strategy because of intrinsic and unresolvable uncertainties in what constitutes a "threshold" level of attack.

(c) Treaties on targeting, intended to lessen the climatic impact of nuclear war, carry with them grave and probably insurmountable verification problems.

(d) A strictly military response to the prospect of nuclear winter might be a conversion of the strategic arsenals to high accuracy, low yield, and perhaps burrowing weapons, a transition under way for other reasons, especially in the American arsenal; but such a conversion has very worrisome implications—e.g., for strategic stability during the transition period.

(e) The already dubious contention that civilian shelters and crisis relocation would be useful in time of nuclear war is made even less credible by the severity and duration of the nuclear winter postwar environment.

(f) Because of inevitable leakage, the Strategic Defense Initiative (Star Wars) cannot prevent nuclear winter, and may even —by encouraging increases in offensive forces—make it more likely.

(g) If comparatively few nuclear weapons can generate global climatic effects, this might provide an incentive for other nations (or even terrorist groups) to acquire nuclear weapons and use them for intimidation, coercion, or revenge. Nuclear winter provides a new dimension to the arguments for halting horizontal proliferation.

(h) Given the high probability, if we wait long enough, of human or machine error in high-technology systems, and the seriousness of even "mild" nuclear winters, the only prudent response is massive reductions in the world strategic arsenals to minimum deterrent levels, at which nuclear winter is much less likely to occur.

These themes have been rediscussed and debated in subsequent commentaries, and are developed further in the present work.

2.4:

D. M. Drew, F. J. Reule, D. S. Papp, D. M. Snow, P. H. B. Godwin, and G. Demack, *Nuclear Winter and National Security: Implications for Future Policy*, Center for Aerospace Doctrine, Research, and Education (CADRE), Maxwell Air Force Base, Alabama (Washington, D.C.: U.S. Government Printing Office, July 1986), Document 635-327, 76 pp. In this combined Air Force/academic study, the policy implications of three possible thresholds, to be determined by future research, for generating nuclear winter are developed: a "high-level" threshold in which a substantial fraction of the strategic inventories are required; a "middle-level" scenario; and a "low-level" scenario in which only some tens or hundreds of nuclear weapons, appropriately targeted, are required to bring about nuclear winter. The high-level scenario is said to have only minor policy consequences. The middle-level scenario is said to challenge the role of NATO's intermediate-range nuclear forces (INF) in nuclear retaliation against a Soviet conventional attack in Europe, to lead possibly to "radical changes in basing modes and [weapons] system characteristics," and to strengthen the opposition to horizontal proliferation; for these and other reasons, mutually "assured destruction [MAD] becomes an unacceptable policy option." (The INF Treaty signed subsequent to this analysis has now removed the NATO forces in question, as well as the corresponding Soviet forces.) The low-level scenario "could result in a radical restructuring of the international system," no less, because the use of any significant number of nuclear weapons would produce unacceptable worldwide consequences. In the present book we argue that the "low-level" scenario may actually describe nuclear reality. Drew and his co-workers also predicted that "nuclear winter could provide the rallying point to pull the arms control community out of its doldrums and back to center stage."

2.5:

R. J. Bee, C. B. Feldbaum, B. N. Garrett, and B. S. Glaser, *Implications of the "Nuclear Winter" Thesis*, Defense Nuclear Agency Report TR-85-29-R-2 (Washington, D.C.: The Palomar

Corporation), June 24, 1985. This study raises the prospect that, because of nuclear winter, "maintenance of a credible deterrent posture by the United States may require reevaluation of U.S. [war] plans." It also argues that "the possibility of a nuclear winter makes the obstacles to survival in a postwar environment appear even more formidable than earlier foreseen," and that coping with "worst case nuclear winter conditions might involve the peacetime expenditure of an unacceptably high level of resources."

2.6:

Other analyses include Thomas Powers, "Nuclear Winter and Nuclear Strategy," *The Atlantic*, November 1984 (the last sentence of which reads, "But to me, recognition of the nuclear-winter problem, awful as it is, seems a piece of immense good fortune at the eleventh hour, and a sign that Providence hasn't given up on us yet."); William J. Broad, "U.S. Weighs Risk That Atomic War Could Bring Fatal Nuclear Winter," *New York Times*, August 5, 1984, 1; J. J. Gertler, "Some Policy Implications of Nuclear Winter," Report P-7045 (Washington, D.C.: The Rand Corporation, 1984); Philip J. Romero, "Nuclear Winter: Implications for U.S. and Soviet Nuclear Strategy," Rand Corporation Graduate Institute Report P-7009-RGI, Santa Monica, Cal. (1984); F. P. Hoeber and R. K. Squire, "The 'Nuclear Winter' Hypothesis: Some Policy Implications," *Strategic Review* (Washington, D.C.), Summer 1985, 39–46 (the last sentence of which reads, "Ill-considered disruption of the basis of . . . deterrence, driven by panic spread by doomsayers, could set the stage for the very catastrophe that they inveigh against"); C. Chagas et al., "Nuclear Winter: A Warning," Document 11 (Vatican City: Pontifical Academy of Sciences, January 23–25, 1984), 15 pp.; Dan Horowitz and Robert J. Lieber, "Nuclear Winter and the Future of Deterrence," *Washington Quarterly*, *8* (3), Summer 1985 ("Even if the original assertions about nuclear winter are no more than marginally correct, the implications for thinking about strategic issues and deterrence remain significant."); *The Consequences of Nuclear War*, Hearings, Joint Economic Committee, 98th Congress (Washington, D.C.: U.S. Government Printing Office, 1986); several papers, including transcripts of the Kennedy/Hatfield Forum on the Worldwide Consequences of Nuclear War, U.S. Senate, in *Dis-*

*armament* (New York: United Nations) 7 (3), 1984, 33–80; B. Weissbourd, "Are Nuclear Weapons Obsolete?," *Bulletin of the Atomic Scientists*, August/September 1984; *The Climatic, Biological, and Strategic Effects of Nuclear War*, Hearings, Subcommittee on Natural Resources, Agricultural Research and Environment, House Committee on Science and Technology, 98th Congress, 2nd Session, Document 126, 39–934-0 (Washington, D.C.: U.S. Government Printing Office, 1985); Michael F. Altfeld and Stephen J. Cimbala, "Targeting for Nuclear Winter: A Speculative Essay," *Parameters: Journal of the U.S. Army War College 15* (3), 1985, 8–15 (a paper devoted to finding a way to fight a nuclear war despite nuclear winter—e.g., "a counter-force war of attrition . . . well below the threshold of self-destruction"); *Nuclear Winter and Its Implications*, Hearings, Senate Armed Services Committee, 99th Congress, 1st Session, Document 99-478 (Washington, D.C.: U.S. Government Printing Office, 1986); F. Solomon and R. Q. Marston, eds., *The Medical Implications of Nuclear War* (Washington, D.C.: National Institute of Medicine, National Academy of Sciences, 1986); Symposium on Nuclear Winter, Royal Institution/British Association for the Advancement of Science, London, December 2, 1986; George H. Quester, "Nuclear Winter: Bad News, No News, or Good News?," in Catherine Kelleher, Frank J. Kerr, and George H. Quester, eds., *Nuclear Deterrence: New Risks, New Opportunities* (Washington, D.C.: Pergamon-Brassey's, 1986); Peter de Leon, "Rethinking Nuclear War: The Strategic Implications of Nuclear Winter," *Defense Analysis, 3* (4), 1987, 319–336; Joseph S. Nye, "Nuclear Winter and Policy Choices," *Survival 28* (2), 1986, 119–127 (in which the author reveals that all previous "policy reactions to the theory of nuclear winter missed the point." He also writes, "The prospect of nuclear winter and the end of our species is so sobering that it must be taken seriously long before the scientists can prove exactly how realistic it is."); Allen Lynch, *Political and Military Implications of the "Nuclear Winter" Theory* (New York: Institute for East-West Security Studies, 1987), which concludes: "The nuclear winter theory, in the way it challenges the very concept of nuclear victory, effectively severs the connection between the operational use of nuclear weapons and foreign and defense policy." We do not list classified analyses. (E.g., "I can imagine in the future that we might in fact want to look at [nuclear winter] scenarios which are, let's say, very close to particular targeting plans, and those might have to be classified."—Richard Wagner, Assistant to the Secretary of Defense for Atomic Energy, Congressional Testimony, July 12, 1984, in *The Consequences of Nuclear War*, cited above, p. 133.) There are also a range of Soviet publications on the subject, some of which are listed in refs. 13.11–13.15, several papers in Volume 18, Number 7, of the Swedish Academy of Sciences journal *Ambio* (1989) and other works referenced below.

2.7:

> It has been a very good thing for the integrity of science, and a sign of courage, that some 40 scientists of high standing have gone public with their considered estimates of the global atmospheric effects and long-term biological consequences of nuclear war. . . . Even allowing for the constraints imposed on scientific opinion in the Soviet Union, it is fair to assume that the same conclusions are held in that quarter. Here, then, is a new basis for dialogue and for an alternative run at restraint. It is a run worth making.
>
> —William D. Carey, "A Run Worth Taking," editorial, *Science* 222, December 23, 1983 (the same issue in which the TTAPS paper was published).

> [Although] apocalyptic visions of the implications of nuclear war have been a feature of popular discussion since the dawn of the nuclear age and although numerous proposals have been made from time to time considering the possible mechanisms for such an extreme result, it appears to me that the nuclear winter study is quite unprecedented in the credibility and explicitness of its apocalyptic speculations. Unless further investigation of the nuclear winter hypothesis convincingly disposes of these speculations, the TTAPS study must be considered to inaugurate a new era in the discussion of nuclear armaments.
>
> In this new era the force of the moral critique of nuclear weapons, based on concerns for the fate of the human species and of other life on the planet as a whole, will be much more widely acknowledged, even in circles where such concerns were formerly dismissed as naive and uninformed.

. . . If the hypothesis should emerge essentially intact from the intense scientific scrutiny that it will surely receive, then the most important question by far is whether the political and military leaderships of the United States and the Soviet Union will acknowledge this reality and accord it appropriate weight as a determinant of their behavior. If they do so, the world will promptly become a great deal safer by virtue of the drastically changed incentives for the initiation of nuclear war by preemptive strike in the context of a severe crisis.

—From "Statement of Sidney G. Winter, Professor of Economics and Management, Yale University," in *The Consequences of Nuclear War: Hearings Before the Subcommittee on International Trade, Finance, and Security Economics of the Joint Economic Committee, Congress of the United States*, 98th Congress, 2nd Session, July 11 and 12, 1984 (Washington, D.C.: U.S. Government Printing Office, 1986), 147–148. Winter is a former Rand Corporation strategic analyst.

Unless [the remaining] uncertainties can be resolved in such a way as to demonstrate rather conclusively that the smoke clouds triggering a nuclear winter will not form, it is difficult to see how the threat to use nuclear weapons can remain a plausible one.

—Thomas M. Donahue (then Chairman, Space Science Board, National Academy of Sciences), "Nuclear Winter," *Michigan Alumnus 91* (4), 1985.

Nuclear winter theory may provide the stimulus for a major paradigm shift in our thinking about national security [Peter C. Sederberg]. . . . If the scientific study . . . should confirm the nuclear winter hypothesis, or even partially confirm it, the implications for strategy and arms reductions should be profound [Martin J. Hillenbrand]. . . . If recent findings prove to be even only remotely accurate, they warrant serious consideration as we face the task of preserving the peace in the years ahead. . . . Should the findings of Turco et al. and others prove accurate, there is a clear requirement to reduce drastically the total megatonnage now available [Robert Kennedy].

—In Peter C. Sederberg, ed., *Nuclear Winter, Deterrence and the Prevention of Nuclear War* (New York:

Praeger, 1986), 9, 98, 151, 160. Kennedy is the General Dwight Eisenhower Professor of National Security at the U.S. Army War College.

2.8:

> [N]uclear winter is not just a theory. It is also a political statement with profound moral implications. If people believe that our nuclear weapons endanger not only our own existence and the existence of our enemies but also the existence of human societies all over the planet, this belief will have practical consequences. . . . It will increase the influence of those who consider nuclear weapons to be an abomination and demand radical changes in present policies. [Freeman J. Dyson, *Infinite in All Directions* (New York: Harper, 1988), 259.]

While holding that 1984 was too early, the strategic analyst Leon Gouré (testimony, House Committee on Science and Technology, September 12, 1984) stated,

> [I]f additional studies and analyses confirm the "Nuclear Winter" hypothesis as constituting a realistic possibility, it is possible that eventually this may significantly influence strategic planning and weapons programs, as well as arms control negotiations.

Nobel Laureate Herbert A. Simon argued,

> The terms of the nuclear standoff have been changed—fundamentally changed. Awakened to that fact, we must proceed at once to an examination of the scientific reality of the nuclear winter and of the implications of this reality for our policies of arming for suicide and our fears that a suicidal weapon might be used by an aggressor against us. ["Mutual Deterrence or Nuclear Suicide," editorial, *Science*, February 24, 1984.]

David F. Emery, then Deputy Director of the U.S. Arms Control and Disarmament Agency, testified before Congress (*The Consequences of Nuclear War*, July 12, 1984, 125, 134, 136, 140, 142; ref. 2.6):

> There's absolutely no doubt that if the results of [a] nuclear exchange have even a small fraction of the implications that Mr. Sagan and others have presented, of course it's going to have an impact on the attitude of leaders of

> nations and public opinion throughout the world. . . . We are trying to convince the Soviet Union to come back to the negotiating table and join us with a comprehensive agreement to eliminate the huge numbers of nuclear weapons and move to a safer, more stable balance at a much lower level. And I think that is exactly the direction that the nuclear winter phenomenon would argue for. . . . The nuclear winter findings reinforce the importance of . . . negotiating very deep reductions, especially in destabilizing systems. That's how I envision the nuclear winter impact, as hopefully a catalyst for both sides to get back to the negotiating table. . . . I don't think we have to wait 2, 5, or 10 years, or a long[er] period of time, to draw the conclusion; I think all of us can draw a conclusion here, and that is, the best way to prevent the disasters that have been outlined is simply to move ahead with nuclear negotiations at the highest possible level. . . . Even if we decided that every horrible effect of nuclear winter were 10 times worse than has been described, it's going to be impossible to solve the problem unless the United States and the Soviet Union work together.

(Shortly after, the U.S./Soviet arms control negotiations resumed, leading eventually to the INF Treaty and the advances in START.)

An internal State Department memorandum to the Secretary of State, dated August 16, 1984, read:

> The implications for United States policy of the nuclear winter theory as it is being argued by Turco, Toon, Ackerman, Pollack and Sagan could be profound if the Administration's policy studies agree with Turco et al.'s conclusions and/or if by default congressional and public attitudes are molded by those results. [Quoted in *Nature*, September 19, 1985, 129.]

What some of those implications might be were restated in a *New York Times* lead editorial:

> There seems at present a solid chance that a nuclear war, besides killing hundreds of millions immediately, could be followed by a nuclear winter that would kill hundreds of millions more. . . . The main message is that deterrence must not be allowed to fail, and long before any meaning-

> ful defense can be achieved, the arsenals need to be reduced to the smallest possible size. Forcing governments to look nuclear weapons in the face may be the best way to sicken their appetite for building ever more. ["Rethinking Nuclear War," *New York Times*, September 29, 1985.]

An anonymous judgment in the journal *Foreign Affairs*: "Clearly the nuclear winter theory, to whatever extent it is valid, has induced a significant debate about man's wisdom in relying upon nuclear weapons and about the entire concept of deterrence" (Winter, 1986/1987, 313).

Lewis Thomas writes,

> The whole problem of nuclear disarmament has been transformed. It is no longer a deep technical puzzle to be waffled over or postponed forever by diplomats and analysts. The discoveries [nuclear winter] . . . mark a turning point not only in the affairs of mankind but also, in Jonathan Schell's prophetic phrase, in the fate of the earth. ["Nuclear Winter, Again," *Discover*, October 1985.]

There are also those who are less impressed with the implications of nuclear winter, as we discuss below.

3.1:

> When the work of Sagan and his associates first appeared, it produced a storm of controversy as well as concern. Conferences were held, official studies were commissioned, special issues of scientific journals were devoted to the topic, television [programs] were produced, and numerous books appeared, all in an effort to understand and determine the veracity of this strange, new, yet compelling vision of the impending apocalypse. [Charles W. Kegley, Jr. and Eugene R. Wittkopf, *The Nuclear Reader: Strategy, Weapons, War*, 2nd ed. (New York: St. Martin's Press, 1989), 254.]

Beyond the comparatively sedate activities described by Kegley and Wittkopf, nuclear winter also engendered comments that, in their vituperation and mean-spiritedness, have gone far beyond the ordinary bounds of polite debate. Sometimes, ostensible scientific and other objections were motivated by a deep unease about nuclear winter's evident policy implications. One of the most revealing class of comments was the contention that nuclear winter was invented (not discovered) *in order* to influence nuclear policy—to impede the deployment of Pershing II missiles in Europe, for example (now wholly removed in accord with the INF Treaty), or to stop Star Wars, or to bring about a global "freeze" in the procurement of nuclear weapons and their delivery systems. Beginning in 1984, it is true, automobile bumper stickers could occasionally be seen bearing the words "Freeze now or freeze later." But taking nuclear winter seriously, we believe, implies major *reductions* in the arsenals, not merely freezing them at levels well above 50,000 nuclear weapons—as was properly argued by people of many political persuasions, including conservatives (e.g., Senator Jake Garn, "Nuclear Winter: The Case for Weapons Reductions and Defense," *Christian Science Monitor*, August 21, 1984, Part II, 15). However, many advocates of the Freeze (e.g., former Presidential Science Adviser George Kistiakowsky) considered it the first step toward major arms reductions. You must first stop the car from hurtling toward the edge of the cliff, they explained, before you can back it up.

Nuclear winter was described as a "deliberately perpetrated hoax," using "political rhetoric and oversimplistic arguments," by Susan G. Long, *High Frontier Newsletter*, March 1988 (reprinted in *Billy James Hargis' Christian Crusade* 35 [8], March 1988). Others have imagined that the nuclear winter hypothesis was "invented" in 1982 for political reasons by "the inner circle of disarmament activists" (Russell Seitz, "The Melting of 'Nuclear Winter' "; this long letter in the editorial pages of *The Wall Street Journal* [Wednesday, November 5, 1986] is notable for its political bias, personal invective, invented quotations, and misunderstanding of the methods and content of science).

As described in Appendix C, the initial motivation for the TTAPS research arose out of our work in the 1970s on dust storms on Mars and volcanic explosions on Earth, and then, beginning around 1980, from our study of the late Cretaceous impact event and the extinction of the dinosaurs. A more proximate cause was a request from the National Academy of Sciences in 1982 for us to consider the environmental consequences of dust generated in a nuclear war. Our expertise and tools for analyzing such problems evolved through a decade of research sponsored chiefly by NASA and the National

Science Foundation. The National Academy study, in which our group participated from the start, was supported by the Defense Nuclear Agency (DNA) of the Department of Defense. It would seem to require a less than firm grip on reality to imagine the National Academy of Sciences and the Department of Defense as constituting an "inner circle of disarmament activists." (Cf. Richard Turco, O. Brian Toon, Thomas Ackerman, James Pollack, and Carl Sagan, "Nuclear Winter Remains a Chilling Prospect," Letter to the Editor, *The Wall Street Journal*, Friday, December 12, 1986.)

In the same vein,

> What is being advertised is not science but a pernicious fantasy that strikes at the very foundations of crisis management, one that attempts to transform the [NATO] Alliance doctrine of flexible response into a dangerous vision. For "Nuclear Winter" does exist—it is the name of a specter, a specter that is haunting Europe. Having failed in their campaign to block deployment of [NATO's] theater weapons, the propagandists of the Warsaw Pact have seized upon "Nuclear Winter" in their efforts to debilitate the political will of the citizens of the Alliance. What more destabilizing fantasy than the equation of theater deterrence with a global *Götterdämerung* could they dream of? What could be more dangerous than to invite the Soviet Union to conclude that the Alliance is self-deterred—and thus at the mercy of those who possess so ominous an advantage in conventional forces? [Russell Seitz, "In From the Cold: 'Nuclear Winter' Melts Down," *The National Interest* 5, Fall 1986, 3–17. This article was reproduced and distributed by the U.S. Department of Defense in the April 2, 1987 number of its internal organ, *Current News*.]

Of the many deficiencies in nuclear winter theory discerned by the *National Review*, then the leading journal of the American right wing, the first was that "the TTAPS group . . . have neglected to explain something: There was no nuclear winter at Hiroshima and Nagasaki." (Cf. this book, box, Chapter 7, "Hiroshima and Nuclear Winter.") The article concludes, "Nuclear winter isn't science. It is propaganda. And the willingness of prominent men of science to debase themselves and their calling for the cheap thrills of political notoriety is a scandal" (Brad Sparks, "The Scandal of Nuclear Winter," *National Re-*

*view*, November 15, 1985, 28–38). A subsequent editorial declared, "Despite the fact that nuclear winter was a fraud from the start . . . the anti-nuclear hucksters largely succeeded in putting one over," and compared the announcement of nuclear winter to the tactics of the Nazis ("Reichstag Fire II," *National Review*, December 19, 1986). The same publication denounced nuclear winter in an article devoted to what it called "scientific lying" (Jeffrey Hart, "The Death of Truth," *National Review*, November 7, 1986).

All of this followed a televised network debate, unmentioned in these articles, between one of us and the *National Review*'s editor, William F. Buckley (*Nightline*, Ted Koppel, moderator, ABC-TV, July 18, 1984), in the course of which Mr. Buckley said,

> That is the critical question: Do we—are we prepared to surrender the Constitution of the United States, the Declaration of Independence in order to avoid this abstract apocalypse by which Dr. Sagan almost seems to be celebrating? Everybody in the world is going to die. . . . Everybody is going to die anyway. And most people, unfortunately, more painfully than they would under a nuclear anaesthetic. So that to the extent that we stand firm, there won't be a nuclear war. To the extent that we question the organic principles behind our defense, we accelerate the likelihood of that war, unless Dr. Sagan can convince the Soviet Union to act rationally, which we devoutly hope he will succeed in doing. [*Nightline*, Show 833; transcript available from Box 234, Ansonia Station, New York, NY 10023.]

Several of Lyndon Larouche's publications condemned nuclear winter as a "fraud" and a "hoax" (e.g., "CFR Journal Confirms Nuclear Winter Hoax," by Carol White, *New Solidarity*, July 14, 1986: "The nuclear winter lie was given wide currency by the Soviets, who used it to foster illusions in U.S. military and policy-making circles that nuclear war was unwinnable, and therefore unthinkable." An earlier issue of the same publication decried research on nuclear winter as fashionable to excess and politically naive—the " 'in thing' among Nuclear Freeze academics and many others, including some at Lawrence Livermore National Labs, who are 'trying to show the Soviets we're sincere' " ["New KGB Road Show Plays in U.S.

Senate," by Paul Gallagher, *New Solidarity*, December 16, 1983]). Larouche publications have also denounced global warming and the greenhouse effect itself as a "fraud" and a "hoax." (E.g., Rogelio Maduro, "The Hoax Behind the 'Greenhouse Effect,'" *21st Century Science and Technology*, January 1989, 34; Maduro, "The Greenhouse Effect Is a Fraud," ibid., March 1989, 14; "The 'Greenhouse Effect' Is a Hoax!," *Executive Intelligence Review*, 1990; etc.)

And in late 1984 a speaker informed the American Civil Defense Association that nuclear winter is "Soviet disinformation"; an American "counter-propaganda, misinformation" campaign was urged to make nuclear winter go away ("Analyst Says Soviets Spend Billions to Spread Misinformation to U.S.," by Ellen Mishkin, *Daytona Beach* [Florida] *News-Journal*, November 18, 1984).

While many far-right publications were incensed by nuclear winter, it was mainly ignored by the left. One exception: *The People* (93 [18], November 26, 1983, 6–7), an organ of the Palo Alto, California–based Socialist Labor Party (and, later, *Socialist Studies*, in 1984) argued that any hope of nuclear winter helping to change strategic policy was illusory. They offered for consideration "that the premise that the governments of the United States and the Soviet Union can be persuaded or otherwise prevailed upon by mass public pressure to abandon the nuclear arms race and the threat of nuclear war is incorrect," and "that both governments by their very nature are committed to maintaining the arms race and the nuclear war threat." These propositions are not very different from those accepted by many mainstream analysts (see Prologue); we shall argue in Chapter 13 that they may well be incorrect.

3.2:

Cf. the physicist and Defense Department consultant Richard Muller (*Nemesis* [New York: Weidenfeld and Nicholson, 1988], 16): "[M]any sober-minded scientists felt that . . . the TTAPS group had shown that the best scientists in the world, in their previous thinking about nuclear war, might possibly have missed the most important, the most damaging effect. . . . [W]e had all learned a nuclear humility." Or the understatement of Undersecretary of Defense Richard DeLauer (in R. Jeffrey Smith, "Nuclear Winter Attracts Additional Scrutiny,"

*Science*, July 6, 1984, 225, 30): "We should all perhaps be a little concerned that we did not recognize a little sooner the importance of smoke to calculations of nuclear effects." Or William J. Broad (*New York Times Book Review*, August 12, 1984, 27): "The question that only the Department of Defense can answer is why it took a group of civilian scientists to drag these issues into the open." As nearly as we can tell, from 1945 until 1983 no meetings were ever held by the Department of Defense to consider the smoke-induced climatic effects of nuclear war. There was no one at a high enough level whose interest (or job description) it was to seek out hitherto undiscovered adverse consequences of nuclear war. This is still the case. Also, nuclear war was already bad enough, some may have thought, and additional horrors would only play into the hands of those who would ban the bomb.

One glancing suggestion that severe climatic consequences might result from the dust raised in a nuclear war—although with no hint of a calculation to back it up—was provided in a 1965 think-tank report to the Department of Defense:

> If a sufficiently large war [more than 10,000 megatons] should occur, and if the scientific uncertainties should turn out to be on the unfavorable side, it is not even inconceivable that a new Ice Age might be triggered. More likely, however, would be the less dramatic climatic consequences such as a temporary cold spell with temperatures averaging a few degrees below normal. [Robert U. Ayres, *Environmental Effects of Nuclear Weapons*, Volume 3, Summary, Hudson Institute Report HI-518-RR, December 1, 1965. Cf. John von Neumann's 1955 assessment, summarized in "Nuclear Winter: Early History and Prehistory," box, Chapter 3.]

Climatic effects were briefly addressed in a study by the U.S. National Academy of Sciences (*Long-Term World-wide Effects of Multiple Nuclear-Weapon Detonations* [Washington, D.C.: National Academy of Sciences, 1975]), which was mainly concerned with ozone depletion caused by nuclear war. The report considered dust only, not smoke, and concluded that climatic effects would be comparable to those produced by volcanoes:

> Stratospheric dust injection from a 10,000 megaton nuclear exchange would be comparable to that from a large vol-

> canic explosion such as Krakatoa in 1883 and therefore might have similar climatic impact. At most, a 0.5°C deviation from the average lasting for a few years might be expected.

This estimate, even for dust alone, may be far too low (ref. 2.2). The study did, however, caution that "climatic changes of a more dramatic nature cannot be ruled out."

In 1978, the U.S. Arms Control and Disarmament Agency issued a report called *Frequently Neglected Effects in Nuclear Attacks*. It neglected nuclear winter.

The physicist Edward Teller, who has played a continuing and central role in U.S. nuclear weapons development since the 1940s, occasionally suggests that he, or scientists working under his direction at the bomb-designing Livermore National Laboratory, had discovered nuclear winter many years earlier. (E.g., *The Climatic, Biological, and Strategic Effects of Nuclear War*, Hearing, Subcommittee on Natural Resources, Agriculture Research and Environment, Committee on Science and Technology, U.S. House of Representatives, September 12, 1984 [Washington, D.C.: U.S. Government Printing Office, 1985]; "Nuclear Winter: Effects of Atomic War," *Nightline*, ABC-TV, November 1, 1983.) If this is true, it is unconscionable—withholding vital information on the consequences of nuclear war from the American people, especially when knowledge of these facts might influence public policy. Even if what Teller says he uncovered was only a *hint* of a climatic catastrophe, insufficiently reliable to justify a scientific paper or a public announcement, it is unconscionable that he did not pursue this line of research, and surprising that the 1975 National Academy report on the long-term consequences of nuclear war—which drew upon the nation's collective scientific wisdom—contains no trace of Teller's purported finding. However, Lowell Wood, a close associate of Teller's, in a 1982 symposium paper estimated a 1°C cooling from dust and a "small" impact from fires after a major nuclear war (Wood, "Concerning the Implications of Large-Scale Nuclear War: The Actual Fate of the Earth," paper prepared for the Second International Conference on the Prevention of Nuclear War, Erice, Italy, August 19–23, 1982; lecture notes kindly provided by Dr. Wood). See also Richard Garwin, in *International Seminar on Nuclear War, 3rd Session: The Technical Basis for Peace* (Frascati, Italy: Servicio Documentazione dei Laboratori Nazionali di Frascati dell'INFN, 1984), 185–186, 298–299. Either way, Teller's comments are curious and troublesome. (See also Teller, ref. 3.3.)

The view of some scientists at Livermore and elsewhere on public disclosure was described by Livermore physicist Joseph B. Knox in 1985:

> At the appropriate time, we should ask the question, what should the populace know about nuclear winter? Prof. [Sergei] Kapitza asked this question last August in Moscow; he said, "When do we take what we've just discussed to the public?" And fortunately I had listened to Dr. Teller often enough to have learned that there are times to zig and times to zag. The response was "I come only as an individual scientist of the United States and I cannot answer that question for our government." Further, I do not want to answer the question at this time as a scientist. [Knox, "Climatic Consequences of Nuclear War: Working Group #1," paper prepared for publication in *Proceedings, International Seminar on Nuclear War*, Erice, Italy, August 19–24, 1985 (Lawrence Livermore National Laboratory, Preprint UCRL-93760, December, 1985).]

These remarks were made nearly two years after extensive worldwide *public* discussion of nuclear winter had begun.

3.3:

This approach is exemplified in the comments of G. Rathjens and R. Siegel (*Issues in Science and Technology 1*, 1985, 123–128) and of E. Teller (*Nature 310*, 1984, 621–624). For a rebuttal, see C. Sagan (*Nature 317*, 1985, 485–488). Cf. also the remark by the strategist Colin Gray, "The burden of proof, to be prudent, should rest on those who would argue that a nuclear war would not trigger a nuclear winter" (in R. Hardin, J. Mearsheimer, G. Dworkin, and R. Goodin, eds., *Nuclear Deterrence: Ethics and Strategy* [Chicago: University of Chicago Press, 1985], 297).

3.4:

E.g., T. Slingo (*Nature 336*, 1988, 421): "The stimulus of nuclear winter led to practical work on the treatment of aerosol transport, scavenging and radiative transfer in numerical models, as well as giving insights into the importance of radiative-convective coupling." Such work has in turn improved our understanding of global greenhouse warming, as well as of the climatic effects of forest fires and volcanic explosions.

3.5:

See, e.g., Bruce Fellman, "Nuclear Winter Comes in From the Cold: Five Years after the Public Was First Stunned by It, the Apocalyptic Vision Emerges as a Useful Tool in the Study of Global Climate Change," *The Scientist 3* (9), May 1, 1989, 1, 18–19; A. B. Pittock, "Climatic Catastrophes: The Local and Global Effects of Greenhouse Gases and Nuclear Winter," in *Natural and Man-made Hazards*, M. I. El-Sabh and T. S. Murphy, eds. (Dordrecht, Holland: Reidel, 1988). The prospect of nuclear winter has also led to a set of regional workshops on climatic change, involving physical and biological scientists and agricultural experts, and branching out lately to the study of climatic changes from other sources, especially greenhouse warming. These regional studies have been organized as the PAN-EARTH project (*P*redictive *A*ssessment *N*etwork for *E*cological and *A*gricultural *R*esponses *to H*uman Activities), under the general direction of Mark Harwell of Cornell University. Among its many virtues, this project calls the attention of scientists and government officials worldwide to problems of climatic change and ozone depletion. Current PAN-EARTH project research studies are under way in China, Japan, Venezuela, and sub-Saharan Africa. Typical technical workshops were held in Saly, Senegal, September 11–14, 1989, and in Maracay, Venezuela, November 13–16, 1989.

3.6:

For example, in editorial comments by J. Maddox (*Nature 307*, 1984, 107 [but see replies, *Nature 311*, 1984, 307–308; *317*, 1985, 21–22]), and S. F. Singer (*Science 227*, 1985, 356 [but see replies, *Science 227*, 1985, 358, 360, 362, 444]). Also see ref. 3.3. Maddox, for many years the editor of the British scientific journal *Nature*, is also the author of a book entitled *The Dooms-*

*day Syndrome* (New York: McGraw-Hill, 1972), which casts light on some of the predispositions he brings to the analysis of nuclear winter; he believes that prophecies of disaster are

> at best pseudoscience. Their most common error is to suppose that the worst will always happen. . . . This book is an attempt to show why these prophecies should not keep people awake at night. . . . The most serious worry about the doomsday syndrome is that it will undermine our spirit.

(Cf. our Chapter 1 and ref. 1.1).

3.7:

There now exists a substantial body of published international scientific work on the subject of nuclear winter, most of which is referred to below (e.g., refs. 3.8, 3.10, 3.11, 3.13, 3.14, 3.16, and Appendix B; see also box, "Wildfires, Martian Dust, and Nuclear Winter," Chapter 3), but also including, e.g., the Royal Society of Canada, "Nuclear Winter and Associated Effects," Ottawa, January 31, 1985, 382 pp. ("We believe the nuclear winter to be a formidable threat. . . . We conclude that the nuclear winter hypothesis does indeed modify the global strategic position. . . . Canada ought to consider at once the military, strategic and social consequences.")

3.8:

S. L. Thompson and S. Schneider, "Nuclear Winter Reappraised," *Foreign Affairs*, Summer 1986, 981–1005; Letter, *Foreign Affairs*, Fall 1986, 171–178. Additional calculations by S. L. Thompson, V. Ramaswamy, and C. Covey ("Atmospheric Effects of Nuclear War Aerosols in General Circulation Model Simulations: Influence of Smoke Optical Properties," *Journal of Geophysical Research* 92, 1987, 10942–10960)—in which smoke is injected according to the recommended profile of the National Research Council (see Figure 2), although lower than in TTAPS and SCOPE—show land temperature drops up to 50% larger than in the original "autumn" calculations. Thompson and Schneider state, "In an attempt to contrast the most soothing of our statements with the most alarming of Carl Sagan's, some analysts misrepresent both of our positions."

3.9:

To compare the TTAPS baseline case with more recent global climate model calculations the following points must be noted: (i) In the TTAPS baseline case (which included a considerable mass of stratospheric dust—produced by high-yield groundbursts—generally neglected in the newer general circulation models), the maximum land temperature drops beneath extended smoke layers were about 35°C; the temperature declines averaged over time were of course smaller. (ii) Based on climatology in the unperturbed atmosphere, TTAPS explicitly estimated that the thermal inertia of the oceans could moderate average midlatitude temperature drops by 30% in continental interiors, and 70% in coastal areas—both below the values for a hypothetical all-land planet. (Oddly, there have been claims that TTAPS made no allowance for the moderating influence of oceans.) Inland and coastal temperature drops would then be about 25°C and 10°C, respectively, with an average land temperature decrease of about 15 to 17°C. (iii) Because a "seasonal average" intensity of sunlight was assumed by TTAPS, the maximum predicted temperature drops can be up to 5 to 10°C larger in summer, implying temperature decreases of roughly 20 to 25°C averaged over land under smoke. (iv) For low-altitude smoke injections, Thompson and Schneider (ref. 3.8) computed midlatitude summer land average temperature drops beneath smoke of about 10 to 12°C. (v) With somewhat higher-altitude smoke injection, Thompson and Schneider computed summer land average temperature drops of about 13 to 17°C. Maximum land temperature decreases as high as 30°C were also predicted by Thompson and Schneider. (vi) For an equivalent TTAPS smoke injection profile, total aerosol thickness, and smoke removal rate, an average surface temperature drop beneath smoke of some 20°C is expected in the Thompson and Schneider model (compared to some 20 to 25°C in the baseline case of the original TTAPS model). This constitutes good agreement. Similar results have been obtained at Livermore National Laboratory (S. J. Ghan, M. C. MacCracken, and J. J. Walton, "The Climatic Response to Large Atmospheric Smoke Injections: Sensitivity Studies with a Tropospheric General Circulation Model," *Journal of Geophysical Research* 93, 1988, 8315–8337).

3.10:

The National Research Council (NRC) of the U.S. National Academy of Sciences, *Effects on the Atmosphere of a Major Nuclear Exchange* (Washington, D.C.: National Academy Press, 1985). This report, which stressed the uncertainties in nuclear winter theory at the time, nevertheless concluded that there was a "clear possibility" of climatic catastrophe following a central exchange.

3.11:

SCOPE (Scientific Committee on Problems of the Environment of the International Council of Scientific Unions) Report 28, *Environmental Effects of Nuclear War*, Volume I, *Physical and Atmospheric Effects*, A. Pittock, T. Ackerman, P. Crutzen, M. MacCracken, C. Shapiro, and R. Turco (1986), 359 pp., and Volume II, *Ecological and Agricultural Effects*, M. Harwell and T. Hutchinson (Chichester: John Wiley, 1985), 523 pp.

3.12:

In several public comments, and in private discussion, Symposium on Global Climate Change, Sundance, Utah, August 24, 1989. See also "Soot Study Puts New Chill Into 'Nuclear Winter,'" by William Booth, *Washington Post*, June 22, 1989. Schneider writes,

> The single most important conclusion, I believe, from the work that has been conducted in the five years since the TTAPS article, is the widespread consensus that has developed that the environmental and societal "indirect" effects of a nuclear war are likely to be extremely serious. . . . [The seriousness] is so substantial that implications for both combatant and non-combatant nations should be considered at the highest policy levels. . . . Those who labored in 1983 to carry the problem to a wider public have performed an important service for both humanity and science. [Stephen H. Schneider, Editorial, *Climatic Change 12*, 1988, 215–219.]

3.13:

R. P. Turco and G. S. Golitsyn, "Global Effects of Nuclear War: A Status Report" (SCOPE Final Summary), *Environment*

*30*, 1988, 8–16; I. Colbeck and Roy M. Harrison, "The Atmospheric Effects of Nuclear War—A Review," *Atmospheric Environment 20*, 1986, 1673–1681 ("All studies show a large potential for major climatic changes as a result of the smoke injected by extensive post-nuclear fires . . . 1-dimensional models were quite correct in establishing the possibility of large-scale climatic perturbations following a nuclear war.")

3.14:

"Climate and Smoke: An Appraisal of Nuclear Winter," R. P. Turco, O. B. Toon, T. P. Ackerman, J. B. Pollack, and C. Sagan, *Science 247*, 1989, 166–176. This work has been called TTAPS II. A press report describing these results as a change from a 1983 estimated temperature drop of 15 to 25°C to a 1989 estimated drop of 10 to 20°C—an insignificant difference considering the nature of the problem—was headlined by the *New York Times* as "Nuclear Winter Theorists Pull Back" (Malcolm W. Browne, January 23, 1990, B5, B9. Cf. C. Sagan and R. Turco, "Don't Relax About Nuclear Winter Just Yet," *New York Times*, March 5, 1990.) See also *idem.*, "Nuclear Winter: Physics and Physical Mechanisms," *Annual Review of Earth and Planetary Sciences* (1991).

3.15:

G. S. Golitsyn and N. A. Phillips, *Possible Climatic Consequences of a Major Nuclear War*, World Climate Program Report WCP-142 (Geneva: World Meteorological Organization, 1986); G. S. Golitsyn and M. C. MacCracken, *Possible Climatic Consequences of a Major Nuclear War*, World Meteorological Organization Technical Document 201 (Geneva: World Meteorological Organization, 1987). Each report was written collaboratively by a Soviet and an American atmospheric scientist.

3.16:

H. A. Nix et al., "Study on the Climatic and Other Effects of Nuclear War: Report of the Secretary-General," United Nations General Assembly *Document A/43/351*, 5 May 1988:

> It appears evident that none would escape the awful consequences of a major nuclear war even if the theatre of conflict was geographically restricted to a small part of the northern hemisphere. . . . The direct effects of a major nu-

clear exchange could kill hundreds of millions; the indirect effects could kill billions.

3.17:

"The Greenhouse Effect: Impacts on Current Global Temperature and Regional Heat Waves," by J. E. Hansen. Testimony, U.S. Senate, Committee on Energy and Natural Resources, June 23, 1988; J. Hansen, I. Fung, A. Lacis, D. Rind, S. Lebedeff, R. Ruedy, G. Russell, and P. Stone, "Global Climatic Changes as Forecast by Goddard Institute for Space Studies: Three-Dimensional Model," *Journal of Geophysical Research* 93, 1988, 9341–9364.

3.18:

Although the greenhouse changes are much more pervasive, affecting even the deep ocean, and would last longer (decades or centuries, rather than months or years).

3.19:

Hugh Clevely, *Famous Fires: Notable Conflagrations on Land, Sea, and in the Air—None of Which Should Ever Have Happened* (New York: The John Day Company, 1957), 141; H. Wexler, "The Great Smoke Pall—September 24–30, 1950," *Weatherwise* 3, 1950, 129–134, 142; E. M. Elsey, "Alberta Forest Fire Smoke—24 September, 1950," *Weather* 6, 1951, 22–25; V. B. Shostakovitch, "Forest Conflagrations in Siberia," *Journal of Forestry* 23, 1925, 365–371; N. N. Veltishchev, A. S. Ginsburg and G. S. Golitsyn, "Climatic Effects of Mass Fires," *Izvestia—Atmospheric and Ocean Physics 24*, 1988, 296–304; Alexander S. Ginzburg, "Some Atmospheric and Climatic Effects of Nuclear War," *Ambio 18* (7), 1989, 384–390; A. W. Brinkman and James McGregor, "Solar Radiation in Dense Saharan Aerosol in Northern Nigeria," *Quarterly Journal of the Royal Meteorological Society 109*, 1983, 831–847; G. S. Golitsyn and A. K. Shukurov, *Doklady* (Proceedings, Soviet Academy of Sciences) 297, 1987, 1334; Y.-S. Chung and H. V. Le, "Detection of Forest-Fire Smoke Plumes by Satellite Imagery," *Atmospheric Environment 18*, 1984, 2143–2151; A. Robock, "Surface Temperature Effects of Forest Fire Smoke Plumes," in *Aerosols and Climate*, P. V. Hobbs and M. P. McCormick, eds. (Hampton, VA: Deepak Publ. [1988]); A. Robock, "Enhancement of Surface Cooling Due to Forest Fire Smoke," *Science 242*, 1988, 911–913; A. Robock, "Forest Fire Smoke Effects on Surface Air Temperatures," Chapman Conference on Global Biomass Burning, Williamsburg, Va., March 19–23, 1990; T. Y. Palmer, "Smoke from the Tropical West Pacific's Burning Forests Results in Tropospheric Heating," *ibid.*; J. B. Pollack, O. B. Toon, C. Sagan, A. Summers, B. Baldwin, and W. Van Camp, "Volcanic Explosions and Climatic Change: A Theoretical Assessment," *Journal of Geophysical Research 81*, 1976, 1071–1083; *idem*, "Stratospheric Aerosols and Climatic Change," *Nature 263*, 1976, 551–555; J. Tillman, University of Washington, Viking data, private communication, 1988; P. M. Anderson et al., "Climatic Changes of the Last 18,000 Years: Observations and Model Simulations," *Science 241*, 1988, 1043–1052.

4.1:

From a multitude of modern examples: "Disco Fire Kills 43 in Spain: Deaths Blamed on Toxic Fumes," Reuters dispatch, *International Herald Tribune*, January 15, 1990, 10:

> Firemen said they found several bodies reclining in armchairs . . . indicating the victims were overwhelmed so quickly they could not even attempt an escape.

Or, the circumstances of the Happy Land Social Club fire in the Bronx, N.Y., on March 25, 1990, as described in "Toxic Smoke Killed Some in Seconds," by Natalie Angier, *New York Times*, March 27, 1990:

> The choking black smoke that the firefighters encountered probably carried aldehydes, cyanides and other toxins from burning wood, plastic and linoleum. Such toxins can trigger an asthma-like attack in a person, causing the bronchial tubes to spasm so wildly that they collapse. . . . [Also,] when the concentrations of carbon monoxide in a room reach a critical level, death usually results in about two minutes.

4.2:

*Newsweek*, January 30, 1984, quoting J. Robert Dille, Manager of the Federal Aviation Administration's Civil Aero-Medical Institute (CAMI), in Oklahoma City.

4.3:

According to the April 5, 1946, testimony of Rudolf Hoess, commander at Auschwitz, Zyklon B was

> dropped into the death chamber from a small opening. It took from 3 to 15 minutes to kill the people in the death chamber depending upon climatic conditions. We knew when the people were dead because their screaming stopped. We usually waited about one-half hour before we opened the doors. . . . ["Closing address for the United States of America by (Supreme Court Justice) Robert H. Jackson, Representative and Chief of Counsel for the United States of America," Nuremberg Tribunal. From Office of United States Chief of Counsel for Prosecution of Axis Criminality, *Nazi Conspiracy and Aggression, Supplement A* (Washington, D.C.: U.S. Government Printing Office, 1947).]

Hoess, not to be confused with Deputy Führer Rudolf Hess, was a precise and methodical bureaucrat, apparently wholly untouched by the horrors he perpetrated.

4.4:

Cf. Barry Commoner, *Making Peace with the Planet* (New York: Pantheon, 1990).

4.5:

The first mention of toxic chemicals being released by fires set in a nuclear war appears to be the important paper by Paul Crutzen and John Birks, "The Atmosphere after a Nuclear War: Twilight at Noon," *Ambio 11* (2–3), 1982, 114–125. However, only oxides of nitrogen and hydrocarbons, "the ingredients of photochemical smog," were mentioned. See also J. W. Birks and S. L. Stephens, "Possible Toxic Environments Following a Nuclear War," in ref. 12.1.

4.6:

Despite the potential seriousness of pyrotoxins in the wake of nuclear war, we know of no serious studies of the phenomenon by any of the world's military establishments.

4.7:

Robert Scheer, *With Enough Shovels* (New York: Random House, 1982).

4.8:

E.g., "Powerful B-53 Bomb Comes Out of Mothballs," *Washington Times*, August 5, 1987, 5. These 9-megaton weapons are needed, it is said, to attack underground shelters for the Soviet leadership. (See box, Chapter 11.)

4.9:

G. Glatzmaier and R. Malone, SCOPE Moscow Workshop, March 21–25, 1988; S. Thompson and P. Crutzen, Defense Nuclear Agency, Global Effects Meeting, Santa Barbara, Cal., April 9–12, 1988; C.-Y. Kao, G. A. Glatzmaier, R. C. Malone and R. P. Turco, "Global Three-Dimensional Simulation of Ozone Depletion Under Post-Nuclear-War Conditions," *Journal of Geophysical Research* (1990). These calculations employ three-dimensional atmospheric circulation computer models with simplified ozone photochemistry. Glatzmaier and Malone also demonstrated quantitatively that the reaction of stratospheric ozone with soot would not significantly reduce the soot abundance (or the resulting climate effect), based on new laboratory chemical reaction data (S. L. Stephens, J. G. Calvert and J. W. Birks, "Ozone as a Sink for Atmospheric Carbon Aerosols: Today and Following Nuclear War," *Aerosol Science and Technology 10*, 1989, 326–331; S. L. Stephens, M. J. Rossi, and D. M. Golden, "The Heterogeneous Reaction of Ozone on Carbonaceous Surfaces," *International Journal of Chemical Kinetics 18*, 1986, 1133–1149).

4.10:

How the flux of dangerous UV-B radiation increases as the ozone layer thins is described by D. Lubin, J. E. Frederick, and A. J. Krueger, "The Ultraviolet Radiation Environment of Antarctica," *Journal of Geophysical Research 94* (1989), 8491–8496. The deleterious effect on marine phytoplankton seems clear. Many other aspects of the effects of increased ultraviolet light on organisms need more research. See, e.g., T. E. Graedel, "Effects of Increased Ultraviolet," *Nature 342*, 1989, 621–622; R. R. Jones and T. Wigley, eds., *Ozone Depletion: Health and Environmental Consequences* (London: Wiley, 1989). (See also ref. 6.3.)

4.11:

After the discovery of nuclear winter, the posture of the Defense Nuclear Agency, the weapons laboratories, and the White

House was an unwillingness to support any major study of biological effects, let alone synergisms. They pleaded uncertainty in our knowledge of the physical outcome of nuclear war. We argued that, on a matter of such importance, the way to deal with any uncertainty in the physical effects is to study the biological effects of a range of assaults on the environment of differing severities. Then we would know how serious various environmental consequences of nuclear war were. This argument was not enthusiastically received. We believe that what is needed is a systematic study of a variety of natural and agricultural ecosystems, into which a wide range of cold, dark, pyrotoxins, radioactivity, and ultraviolet irradiation are introduced. The investigations could be done under carefully monitored conditions in the field, and also in large terraria and aquaria. There is probably much fundamental ecological knowledge to be gained, since such systematic experiments, at first changing one variable at a time, have never been performed. But even more important, such studies would provide a much better understanding of the biological consequences of nuclear war. The foregoing remarks are critical of the American nuclear winter research program, which we know best, but we believe it applies equally well to the research programs of the other nuclear weapons states. (See also ref. 6.5.)

5.1:

J. John Sepkoski, Jr., "Phanerozoic Overview of Mass Extinction," in D. M. Raup and D. Jablonski, eds., *Pattern and Process in the History of Life* (Berlin: Springer Verlag, 1986); and private communication, 1985.

5.2:

An early example:

> A chemist decried loose talk of destroying the human race by atomic energy: Daily [world] increases in population equaled the casualties produced by the Nagasaki bomb. Fissile materials were so rare that it seemed most unlikely that they could ever be used to alter climates. [Summary of a scientific meeting, *Journal of the British Interplanetary Society* 7 (6), November, 1948, 233.]

Similar concerns about "loose talk" were voiced in the debate in the 1950s and early 1960s about radioactive fallout, about accidents at nuclear weapons plants and facilities, and about the threat nuclear war poses to the ozone layer.

5.3:

Herman Kahn, *On Thermonuclear War*, 2nd ed. (Westport, Conn: Greenwood Press, 1961), 96, 149.

5.4:

E.g., "many of us have felt that even a much more confined thermonuclear war is already an unacceptable disaster, morally so bad that nothing could really be meaningfully worse." (George Quester, in ref. 2.6.)

5.5:

One attempt to address the policy implications of extinction has been made by Joseph S. Nye, Jr., *Nuclear Ethics* (New York: Free Press, 1986): "It does not follow from the fact that extinction is an unlimited consequence that even a tiny probability is intolerable and that our generation has no right to take risks" (p. 64). Nye is paraphrased as arguing that the "correct" question is not "What is worth defending at the cost of the survival of our species?," but "Is defending our way of life worth raising the risk that the species will be destroyed from one in ten thousand to one in one thousand, for a certain period of time?" (Peter Grier, "Is Proving the Terror of Global Nuclear War a Waste of Time?," *Christian Science Monitor*, November 10, 1986.) Not least of the problems with this argument is the fact that we don't have a ghost of a chance of making such probability estimates with adequate reliability.

5.6:

> I first became involved in nuclear targeting in 1954. What came home clear to me at that time was that the easiest thing to do was to destroy cities. . . . Conversely, the hardest thing to do was to protect cities from attack. . . . I estimate that 2% of the force of either side, impacting on the cities of the other side, would do catastrophic damage to most of the urban-industrial area. It takes only a few hundred weapons, and probably not more than a couple hundred, to impact and make very, very great damage. [Gen. David C. Jones, former Chairman of the Joint Chiefs

of Staff, in testimony before the Senate Armed Services Committee, March 30, 1987.]

5.7:

Desmond Ball, *Déjà Vu: The Return to Counterforce in the Nixon Administration* (Santa Monica: California Seminar on Arms Control and Foreign Policy, 1974, 11); also *International Herald Tribune*, May 9, 1978.

5.8:

Anthony Cave Brown, ed., *Operation World War III: The Secret American Plan "Dropshot" for War with the Soviet Union* (London: Arms and Armour Press, 1978) [U.S. edition, *Dropshot* (New York: Dial, 1978)]; David Alan Rosenberg, "American Atomic Strategy and the Hydrogen Bomb Decision," *Journal of American History 46*, 1979, 62–87.

5.9:

R. P. Turco, "Synthesis of Fallout Hazards in a Nuclear War," *Ambio 18* (7), 1989, 391–394. See also TTAPS (ref. 2.2) and ref. 3.11.

5.10:

For smaller-scale nuclear wars, if such are possible, see W. H. Daugherty, B. G. Levi, and F. N. von Hippel, "The Consequences of 'Limited' Nuclear Attacks on the United States," *International Security 10*, 1986, 3–45; B. G. Levi, F. N. von Hippel, and W. H. Daugherty, "Civilian Casualties from 'Limited' Nuclear Attacks on the U.S.S.R.," *ibid. 12*, 1987, 168–189. See also W. M. Arkin, B. G. Levi, and F. von Hippel, "The Consequences of a 'Limited' Nuclear War in East and West Germany," *Ambio* 9 (2–3), 1982, 163–173, and addendum, *ibid. 12* (1), 1983, 57.

5.11:

*Ambio 11* (2–3), 1983; S. Bergstrom et al., "Effects of Nuclear War on Health and Health Services" (Rome: World Health Organization Publication A 36.12, 1983); World Health Organization, "Effects of Nuclear War on Health and Health Services" (Geneva: 1988). The 1983 WHO estimate of fatalities from prompt nuclear war effects—1.1 billion killed outright, and another 1.1 billion dying later—assumes heavy targeting in

China, India, and Southeast Asia, which some analysts consider unlikely or even preposterous. We find it hard to believe that China would remain uninvolved in a major U.S./U.S.S.R. strategic exchange. India's development of nuclear weapons is in part intended to deter China, suggesting possible Indian targets in a future global war. See also Chapter 12.

5.12:

Andrei Sakharov and Ernst Henry, "Scientists and Nuclear War," in Stephen F. Cohen, ed., *An End to Silence: Uncensored Opinion in the Soviet Union* (New York: Norton, 1982), 230.

5.13:

P. R. Ehrlich, J. Harte, M. A. Harwell, P. H. Raven, C. Sagan, G. M. Woodwell, J. Berry, E. S. Ayensu, A. H. Ehrlich, T. Eisner, S. J. Gould, H. D. Grover, R. Herrera, R. M. May, E. Mayr, C. P. McKay, H. A. Mooney, N. Myers, D. Pimentel, and J. M. Teal, "Long-Term Biological Consequences of Nuclear War," *Science* 222, 1983, 1293–1300.

5.14:

For maladaptive behavior induced by crisis—especially in military and civilian leaders—see, e.g., Jerome D. Frank, *Sanity and Survival: Psychological Aspects of War and Peace* (New York: Random House, 1967); George V. Coelho, David Hamburg, and John E. Adams, eds., *Coping and Adaptation* (New York: Basic Books, 1974); Lester Grinspoon, "Crisis Behavior," *Bulletin of the Atomic Scientists*, April 1984, 25–28; Richard A. Gabriel, *No More Heroes: Madness and Psychiatry in War* (New York: Hill and Wang, 1987); Richard Ned Lebow, *Nuclear Crisis Management: A Dangerous Illusion* (Ithaca, N.Y.: Cornell University Press, 1987); *idem., Between Peace and War* (Baltimore: Johns Hopkins University Press, 1981). The psychological effects of nuclear war and nuclear winter on survival have been given inadequate attention. Some preliminary studies appear in L. Grinspoon, ed., *The Long Darkness: Psychological and Moral Perspectives on Nuclear Winter* (New Haven: Yale University Press, 1986); James Thompson, *Psychological Aspects of Nuclear War* (New York: John Wiley, 1985); M. Pamela Bumsted, ed., *Nuclear Winter: The Anthropology of Human Survival* (Proceedings of a session at the 84th annual meeting, American Anthropological Association, December 6, 1985, Washington, D.C.), Los Alamos National Laboratory Document LA-UR-86-370, 1986; F. Solomon and R. Q. Marston, eds., *The Medical Implications of Nuclear War* (Washington, D.C.: Institute of Medicine, National Academy of Sciences, 1986).

5.15:

C. D. Laughlin and I. A. Brady, eds., *Extinction and Survival in Human Populations* (New York: Columbia University Press, 1978).

5.16:

We know of no scientific work on nuclear winter that concludes human extinction to be thereby inevitable or even likely. Most formulations state that under some (extreme) circumstances extinction cannot be excluded, or words to that effect. Even consideration of a 10,000-megaton exchange in which all uncertain parameters were assumed to take their most adverse possible values concluded no more than this (ref. 5.13). Nevertheless, a number of commentaries have assumed that TTAPS or others drew the conclusion that human extinction was probable. There is a wide gulf separating "probable" from "not impossible." But if the stakes are high enough, even "not impossible" must be taken very seriously.

The assertion by Thompson and Schneider that "global apocalyptic conclusions of the initial nuclear winter hypothesis can now be relegated to a vanishingly low level of probability" (ref. 3.8) was based on a nuclear winter model that may have been too mild (refs. 3.8, 3.9, 3.13, 3.14). (More recently, "vanishingly low" has been softened to "highly remote" [ref. 3.12].) Such a confident dismissal requires much better evidence than has been offered at least thus far: "We have here a tension between the usual standards of scientific caution and the usual standards of military prudence. This is not a debate on some arcane point of theoretical physics in which errors, if any, will eventually be corrected by the traditional, tested methods of scientific criticism and debate. Here, if we make a mistake, the consequence may be irreversible." (C. Sagan, *Foreign Affairs* 65, 1986, 163–168.)

Because of the unprecedentedly high stakes, it seems to us

that two rules of evidence apply: (1) in any debate in which human extinction following nuclear war is dismissed, the burden of proof must fall on those doing the dismissing; and (2) if we acknowledge that the issue cannot be decided unambiguously, it is prudent to err on the side of the argument that holds extinction to be possible—even if we were sure the probability was small.

On the other hand, some conclusions clearly go too far in the opposite direction—e.g., the comment by Marshal Akhromeyev, then Chief of the Soviet General Staff, that *any* use of nuclear weapons would mean "the entire humanity and the whole life on our planet would be annihilated" (S. F. Akhromeyev, interview by Robert Scheer, "Then Came Gorbachev," *Playboy*, August 1988). This represents a considerable change in Soviet thinking. Consider, in contrast, these representative earlier Soviet statements: "In the West, for instance, it is claimed that humanity, world civilization, would perish in the event of such a war. . . . Marxist-Leninists resolutely reject these attempts. They have always considered and still consider war, all the more so with thermonuclear war, as the greatest calamity for the people. But Communists harbor no sentiments of hopelessness or pessimism." [Rear Adm. V. Shelyug, "Two Ideologies, Two Views of War," *Krasnaya Zvezda*, February 7, 1974.] Or, "However grievous the consequences of atomic war might be, it must not be identified with the 'destruction of world civilization.' Such an identification willy-nilly brings grist to the American imperialist mill." [*Kommunist 4*, March 1955, 12–23; cited in H. S. Dinerstein, *War and the Soviet Union* (New York: Praeger, 1959), 77.]

5.17:

In his widely read and influential book *The Fate of the Earth* (New York: Knopf, 1982), Jonathan Schell does argue that extinction is not only a possible but perhaps even a likely consequence of nuclear war. However, the argument is built upon blast, fire, prompt radioactive fallout, and depletion of the ozone layer, without any climatic effects included. The book was published just as nuclear winter was being discovered. Although Schell could not have discussed nuclear winter, in the broadest sense he anticipated it: "Given the incomplete state of our knowledge of the Earth, it seems unjustified at this point to assume that further developments in science will not bring forth further surprises."

This last point had a few earlier advocates—e.g., the Director of the U.S. Arms Control and Disarmament Agency in 1974:

> The damage from nuclear explosions to the fabric of nature and the sphere of living things cascades from one effect to another in ways too complex for our scientists to predict. Indeed, the more we know, the more we know how little we know. [Fred Iklé, "Nuclear Disarmament Without Secrecy," *U.S. Department of State Bulletin*, September 30, 1974, 454–458.]

5.18:

L. W. Alvarez, W. Alvarez, F. Asaro, and H. V. Michel, "Extraterrestrial Cause for the Cretaceous-Tertiary Extinction," *Science 208*, 1980, 1095; W. Alvarez, F. Asaro, H. V. Michel, and L. W. Alvarez, "Iridium Anomaly Approximately Synchronous with Terminal Eocene Extinctions," *Science 216*, 1982, 86; O. B. Toon, J. B. Pollack, T. P. Ackerman, R. P. Turco, C. P. McKay, and M. S. Liu, "Evolution of an Impact-Generated Dust Cloud and Its Effects on the Atmosphere," in *Geological Implications of Impacts of Large Asteroids and Comets on the Earth*, Leon Silver and Peter Schultz, eds., Geological Society of America Special Paper No. 190, 1982, 187–199; J. B. Pollack, O. B. Toon, T. P. Ackerman, C. P. McKay, and R. P. Turco, "Environmental Effects of an Impact-Generated Dust Cloud: Implications for the Cretaceous-Tertiary Extinctions," *Science 219*, 1983, 287–289; W. S. Wolbach, R. S. Lewis, and E. Anders, "Cretaceous Extinctions: Evidence for Wildfires and Search for Meteoritic Material," *Science 230*, 1985, 167–170; E. Argyle, "Cretaceous Extinctions and Wildfires," *Science 234*, 1986, 261–264; W. S. Wolbach, I. Gilmour, E. Anders, C. J. Orth, and R. R. Brooks, "Global Fire at the Cretaceous-Tertiary Boundary," *Nature 334*, 1988, 665–669; H. J. Melosh, N. M. Schneider, K. J. Zahnle, and D. Latham, "Ignition of Global Wildfires at the Cretaceous/Tertiary Boundary," *Nature 343*, 1990, 251–254; A. Hallam, "End-Cretaceous Mass Extinction Event: Argument for Terrestrial Causation," *Science 238*, 1987, 1237–1242, and C. B. Officer, A. Hallam, C. L. Drake, and J. D. Devine, "Late Cretaceous and Paroxysmal Cretaceous/Tertiary Extinctions," *Nature 326*, 1987, 143–149; L. W. Alvarez, "Mass

Extinctions Caused by Large Bolide Impacts," *Physics Today 40*, 1987, 24–33 (a rebuttal of the preceding two articles); H. C. Urey, "Cometary Collisions and Geological Periods," *Nature 242*, 1973, 32; F. Hoyle and C. Wickramasinghe, "Comets, Ice Ages, and Ecological Catastrophes," *Astrophysics and Space Science 53*, 1978, 523–526; E. C. Prosh and A. D. McCracken, "Postapocalypse Stratigraphy: Some Considerations and Proposals," *Geology 13* (1), 1985, 4–5. See also *Comet*, by Carl Sagan and Ann Druyan (New York: Random House, 1985).

6.1:

For instance, Herman Kahn, *On Thermonuclear War*, 2nd ed. (Westport, Conn.: Greenwood Press, 1961; reprinted 1978). The necessity to consider simultaneously both the probability and the severity of an event, and at the same time to distinguish between them, suffuses Kahn's book. One of many examples: "It is not that any of these possibilities has a very high probability of occurring. The point is that the results would be terribly serious if they did occur" (p. 154). A similar point is made in the Einstein/Russell Mainfesto: "Many warnings have been uttered by eminent men of science, by authorities in military strategy. None of them will say that the worst results are certain. What they do say is that these [apocalyptic] results are possible, and no one can be sure that they will not be realized."

6.2:

For Bhopal, see "Bhopal Payments Set at $470 Million by Union Carbide," by Sanjoy Hazarika, *New York Times*, February 15, 1989, 1. More than 3,500 died, and more than 200,000 were injured. If no compensation were available for the survivors of the people killed, a little more than $2,000 would be available for each injured person; if nothing were available for the injured, a little more than $100,000 would be earmarked for the survivors of those killed. The average compensation lies between these values. The total number of claims for compensation was over 500,000. The average settlement paid by the Johns Manville Corporation to compensate American victims of the usually lethal mesothelioma cancer caused in schools and elsewhere by its asbestos products was $38,000. ("Manville Trust Fund in Trouble," by Stephen Labaton, *New York Times*, February 7, 1989, D1.) The usual liability offered by U.S. airlines to international passengers ranges from $10,000 to $75,000 per passenger death. A bill proposed in the Senate would provide $100,000 to former uranium miners (or their survivors)—many of whom are Navajos—who suffered serious radiation damage without having been advised of the dangers of their work; those residents of Nevada and Utah endangered by radioactive fallout from aboveground nuclear testing would receive $50,000 each. ("Uranium Miners Tell Panel Radiation Caused Ailments," *New York Times*, March 14, 1990, A20.) On the other hand, life in other parts of the world is often judged more cheaply by citizens of affluent nations. After the accidental B-52 drop of 36 tons of high explosives on the main street of the "friendly" Cambodian town of Neak Luong, the United States paid survivors, next-of-kin, and others about $100 each. (William Shawcross, *Sideshow: Kissinger, Nixon and the Destruction of Cambodia* [New York: Pocket Books, 1975], 294.)

6.3:

The ozone problem is described in a long series of reports, of which the assessments by the World Meteorological Organization ("Atmospheric Ozone, 1985," WMO Report 16, 3 vols., Geneva, 1985) and by NASA (*Present State of Knowledge of the Upper Atmosphere: An Assessment Report*, 201 pp., 1988) are exemplary. There is a possibility that even small increases in the near-ultraviolet solar flux reaching the Earth's surface could have "profound consequences" for oceanic phytoplankton, and therefore, through the food chain, for the entire marine ecosystem (C. H. Kruger et al. and R. B. Setlow et al., *Causes and Effects of Stratospheric Ozone Reduction: An Update* (Washington, D.C.: National Academy of Sciences, 1982). In the last few years the ozone depletion issue has become more clear (especially because of the discovery of the hole in the Antarctic ozonosphere), and the dangers are now recognized as potentially much more serious—with indirect effects possibly including a compromise of the immune systems of exposed humans and even a global ecological catastrophe caused by the deaths of primary photosynthetic producers. (See also ref. 4.10.)

6.4:

The $CO_2$ greenhouse effect (see Appendix A) may result in a warming of the Earth by an average of several degrees Centigrade over the next century (cf. ref. 3.17); during this same epoch, a much smaller general background cooling trend would otherwise be expected as the Earth's climate drifts out of the present interglacial period. Variations in temperature, sea level, and weather over time due to the $CO_2$ warming could lead to massive deforestation, agricultural failures, inundation of coastal cities and low-lying land areas, drought, crop and human migration, and economic disruptions. It is likely that these disturbances would occur slowly enough, however, to avoid anything like the casualties and hardships of nuclear war, although the face of global society could be markedly altered. For a discussion of greenhouse warming, its uncertainties, and possible mitigating steps, see, e.g., C. Sagan, *American Journal of Physics*, 58, 1990, 721–730; Michael C. MacCracken *et al.*, *Energy and Climate Change: Report of the DOE Multi-Laboratory Climate Change Committee* (Chelsea, MI: Lewis, 1990).

6.5:

By contrast, the research budget for understanding the consequences of nuclear war seems profoundly disproportionate to the dangers. At its peak, total research expenditures on nuclear winter in the United States were roughly $5.5 million a year, the bulk of which was available only from or to the nuclear weapons establishment itself, either disbursed by the Defense Nuclear Agency or given to the national weapons laboratories. Apparently, no funding from Department of Energy headquarters was ever granted to the weapons laboratories explicitly to study nuclear winter. We are told that to study nuclear winter Livermore and Los Alamos were obliged to use funds intended for general weapons development and taxes internally levied on all other programs. These funds and taxes amounted to about half of the $5.5 million per year. Today the funding is not much more than $1 million per year. Only about $0.5 million per annum was ever available from the National Science Foundation—and less is available today. Despite the recommendations of a review ordered by the Executive Office of the President (see box, "Would Billions of People Really Be Killed by Nuclear Winter?," Chapter 5), virtually no research is being done on simulations of nuclear winter effects on agriculture. The maximum annual research budget for studying nuclear winter was less than the cost of a single attack helicopter and around one-thousandth the annual budget for researching and developing the Strategic Defense Initiative (Star Wars). The budgets for nuclear winter research in the U.S.S.R. and other nations appear to be even smaller.

6.6:

Cf. Paul C. Warnke, former chief U.S. arms control negotiator and former Assistant Secretary of Defense:

> Pascal said that even the remote prospect of eternal damnation should lead to every effort to avoid it. I don't need to be totally convinced of the inevitability of nuclear winter to feel that no effort is too great to see to it that a strategic nuclear exchange does not occur. [In Peter C. Sederberg, ed., *Nuclear Winter, Deterrence and the Prevention of Nuclear War* (New York: Praeger, 1986), 32.]

6.7:

See Peter Stein and Peter Feaver, *Assuring Control of Nuclear Weapons* (Lanham, Md.: University Press of America, 1987); G. E. Miller, "Who Needs Pals?," *Proceedings, U.S. Naval Academy*, July, 1988, 50–56; and rebuttal by Feaver and Stein, *ibid.*, October 1988, 35.

7.1:

Sagan (ref. 2.3) discussed a broad "threshold" regime in which severe effects become possible, while Thompson and Schneider (ref. 3.8) held that even such a generalized concept may be misleading. There is no doubt that various sorts of confusion, both on the science and the policy, have attached themselves to the threshold concept. A prominent American nuclear strategist proposed to us that nuclear winter could not happen, because once our arguments for a threshold were published no one would ever explode the corresponding number of nuclear weapons. He then went on to argue, on these very grounds, that our analysis should not be published.

7.2:

Other properties of the smoke are also important, including its height of injection, particle size distribution, composition (a larger fraction of black soot has a greater effect), and optical properties (e.g., soot from oil or plastics is much darker and absorbs much more light than smoke from burning vegetation).

The atmospheric dispersal and removal rates and the geographic distribution of the initial injection also affect the climatic impact. Dust from high yield groundbursts adds to the overall obscuration, and may in some cases be climatically significant.

7.3:

Existing climate models do not provide a high-resolution description of weather variability, and many of the models average the diurnal solar heating; so in the scientific literature the temperature *extremes* over a day/night cycle are not always calculated, and, when calculated, are not always stated. The potential severity for plants and ecosystems of the extreme of the diurnal temperature variations is therefore not obvious in these results; e.g., rice crops are imperiled if nighttime temperatures ever drop below the freezing point, but diurnal averages might miss a predawn freeze.

7.4:

J. B. Hoyt, "The Cold Summer of 1816," *Annals of the Association of American Geographers 48*, 1958, 118–131; H. Stommel and E. Stommel, "The Year Without a Summer," *Scientific American 240*, 1979, 176; *idem, Volcano Weather: The Story of 1816, the Year Without a Summer* (Newport, R.I.: Seven Seas Press, 1983). Our summary of the history of the 1816/1817 climate anomaly is taken mainly from the book by Stommel and Stommel and ref. 7.5. The connection between the Tambora explosion and widespread Northern Hemisphere summer frosts is based on strong circumstantial evidence—it was one of the most violent volcanic explosions in hundreds of years—but the connection is not beyond question. Consider, however, other very similar cases described in the box, "Volcanic Winter," in this chapter.

7.5:

J. D. Post, *The Last Great Subsistence Crisis in the Western World* (Baltimore: Johns Hopkins University Press, 1977).

7.6:

The global average temperature decrease of about 1°C applies to land and oceans, taken together, and implies an average land temperature drop of perhaps 2°C. Even with such a relatively small average temperature decrease, however, frost damage was recorded (although the more hardy trees were not killed) in bristlecone pines in the Western U.S. [V. C. LaMarche and K. K. Hirschboeck, "Frost Rings in Trees as Records of Major Volcanic Eruptions," *Nature 307*, 1984, 121] and, as we have mentioned, crop failures were widespread in the Northeast U.S. and Europe during the spring, summer, and fall of 1816 (see ref. 7.4).

7.7:

During the "Little Ice Age," between about 1450 and 1850, global temperatures were also about 1°C colder than they are today, but the changes occurred more slowly. In winters people skated on the Thames, the Seine, and the canals of Holland, and in 1780 people walked the five miles from Staten Island to Manhattan over the ice. (G. Parker, *Europe in Crisis, 1598–1648* [Sussex: Harvester, 1980]; D. Ludlum, *Early American Winters: 1604–1820* [Boston: American Meteorological Society, 1966].)

7.8:

John Imbrie and Katherine Palmer Imbrie, *Ice Ages: Solving the Mystery* (Short Hills, N.J.: Enslow Publishers, 1979).

7.9:

S. J. McNaughton, R. W. Ruess, and M. B. Coughenour, "Ecological Consequences of Nuclear War," *Nature 321*, 1986, 483–487.

7.10:

What follows is a pocket primer, if you've taken a high-school algebra course or two, on how to calculate with optical depths. This is just to satisfy your curiosity; the discussion in the body of the book stands on its own and should be understandable if you choose to skip this note altogether.

The influence of smoke or dust or other aerosols on light is usually measured by a quantity called the optical depth. It has two components: one due to scattering or reflection of light off the fine aerosol particles, and the other due to absorption of light inside those same particles. The intensity of sunlight entering a layer of such particles might be reduced by being bounced around by the suspended particles (and eventually emerging from the layer): that's scattering. Or the sunlight might be gobbled up by the particles (if they're dark in color), heating them: that's absorption. In smoke, particularly smoke generated from burning cities or petroleum depots, absorption is generally more effective than scattering in attenuating sunlight. So the simplest measure of smoke optical effects is the optical depth due to absorption alone (indicated by the subscript a). The average absorption optical depth, $\tau_a$, may be estimated for a given mass of smoke, m—assumed to be distributed uniformly over some area A—as

$$\tau_a = m\, \sigma_a / A,$$

where $\sigma_a$ (Greek lowercase sigma) is the smoke absorption coefficient. Here m can be measured in grams (g), A in square centimeters ($cm^2$), and $\sigma_a$ in $cm^2/g$. $\tau_a$ (pronounced "tau-sub-a") is called a dimensionless number and doesn't have grams or square centimeters trailing after it.

We define $I_0$ to be the intensity of sunlight falling on the top of the smoke clouds. I is how much sunlight actually makes it all the way through the clouds to the surface. $I/I_0$, therefore, measures what fraction of the incident sunlight isn't absorbed by the smoke. The average reduction in sunlight caused by absorption in smoke may be roughly estimated using Beer's law:

$$I/I_0 = \exp(-\tau_a/\mu) = e^{-\tau_a/\mu}.$$

$\mu$ (Greek lower case mu) is a measure of how much longer the path of sunlight is through the atmosphere than when the Sun is at the zenith (i.e., directly overhead); it allows for the fact that the Sun rises and sets. The average value of $\mu$ turns out to be roughly 0.58. And "exp," for exponential, means that what follows in parentheses is the exponent or power to which a transcendental number called the base of natural logarithms ($e = 2.718\ldots$) is to be raised. All this can be worked out almost instantly on standard scientific pocket calculators or from tables of exponents or logarithms.

The bigger the optical depth, the less sunlight makes it through the layer. Thus, if $\tau_a = 0$, $I/I_0 = \exp(0) = 1$. Then $I = I_0$ and all of the incident light is transmitted by the layer—the layer is transparent. If $\tau_a = 0.05$, $I/I_0 = 0.92$ and most of the light still makes it through; if $\tau_a = 1$, $I/I_0 = 0.18$ and most doesn't; and if $\tau_a = 2$, $I/I_0 = 0.03$ and hardly any of it does (cf.

Fig. 4). If $\tau_a$ is 2 or more, it's getting dark. If $\tau_a = \infty$, $I/I_0 = \exp(-\infty) = 0$, so $I = 0$; with an infinite optical depth, no light gets through at all—the Sun is invisible at noon, and the sky is black. But because of the nature of exponentials, by the time $\tau_a$ gets much larger than 2, it's not much different from the never-realizable case of $\tau_a = \infty$.

The absorption coefficient for urban smoke is measured to range from about 30,000 to about 120,000 $cm^2/g$ (square centimeters per gram). The greatest absorption occurs in black sooty smoke. If 5 teragrams (1 Tg = $10^{12}$ g = 1 trillion grams = 1 million metric tons = 1 megaton) of very sooty smoke were evenly distributed over an entire hemisphere of the Earth ($A = 2.5 \times 10^{18}$ $cm^2$), then $\tau_a = 0.2$ and $I/I_0 = 0.71$. (This is for $\sigma_a = 100{,}000$ $cm^2/g$). That is, 5 Tg of soot is capable of absorbing up to 30% of all solar energy incident on the hemisphere. This would cause major climate anomalies. If you like, you might try working out other cases for yourself.

Nuclear weapons are very good at burning cities: An airburst of a run-of-the-mill 400-kiloton nuclear weapon can burn an area between 300 and 500 square kilometers (refs. 3.10, 3.11). Richard D. Small ("Atmospheric Smoke Loading from a Nuclear Attack on the United States," *Ambio 18* [7], 1989, 377–383) estimates the amount of flammable materials in American cities and elsewhere in the U.S. and concludes that a total of 37 Tg of sooty smoke (almost two-thirds of it from burning buildings) might be released into the atmosphere from a major Soviet attack—neglecting the spread of fires. Initially the soot would be patchy, since different targets release different quantities of smoke, but it would soon become nearly uniform (ref. 7.11). Very little of this would be rained out in the days and weeks following the war. If spread uniformly over the Northern Hemisphere, this would correspond to $\tau_a = 0.62$. If we now add smoke from Europe, the Soviet Union, and elsewhere, optical depths of 2 or more become possible, corresponding well to our TTAPS II baseline value of $\tau_a = 2.3$ (ref. 3.14).

There are some differences between Small's estimates and our own, but they concern wood and lumber—the sources of low-soot cellulosic smoke. Small's estimate for the U.S. inventory of noncellulosic highly sooting materials (petroleum, plastics, asphalt) is about 500 Tg. (Of course, only a fraction of this would be burned and the soot injected into the atmosphere in a nuclear war.) Extrapolating to Europe and the Soviet Union gives roughly 1500 Tg. The TTAPS II estimate in this highly sooting category of combustibles (see Table 2, ref. 3.14) is 925 Tg, in good agreement. When care is taken to compare the same target zones and to give appropriate weight to sootier smoke, the two estimates give mutually compatible answers.

Our detailed inventories show about 7,000 to 13,500 Tg of flammable material, mostly urban and suburban, in NATO and WTO—mainly wood and lumber, primary and secondary petroleum products, plastics, and asphalt roofing (ref. 3.14); values for the entire developed world are not much more. Some of these inventories—especially plastics—are increasing rapidly. The total amount of this material in flames in a central exchange is estimated to be between 2500 and 8500 Tg, but only 20 to 300 Tg of that would be emitted to the atmosphere as soot. Burning plastics are good at emitting soot; burning wood is much less efficient. Allowing for uncertainties in $\sigma_a$, we find the corresponding range in $\tau_a$ to be between 0.2 and 10, with an average value of 2.3 (ref. 3.14). For comparison, the average amount of soot in the present atmosphere—chiefly from forest fires—is less than 1 Tg. This corresponds to an ordinary value of $\tau_a$—in the absence of nuclear war—of less than 0.02, averaged over the globe.

7.11:

The initial smoke cloud produced by a large fire is extremely dense and localized. The cloud structure, although very complex, is not open, like a field of cumulus clouds, but blanketing and continuous. Within hours, this localized smoke can spread over hundreds of kilometers in the prevailing winds, and in a matter of roughly a week the smoke can circumnavigate the globe. The smoke pall generated by many widespread fires, although initially patchy, tends to become homogeneous over continental scales with amazing rapidity. Anyone who has stood beneath the smoke plume from a distant forest fire can attest to the fact that such a pall is continuous and, even though variable in thickness, relatively uniform in its ability to block sunlight. Satellite images of continent-sized smoke clouds created by localized forest fires show this blanketing effect of smoke. The patchiness of smoke plumes on scales greater than hundreds of kilometers has been simulated in computer models of the global atmosphere; it is found that smoke from dispersed nuclear war fires would merge into a hemispheric-wide pall within a week or two following the war.

Ordinary clouds are generally very patchy in appearance because they are composed of water, which can easily condense or evaporate with even a small change in air temperature. So cumulus clouds form in regions where the air is rising and cooling, and are absent in regions where the air is subsiding and warming. This sensitivity to the local state of the atmosphere leads to highly variable fields of water clouds, which we notice nearly every day. But smoke is not water; smoke contains *nonvolatile** particles that do not care whether the air is slightly cooler or warmer. Indeed, air motions, particularly small-scale turbulent motions that exist throughout the atmosphere, have the effect of mixing the smoke into larger air masses, thus making the cloud larger and generally more uniform. You could once see the same effect when the smoke from a few cigarettes quickly dispersed to fill a room (or crowded airliner) with a uniform haze.

Initially the smoke clouds from a nuclear war would be patchy; one country might for a day or two experience more cold and dark than the average, and an adjacent country less. But after a week or two most of the patchiness and the attendant climate roulette would disappear, replaced by a uniform smoke pall.

7.12:

Michael R. Rampino, Stephen Self, and Richard B. Stothers, "Volcanic Winters," *Annual Review of Earth and Planetary Sciences 16*, 1988, 73–99; Richard B. Stothers and Michael R. Rampino, "Historic Volcanism, European Dry Fogs, and Greenland Acid Precipitation, 1500 B.C. to A.D. 1500," *Science 221*, 1983, 411–413; Stefi Weisburd, "Excavating Words: A Geological Tool," *Science News 127* (8), February 9, 1985, 91–94; K. D. Pang, S. K. Srivastava, and H.-h. Chou, "Climatic Impacts of Past Volcanic Eruptions: Inferences from Ice Core,

* By nonvolatile we mean here that the smoke particles will not evaporate if the air holding the smoke remains below some reasonable temperature—e.g., 47°C, which is the warmest temperature reached in the lower atmosphere.

Tree Ring and Historical Data," *EOS 69*, 1988, 1062; G. J. Symons, ed., *The Eruption of Krakatoa and Subsequent Phenomena* (London: Harrison and Sons, 1888) (This is the classic study, performed for the Royal Society of London, by many scientists—including Bertrand Russell's uncle Rollo, who calculated that the stratospheric dust had traveled around the world at about 75 miles per hour); Tom Simkin and Richard S. Fiske, *Krakatau 1883: The Volcanic Eruption and Its Effects* (Washington, D.C.: Smithsonian Institute Press, 1983); C. G. Abbot, "Do Volcanic Explosions Affect Our Climate?" *National Geographic 24* (2), 1913, 181–197; Special issue on El Chichón, J. B. Pollack, ed., *Geophysical Research Letters, 10* (1), November 1983; Frans J. M. Rietmeijer, "El Chichón Dust a Persistent Problem," *Nature 344*, 1990, 114–115; P. M. Kelly and C. B. Sear, "Climate Impact of Explosive Volcanic Eruptions," *Nature 311*, 1984, 740–743.

8.1:

Cf. W. M. Arkin and R. W. Fieldhouse, *Nuclear Battlefields* (Cambridge, Mass.: Ballinger, 1985); T. B. Cochran, W. M. Arkin and M. M. Hoenig, *Nuclear Weapons Databook*, Volume I, *U.S. Nuclear Forces and Capabilities* (Cambridge, Mass.: Ballinger, 1984); *The Military Balance* (London: International Institute of Strategic Studies, 1987/1988 and later editions); ref. 17.26.

8.2:

Cf. George M. Seignious III and J. P. Yates, "Europe's Nuclear Superpowers," *Foreign Policy 55*, 1984, 40–53; R. D. Small, B. W. Bush and M. A. Dore, "Initial Smoke Distribution for Nuclear Winter Calculations," *Aerosol Science and Technology 10*, 1989, 37–50; *ibid.*, Pacific Sierra Research Corp. Report 1761, November, 1987; Small, ref. 7.10.

8.3:

This is immediately implied by the comparatively low accuracy of Chinese strategic rockets, as well as other evidence. Zhang Aiping, one of those principally responsible for Chinese nuclear weapons, is quoted as indicating that "the ability to destroy urban areas or 'soft' military targets in a retaliatory strike" is what matters. (John Wilson Lewis and Xue Litai, *China Builds the Bomb* [Stanford, Cal.: Stanford University Press, 1988], 214.) See also Dingli Shen, "The Current Status of Chinese Nuclear Forces and Nuclear Policies," Princeton University Center for Energy and Environmental Studies Report 247, 1990.

8.4:

Desmond Ball, *Targeting for Strategic Deterrence*, Adelphi Paper 185 (London: International Institute for Strategic Studies, 1983).

8.5:

Noel Gayler, Testimony, Joint Economic Committee, July 11–12, 1984, published in *The Consequences of Nuclear War*, Hearings, 98th Congress (Washington, D.C.: U.S. Government Printing Office, 1986).

8.6:

Indeed, all major analyses of possible nuclear war scenarios include oil refining and storage capacity as principal targets (e.g., *The Effects of Nuclear War*, Office of Technology Assessment, U.S. Congress [Washington, D.C.: U.S. Government Printing Office, 1979], 151 pp.; refs. 8.1, 8.19, 8.20, 8.21). Some analysts have gone so far as to suggest that oil should be intentionally stored in large quantities near the highest priority strategic targets—such as missile silos—to guarantee soot generation and nuclear winter after a first strike, thereby strengthening deterrence (e.g., Donald Bates, "The Ultimate Deterrent," *Thoughts on Peace and Security* 2, March/April, 1986). We believe this provocative idea has serious technical and, especially, political drawbacks.

8.7:

Some 450 oil refineries (and 3,000 oil pipelines) are on hypothetical, nearly identical (ref. 8.16) target lists published both in the United States (ref. 8.20) and in the Soviet Union (ref. 8.21). However, most storage capacity is concentrated in a small number of large refineries and depots (see Fig. 5).

8.8:

"Evaluation of Current Strategic Air Offensive Plans," Joint Chiefs of Staff JCS 1952/1 (Top Secret), December 21, 1948. This was the JCS approval of Strategic Air Command Emergency War Plan SAC EWP 1-49. Reprinted in T. H. Etzold and J. L. Gaddis, eds., *Containment: Documents on American Policy and Strategy, 1945–1950* (New York: Columbia University Press, 1978).

8.9:

Around 1950, Gen. Curtis LeMay, Commander of the Strategic Air Command, instructed Sam Cohen, later inventor of the neutron bomb, to "Give me a bomb that can wipe out all of Russia." (Michael Kepp, "Much Ado About Nothing [Bombs]" [Profile of Sam Cohen], *Oui*, April 1982, 93 ff. See also Sam Cohen, *The Truth About the Neutron Bomb: The Inventor of the Bomb Speaks Out* [New York: William Morrow, 1983], 30.) Unknowingly, LeMay may already have had this capability—and a great deal more besides.

8.10:

Scott D. Sagan, *Moving Targets: Nuclear Strategy and National Security* (Princeton: Princeton University Press, 1989), 12, 44–45, 47–48.

8.11:

*Department of Defense Authorization for Appropriations for Fiscal Year 1981*, Hearings, Senate Armed Services Committee, Part 5, 2721 (Washington, D.C.: U.S. Government Printing Office, 1980).

8.12:

U.S. Arms Control and Disarmament Agency, *Effectiveness of Soviet Civil Defense in Limiting Damage to Population* (ACDA Civil Defense Study Report. No. 1, November 16, 1977), 18–20.

8.13:

For possible asymmetries in U.S./Soviet economic targeting, see "Economic Targeting in Nuclear War: U.S. and Soviet Approaches," by B. S. Lambeth and K. N. Lewis, *Orbis*, Spring

1983, 127–149. Both the co-location of cities with industrial targets and the urban population densities are greater in the U.S.S.R. than in the U.S. (U.S. Arms Control and Disarmament Agency, *An Analysis of Civil Defense in Nuclear War*, December 1978).

8.14:

U.S. Arms Control and Disarmament Agency, *The Effects of Nuclear War*, April 1977.

8.15:

In public statements, the United States often declares that civilian populations are not targeted "*per se*." This new locution was first announced

> by Secretary of Defense Elliot Richardson, who testified in April 1973 that: "We do not in our strategic planning target civilian populations *per se*." And the Chairman of the JCS explained in 1976: "We do not target population *per se* any longer. We used to. What we are now doing is targeting a war recovery capability." . . . [Assistant Defense Secretary] Richard Perle . . . testified in March 1982, "as a conscious matter of policy, we have not planned for the deliberate destruction of the population" (ref. 8.4, p. 32).

Occasionally even these circumlocutions are dropped, as in "Our strategy does not target population" (Caspar N. Weinberger, *The Potential Effects of Nuclear War on the Climate: A Report to the United States Congress*, Department of Defense, March 1985). Or, "Regrettably, some analyses of nuclear war's climatic effects have assumed targeting of cities. If one were to regard this as an inevitable result of nuclear attack, or as U.S. policy, it would completely distort these analyses" (Caspar N. Weinberger, "Nuclear Winter," Hearings, Defense Appropriations Subcommittee, House Armed Services Committee, *Department of Defense Appropriations for 1986* [Washington, D.C.: U.S. Government Printing Office, 1985], 256–257). In fact, populations are heavily targeted (see refs. 8.2, 8.5), and whether it's *per se* or not provides comfort only to those who build, launch, and justify the nuclear weapons, not to the multitudes of men, women, and children in the targeted cities.

8.16:

> Soviet military thought did not then [1958] and does not now [1983] draw a sharp distinction between weapons intended to strike military targets in the event of war and weapons designed to deter through the threat of destroying cities. [David Holloway, *The Soviet Union and the Arms Race* (New Haven: Yale University Press, 1983), 68.]

Here, for example, is a list (taken from William M. Arkin and Richard W. Fieldhouse, *Nuclear Battlefields: Global Links in the Arms Race* [Cambridge, Mass.: Ballinger, 1985]) of American cities with populations over 100,000 people that contain strictly military targets and so are likely to be targeted in a "counterforce" nuclear war: Phoenix and Tucson, Arizona; Little Rock, Arkansas; San Diego, San Francisco, Long Beach, San Jose, Sunnyvale, Stockton, Oxnard, Sacramento, San Bernardino, and Concord, California; Denver and Colorado Springs, Colorado; Washington, D.C.; Jacksonville, Tampa, Miami, and Fort Lauderdale, Florida; Honolulu, Hawaii; Chicago, Illinois; Fort Wayne, Indiana; Des Moines, Iowa; Wichita, Kansas; New Orleans, Louisiana; Springfield, Massachusetts; Detroit, Michigan; Duluth, Minnesota; Kansas City and St. Louis, Missouri; Omaha, Nebraska; Las Vegas, Nevada; Albuquerque, New Mexico; Columbus, Dayton, and Cincinnati, Ohio; Oklahoma City, Oklahoma; Pittsburgh, Pennsylvania; Charleston, South Carolina; Knoxville, Tennessee; Amarillo, San Antonio, and Fort Worth, Texas; Salt Lake City, Utah; Arlington, Chesapeake, Newport News, Norfolk, and Alexandria, Virginia; Seattle/Tacoma, Washington; and Milwaukee, Wisconsin. The list is incomplete. If we go to smaller cities, much larger numbers —among them, Bangor, Maine, Grand Forks and Minot, North Dakota, and Plattsburg, New York—would be added. Richard D. Small ("Atmospheric Smoke Loading from a Nuclear Attack on the United States," *Ambio 18* [7], 1989, 377–383) estimates that fully half of the urban-suburban areas of the U.S. are likely Soviet targets in a major nuclear war. A similar set of Soviet cities is of course on American "counterforce" targeting lists.

The only difference between published U.S. and Soviet target lists (besides 100 "satellite command and measuring complexes," added to both U.S. and Soviet targets) is the Soviet addition (ref. 8.21) of 100 U.S. "leadership targets"—where the

"National Command Authority" might be hiding—for Soviet planners, but 10,000 Soviet leadership targets on the National Strategic Target List (NSTL) for U.S. planners. The latter number is judged "unreasonable" by Bing (ref. 9.8), although it is just the number given by Ball (ref. 8.4) for SIOP-6. So large a number of partly or mainly urban leadership targets has serious implications for nuclear winter—assuming that the bulk of the attack is accomplished by air and surface bursts. However, only a small fraction of these targets would be attacked in a real war, because there are always far more targets than weapons. Ball (private communication, 1989) estimates about 850 Soviet leadership targets in the American SIOP for a "generated" war, and about 600 for a nongenerated war. These numbers of cities alone—apart from urban areas ignited (although, doubtless, not "*per se*") and apart from economic/industrial targeting—look to be more than enough to generate nuclear winter. By October 1, 1987, when SIOP-6D took effect, the NSTL had been pruned to a total of some 15,000 targets, of which 3,000 to 4,000 were in the "leadership" and $C^3I$ (command, control, communications, and intelligence) categories (*ibid.*). See also ref. 8.19.

8.17:
Pierre Salinger, "Gaps in the Cuban Missile Crisis Story," *New York Times Magazine*, February 5, 1989.

8.18:
Andrei Sakharov, *Memoirs* (New York: Knopf, 1990).

8.19:
Desmond Ball and Robert C. Toth, "The New SIOP: Taking War-Fighting to Dangerous Extremes," Reference Paper 173, Strategic and Defence Studies Centre, Australian National University, Canberra, December 1989. See also ref. 8.10.

8.20:
Ronald Siegel, *Strategic Targeting Options* (Cambridge, Mass.: MIT Press, 1981).

8.21:
R. Sagdeev, A. Kokoshin, et al., *Strategic Stability Under Conditions of Radical Nuclear Arms Reductions* (Moscow: No-

vosti, 1987). Reprinted in Sanford Lakoff, ed., *Beyond START?* (San Diego: University of California Institute on Global Conflict and Cooperation, Policy Paper No. 7, 1988).

8.22:

C. Sagan, "On Minimizing the Consequences of Nuclear War," *Nature 317*, 1985, 485–488.

9.1:

Sam Cohen, credited with the invention of the neutron bomb, has argued that what is publicly known about U.S. and Soviet targeting plans guarantees—at a level of reliability comfortable at any rate for him—that cities would not be attacked in a major nuclear war, and smoke would not be generated. Nuclear winter therefore is a nonproblem. And if the superpowers once had plans to target cities, they would abandon them now that they know about nuclear winter: "Neither the United States nor the U.S.S.R. wants to put the world into a nuclear deep freeze. . . . Why would either of these nations want to wage war so insanely?" (Cohen, "Nuclear Winter and Nuclear Reality," *Washington Times*, June 28, 1984, 1C, 2C, and private communication.) We agree that knowledge of nuclear winter improves deterrence and discourages attacks on cities, but too many strategic targets are in or near cities, and nuclear war is unlikely to bring out the most sober and rational tendencies of world military and civilian leaders. Cohen's views represent one of several attempts to make nuclear winter go away by thinking at it in the right way.

9.2:

Theodore Postol, "Strategic Confusion—With or Without Nuclear Winter," *Bulletin of the Atomic Scientists 41* (2), February, 1985, 14–17.

9.3:

E.g., Desmond Ball, "Can Nuclear War Be Controlled?," *Adelphi Paper 169* (London: International Institute of Strategic Studies, 1981); Paul Bracken and Martin Shubik, "Strategic War: What Are the Questions and Who Should Ask Them?," *Technology and Society 4*, 1982, 155–179; Paul Bracken, *The Command and Control of Nuclear Forces* (New Haven: Yale

University Press, 1983); Bruce G. Blair, *Strategic Command and Control* (Washington, D.C.: The Brookings Institute, 1985); ref. 10.6. Also, former Defense Secretary Harold Brown: "We know that what might start as a supposedly controlled, limited strike could well, in my view very likely, escalate to a full-scale nuclear war" (Convocation Address, Naval War College, Newport, R.I., August 20, 1980); former Communist Party General Secretary (and Marshal of the Soviet Union) Leonid Brezhnev: "A 'limited' nuclear war, as conceived by the Americans in, say, Europe, would from the outset mean certain destruction of European civilization. And, of course, the United States too would be unable to escape the flames of war" (Speech, 26th Party Congress, reported in *Pravda*, February 24, 1981); and Soviet Defense Minister N. V. Ogarkov: "The calculation of the strategists across the ocean, based on the possibility of waging a so-called "limited" nuclear war, now has no foundation whatever. . . . Any so-called limited use of nuclear forces will inevitably lead to the immediate use of the whole of [both] sides' nuclear arsenals. This is the terrible logic of war." [Quoted in Leon Gouré, " 'Nuclear Winter' in Soviet Mirrors," *Strategic Review*, Summer 1985, 22–38.]

9.4:

George F. Kennan, Albert Einstein Peace Prize Acceptance Speech, May 19, 1981. *Manchester Guardian Weekly*, May 31, 1981. Also Kennan, *The Nuclear Delusion: Soviet-American Relations in the Atomic Age* (New York: Pantheon, 1982), 180.

9.5:

This interpretation remains valid no matter which of the several mutually contradictory accounts of what was agreed to by Ronald Reagan and Mikhail Gorbachev at the Reykjavik summit one cares to believe (including the U.S. position that all strategic *ballistic missiles* would be destroyed in a decade, and the Soviet position that all strategic nuclear *weapons* would be destroyed in a decade).

9.6:

Strobe Talbott, "Why START Stopped," *Foreign Affairs* 67 (1), Fall, 1988, 49–69.

9.7:
Wolfgang K. H. Panofsky et al., *Reykjavik and Beyond: Deep Reductions in Strategic Nuclear Arsenals and the Future Direction of Arms Control* (Washington, D.C.: National Academy of Sciences, 1988).

9.8:
G. F. Bing, P. Garrity, W. F. Hanrieder, M. D. Intrilligator, R. Kolkowicz, S. Prowse, A. Wohlstetter, and K. Waltz, in S. Lakoff, ed., *Beyond START?* (San Diego: University of California Institute on Global Conflict and Cooperation, Policy Paper No. 7, 1988).

9.9:
Neil Kinnock, Labour Party Defence Policy Statement, *Times* (London), December 11, 1986.

9.10:
The Labour Party's National Executive Committee, "Peace in Our Changing Times," *The Guardian*, May 8, 1989, 17.

9.11:
David Owen, private communication, December 3, 1986.

9.12:
In repeated statements at the United Nations. At the second Special Session on Disarmament, on June 21, 1982, the Foreign Minister, Huang Hua, stated that a 50% cutback in arsenals by the U.S. and U.S.S.R. was a prerequisite for China to enter into negotiations on nuclear disarmament. (Reprinted in Ken Coates, ed., *China and the Bomb* [Nottingham: Spokesman Books, 1986], 71, 73.) Liang Yufan, at the U.N. General Assembly, 38th session, 1983, gave more details:

> After the Soviet Union and the United States had taken practical action to stop testing, improving and manufacturing nuclear weapons and agreed on reducing by half their nuclear weapons and means of delivery of all types, a widely representative international conference should be convened with the participation of all nuclear weapons states to negotiate the general reduction of nuclear weapons by all nuclear weapons states.

See also, Qian Jiadong, "China's Position on Nuclear Non-Proliferation," in *Nuclear War, Nuclear Proliferation and Their Consequences*, Sadruddin Aga Khan, ed. (Oxford: Oxford University Press, 1985). The proposed START treaty would go far to meet the Chinese conditions. In an address to the United Nations on June 2, 1988, the Foreign Minister, Qian Qichen, offered—in exchange for "drastic" reductions in all types of nuclear weapons and an end to the manufacture, testing, and deployment of nuclear weapons by the superpowers—to call an international conference to discuss "a thorough destruction of nuclear armaments." The admission fee has apparently gone up in the interim (from "half" to "drastic"), but the possible payoff has also increased (from "reductions" to "thorough destruction").

9.13:

In a series of statements (for example, his interview with the Paris correspondent of the Japanese newspaper *Asahi Shimbun*, July 12, 1989), President François Mitterrand has reiterated the conditions of his 1983 address to the United Nations General Assembly: "We cannot reject the idea—and I do not—that the five nuclear Powers should together debate, when the time comes, a permanent limitation of their strategic systems. We must therefore set out clearly the conditions for progress in this field." (*Official Records of the General Assembly*, Thirty-eighth Session, *Plenary Meetings*, Volume 1 [New York: United Nations, 1988], 109.) The conditions included major divestments in U.S. and Soviet strategic weaponry, reduction in Soviet conventional weaponry, and restrictions on the development of antisatellite and space-based weaponry as well as antisubmarine warfare. When asked, at a May 18, 1989, press conference in Paris, why France did not unilaterally end nuclear testing and the procurement of new strategic weaponry, Mr. Mitterrand answered: "I am going to give you a very simple answer. If the United States of America and the Soviet Union renounce [such programs], and Great Britain also, we will follow that path." In a press conference in Sofia, Bulgaria, on January 19, 1989, Mr. Mitterrand said, "France would take part in nuclear disarmament from the moment when disarmament by the two great powers will have attained to a level which still remains rather distant. . . . We are on the right path. Let us encourage the Americans, encourage Mr. Gorbachev, and every time we can take positions favorable to disarmament, let us do so."

A Paris-datelined report by William Echikson ("France Shifts Arms-Control Stance," *Christian Science Monitor*, September 28, 1988, 7) quoted an official of the French Defense Ministry as saying, "Our policy has shifted because the realities around us have shifted." The story continues,

> If Mr. Gorbachev were to make an acceptable offer on conventional arms, a top-ranking adviser to Mr. Mitterrand told some Western reporters privately last week, France might respond by not developing neutron bombs and the Hades tactical missile, a sophisticated land-based missile with a 200-mile range.
>
> The offer, if confirmed, would represent a major reversal from the previous French position. Although the adviser reportedly ruled out any reduction to long-range strategic land-based missiles and submarines, the French previously have refused to cut their small nuclear forces until the superpowers eliminate almost all of their own missiles.
>
> After the story broke, the Élysée Palace labeled it "unfounded." But analysts here say Mitterrand is rethinking France's strategic policy. In his campaign earlier this year, he said arms control would be a high priority for his second term.

Public awareness and debate on the nature of nuclear war and related policy matters is much less evident in France than, say, in the U.S., the U.S.S.R., Britain, and many nonnuclear nations (e.g., Michel Haag, "France's Low Awareness of Accidental Nuclear War Dangers," Technical Report 5 (Santa Barbara, Cal.: Nuclear Age Peace Foundation, 1987]). Alone among major European nations, so far as we know, France has had no research efforts and no public or scientific symposia on nuclear winter.

After a brief description by one of us (C.S.) of how France's *force de frappe*, if used as planned against cities, might by itself bring about nuclear winter, President Mitterand's response was an assurance that French nuclear forces were in "safe hands" (private communication, 1984).

9.14:

"Statement of Robert S. McNamara before the House Armed Services Committee on the Fiscal Year 1966–1970 Defense Program and 1966 Defense Budget," February 18, 1965, 39. See also A. C. Enthoven and K. W. Smith, *How Much Is Enough?: Shaping the Defense Program, 1961–1969* (New York: Harper & Row, 1971), 207. Near the end of the Eisenhower Administration, 232 single-warhead Polaris missiles on invulnerable submarines were declared by the U.S. Navy to be "sufficient to destroy all of Russia" (ref. 18.8). Even smaller arsenals, targeted on cities, are adequate for what McGeorge Bundy has called "existential deterrence" ("To Cap the Volcano," *Foreign Affairs 48* [2], October 1969, 1–20). A recent Soviet report estimates that 400 1-megaton explosions are sufficient to destroy targets, on either side, comprising "25–30% of the population and up to 70% of the industrial capacity" (ref. 8.21).

9.15:

Vera Rich, "Nuclear Winter Expert Vanishes Without Trace," *Nature 316*, July 4, 1985, 3; Keith Rogers, "Soviet Researcher Vanishes," *The Valley Times* (serving Livermore, California), July 13, 1985; *idem*, "No Word on Missing Soviet," *The Valley Times*, July 17, 1985; Philip M. Boffey, "Russian Scientist Vanishes in Spain," *New York Times*, July 16, 1985, A4; Ralph de Toledano, "Is That Missing Physicist Crooning to the CIA?," *Washington Times*, October 29, 1985; *idem*, "Incredible Vanishing Scientist," *Washington Times*, May 13, 1986; *idem*, "Soviets Embarrassed the CIA," Copley News Service, May 23, 1987; Andrew C. Revkin, "Why Is This Top Soviet Scientist Missing?: The Curious Case of Vladimir Alexandrov," *Science Digest 94* (7), July 1986, 32–43; Iona Andronov, "Where Is Vladimir Alexandrov?," *Literaturnaya Gazeta*, July 23, 1986, translated by Hugh W. Ellsaesser, UCRL-Trans-12103, Lawrence Livermore National Laboratory.

10.1:

Bernard Brodie, "Implications for Military Policy," in *The Absolute Weapon: Atomic Power and World Order*, Bernard Brodie, ed., (New York: Harcourt Brace, 1946). Brodie continues,

Thus, the first and most vital step in any American security program for the age of atomic bombs is to take measures to guarantee to ourselves in case of attack the possibility of retaliation in kind. The writer in making that statement is not for the moment concerned about who will *win* the next war in which atomic bombs are used. Thus far the chief purpose of our military establishment has been to win wars. From now on its chief purpose must be to avert them. It can have almost no other useful purpose.

10.2:

Another reason that first strike is an unreliable and dangerous doctrine was given in 1974 Congressional testimony by former Secretary of Defense James Schlesinger: "[If] you have any degradation in operational accuracy, American counter-force capability goes to the dogs very quickly. We know that, and the Soviets should know it, and that is one of the reasons that I can publicly state that neither side can acquire a high-confidence first-strike capability." (James Fallows, *National Defense* [New York: Vintage Books, 1982], 155–156.) Nevertheless, in military doctrine, government procurement, and public perception, preparing for and deterring a massive first strike has played a central role in the nuclear arms race.

Standard discussions of deterrence include Thomas Schelling, *Strategy of Conflict* (New York: Oxford University Press, 1960); ref. 6.1; Lawrence Freedman, *The Evolution of Nuclear Strategy* (New York: St. Martin's, 1983); ref. 17.3; Robert Jervis, *The Illogic of American Nuclear Strategy* (Ithaca, N.Y.: Cornell University Press, 1984); R. Hardin, J. J. Mearshimer, G. Dworkin, and R. E. Goodin, eds., *Nuclear Deterrence: Ethics and Strategy* (Chicago: University of Chicago Press, 1985); and ref. 17.8. For an excellent short summary by a firsthand participant, see Robert S. McNamara, *Blundering into Disaster* (New York: Pantheon, 1986).

10.3:

A concern expressed, perhaps a little unfairly, by Douglas Lackey: "The thought that an American President may lack the nerve to destroy civilization depresses the military mind." ("Missiles and Morals," in James P. Sterba, ed., *The Ethics of War and Nuclear Deterrence* [Belmont, Cal.: Wadsworth, 1985].)

10.4:

Maxwell Taylor, *Precarious Security* (New York: Norton, 1976), 68–69.

10.5:

Commercial airliners have experienced simultaneous failure of all jet engines when flying in the middle troposphere through volcanic dust clouds following the Mt. Galunggung, Indonesia (1981), and the Mt. Redoubt, Alaska (1989), eruptions. (E.g., Richard Witkin, "Jet Lands Safely After Volcanic Ash Stops Engines," *New York Times*, December 17, 1989, 47. Also, William J. Broad, "Threat to U.S. Air Power: The Dust Factor," *Science 213*, 1981, 1475–1477, which concludes, "Considering the ease with which the dust factor was overlooked, one wonders whether there are other unforeseen impediments that might considerably complicate the idealized war scenarios the military has in mind.")

10.6:

*Crisis Stability and Nuclear War*, Kurt Gottfried and Bruce G. Blair, eds. (Oxford: Oxford University Press, 1988).

10.7:

If we are seriously concerned about an asymmetry of perception, one remedy is to enter into joint U.S./Soviet deliberations on the science and policy implications of nuclear winter—as proposed in Senate Concurrent Resolution 36, "To Establish a Joint Commission for Joint Study by the United States and the Soviet Union on Nuclear Winter," 99th Congress (bill died in formation April 2, 1985), and in "sense of the Congress" provisions attached to the fiscal year 1986 and 1987 Foreign Relations Authorization Acts:

> The United States and the Soviet Union should jointly study . . . "nuclear winter," and the impact that nuclear winter would have on the national security of both nations; such a joint study should include the sharing and exchange of information and findings on the nuclear winter phenomenon, and make recommendations on possible joint research projects that would benefit both nations; and at an appropriate time the other nuclear weapon states . . . should be involved in the study.

Despite these resolutions, no such governmentally sponsored joint study has been undertaken. There have been and likely still are influential figures in American nuclear policy who are unhappy with the prospect of international discussions of the policy implications of nuclear winter. For example, in 1984 a resolution was offered in the U.N. General Assembly merely to compile excerpts from existing scientific studies on the climatic effects of nuclear war to distribute to member nations for their information. The vote (Resolution 39/148 F, December 17, 1984) was 130 in favor, with 11 nations, including the United States, abstaining. A subsequent resolution (40/152 G, December 16, 1985) proposed a more systematic United Nations study of the subject. It passed with 141 in favor. There were 10 abstentions, and only one nation opposed—the United States. This resolution led to the report cited in ref. 3.16. In both cases all the other abstaining nations were—like Israel and Grenada—strongly beholden to the United States and/or NATO allies of the United States.

Scientists have themselves arranged for joint meetings, beginning in 1983—the first in April in Cambridge, Massachusetts, the second in August in Erice, Sicily, and the third in October in Washington, D.C. (with a link to Moscow by satellite). Since 1983 a large number of multilateral meetings have been held, prominently under SCOPE auspices (ref. 3.11 and SCOPE box, Chapter 3). But this is a very different matter from meetings that include discussions of policy issues and that are held under official U.S./U.S.S.R. auspices (perhaps with other nations as well).

10.8:

A. B. Pittock, "The Environmental Impact of Nuclear War: Policy Implications," *Ambio 18* (7), 1989, 367–371.

11.1:

For example, W. H. Daugherty et al. (ref. 5.10) calculate that an attack on the U.S. strategic nuclear forces by the Soviet Union could produce up to 43 million casualties from prompt fallout alone (an average of 23 million over a range of cases, and a lower limit of 12 million). In Europe, even greater relative casualties would be expected because of the large number and wide dispersal of military targets, as well as the high population density (J. Duffield and F. von Hippel, *The Short-Term Consequences of Nuclear War for Civilians*, Symposium on the Environmental Effects of Thermonuclear War, American Association for the Advancement of Science, Detroit, May 1983 [New York: Macmillan, 1984]; ref. 5.10). A recent estimate of casualties from all sources of radioactivity in a major conflict is around 180 million people (ref. 5.9)—mainly Americans, Soviets, and Europeans, with an upper range close to 300 million.

11.2:

Optimistic forecasts have been forthcoming from the Federal Emergency Management Agency (*Nuclear Attack Planning Base—1990: Final Project Report*, Federal Emergency Management Agency Report NAPB-90, Washington, D.C., April 1987), the Defense Department ("Sensitivity of Collateral Damage Calculations to Limited Nuclear War Scenarios," in *Analyses of Effects of Limited Nuclear War* [Washington, D.C.: U.S. Department of Defense, 1975], 14), and from various civil defense advocates. Even some of the less optimistic assessments still foresee a steady recovery of U.S. society over several decades (A. Katz, *Life After Nuclear War* [Cambridge, Mass.: Ballinger, 1982]). None of these studies takes account of nuclear winter. An early attempt by a FEMA official to explain why nuclear winter is not considered in FEMA's plans for post-nuclear-war America is given in Congressional testimony by David McLoughlin (*The Consequences of Nuclear War*, Hearings, July 12, 1984, Joint Economic Committee [ref. 2.6]).

As a kind of self-parody of FEMA's penchant for making nuclear war seem readily survivable, consider the following advice:

> "If a weapon detonates nearby:
> a. Extinguish fires
> b. Repair damage."
>
> [From Federal Emergency Management Agency, *Shelter Management Handbook* (Washington, D.C.: U.S. Government Printing Office, May 1984). Reprinted in Donna Uthus Gregory, ed., *The Nuclear Predicament: A Sourcebook* (New York: St. Martin's Press, 1986), 237.]

This official tradition of minimizing the dangers of nuclear weapons goes back to Alexander de Seversky, author of *Victory*

*Through Air Power* (which was made into a stirring, patriotic, and now nearly forgotten Walt Disney film that urged almost total reliance on intercontinental bombers to win World War II). As "Special Consultant to the Secretary of War," Major de Seversky wrote in *Reader's Digest* six months after Hiroshima and Nagasaki that the effects of those explosions "had been wildly exaggerated. . . . The same bombs dropped on New York or Chicago, Pittsburgh or Detroit, would have exacted no more toll in life than one of our big blockbusters, and the property damage might have been limited to broken window glass over a wide area" (de Seversky, "Atomic Bomb Hysteria," *Reader's Digest*, February 1946).

11.3:

Federal Emergency Management Agency reports analyzed and criticized in Jennifer Leaning and L. Keyes, eds., *The Counterfeit Ark: Crisis Relocation for Nuclear War* (Cambridge, Mass.: Ballinger, 1984). See also ref. 11.2.

11.4:

A distinction between shorter-term "acute" and longer-term "chronic" nuclear winter effects was first suggested by Donald Kennedy (in Paul R. Ehrlich, Carl Sagan, Donald Kennedy, and Walter Orr Roberts, *The Cold and the Dark: The World After Nuclear War* [New York: Norton, 1984], xxvii). The acute phase is currently taken as one to three months, the chronic phase as one to three years.

The duration of nuclear winter effects is among the most poorly understood aspects of the subject—in part because of our lack of real-world experience with the highly perturbed and anomalous state of the atmosphere following a nuclear war. Our original TTAPS report mentioned the likelihood both of self-lofting of smoke by solar heating (and therefore prolonged duration of nuclear winter) and of climatic feedbacks in the world weather system. But we calculated neither. We predicted significant departures from the ordinary climate to last for periods ranging between a few months and a year or two. Subsequent work has confirmed that both self-lofting and feedback should occur (refs. 3.9, 3.11; also, R. M. Haberle, T. P. Ackerman, O. B. Toon and J. L. Hollingsworth, "Global Transport of Atmospheric Smoke Following a Major Nuclear Exchange," *Geo-*

*physical Research Letters 12* (1985), 405–408; R. C. Malone, L. H. Auer, G. A. Glatzmaier, M. C. Wood and O. B. Toon, "Nuclear Winter: Three-Dimensional Simulations Including Interactive Transport, Scavenging and Solar Heating of Smoke," *Journal of Geophysical Research 91*, 1986, 1039–1053). It seems possible that a central exchange could lead to climatic anomalies lasting for years after the nuclear war was "over." (C. Covey, "Protracted Climatic Effects of Massive Smoke Injection into the Atmosphere," *Nature 325*, 1987, 701–703; A. Robock, "Snow and Ice Feedbacks Prolong Effects of Nuclear Winter," *Nature 310*, 1984, 667 [which predicts chronic *ocean* surface temperature decreases of 2 to 6°C, and widespread land temperatures that could compromise agriculture at least through the second postwar summer]; T. P. Ackerman, R. P. Turco, and O. B. Toon, "Persistent Effects of Residual Smoke Layers," in P. V. Hobbs and M. P. McCormick, eds., *Aerosols and Climate* [Hampton, Va.: Deepak Publ., 1988], 443–458.) But much more work is needed on this issue. (Cf. also ref. 12.7.)

11.5:

There are a few possible exceptions for small, wealthy, ethnically more homogeneous nations such as Switzerland and Sweden.

11.6:

Federal Emergency Management Agency, *Nuclear Attack Planning Base—1990: Final Project Report*, Report NAPB-90 (Washington, D.C.: U.S. Government Printing Office, April 1987). We know FEMA is officially aware of nuclear winter from Congressional testimony by FEMA officials (cf. *The Consequences of Nuclear War*, 174–198, 297, ref. 2.6), and from the fact that it has itself commissioned a number of contractor reports on nuclear winter (the first of which seems to be C. V. Chester, F. C. Kornegay, and A. M. Perry, "A Preliminary Review of the TTAPS Nuclear Winter Scenario," Oak Ridge National Laboratory/Martin Marietta Corp. Report ORNL/TM-1223 to FEMA, July 1984).

"Survivalists" and others who hope to emerge intact from shelters after a nuclear war have special reasons to be unhappy about the threat of nuclear winter. According to an article in the *Portland Oregonian* ("N-attack Shelter Plan Pushed with

Surviving in Mind," by John Darling, January 29, 1988), an Oregon manufacturer of shelters also sells a book that offers this assessment: "The myth of nuclear winter is still spread by uninformed people and by people who want to keep America defenseless." Among objections to nuclear winter voiced at a 1985 meeting of the American Civil Defense Association were that the Soviets would not target American cities; that if they did, the fires would be minor ("There will be small fires that can be beat out with wet towels or stamped on with boots"); and that "Nuclear winter is a snow job. . . . These scientists are creating war horror thoughts to undermine America's will to resist." But civil defense also has, it turns out, a political purpose—at least in the minds of some: "If a nation thought it could protect a large part of its people, would it not mean that the country could start [and survive] a nuclear war? . . . The nation who has the most survivors will be a nation able to rebuild its economy." ("Civil Defense Group Has Own Ideas on Aftermath of Nuclear War," by Karla Tipton, *Palmdale* [California] *Antelope Valley Press*, November 19, 1985.)

11.7:

What little public evidence there is suggests that nuclear winter has not been explicitly taken into account in the U.S. SIOP—at least until 1986 (ref. 13.26) and, more likely, never. Nothing is publicly known about the influence of nuclear winter on the Soviet SIOP. Powers (ref. 2.6) writes:

> According to many sources, there is no independent review [e.g., by the Executive Office of the President] of the SIOP at any stage in the planning process; the size of the U.S. arsenal provides the only limit to the size of a major projected war; and no one involved in drawing up the SIOP is authorized to consider the gross environmental effects of carrying out the plan.

Nevertheless, beginning in 1987, U.S. targeting doctrine began slowly to move in a direction consonant with the implications of nuclear winter. (See Chapters 8 and 13.)

12.1:

*The Medical Implications of Nuclear War*, F. S. Solomon and R. Q. Marston, eds., U.S. National Institute of Medicine (Washington, D.C.: National Academy of Sciences, 1986), 619 pp.

12.2:
D. S. Greer and L. S. Rifkin, "The Immunological Impact of Nuclear Weapons," in ref. 12.1.

12.3:
Particularly Nevil Shute's novel, *On the Beach* (New York: Morrow, 1957), in which the radioactivity from a cobalt bomb Doomsday Machine renders the human species extinct.

12.4:
Cao Hongxing and Liu Yuhe, "The Climatic Effects of Nuclear War," Pan-Earth Workshop report, Beijing, August 25–31, 1988, Academia Sinica, China.

12.5:
S. H. Schneider and S. L. Thompson, "Simulating the Climatic Effects of Nuclear War," *Nature* 333, 1988, 221–227; J. F. B. Mitchell and A. Slingo, "Climatic Effects of Nuclear War: The Role of Atmospheric Stability and Ground Heat Fluxes," *Journal of Geophysical Research* 93 (D6), 1988, 7037–7045; S. J. Ghan, M. C. MacCracken, and J. J. Walton, "The Climatic Response to Large Atmospheric Smoke Injections: Sensitivity Studies with a Tropospheric General Circulation Model," *ibid.* 93 (D7), 1988, 8315–8338.

12.6:
A. B. Pittock, *Beyond Darkness: Nuclear Winter in New Zealand and Australia* (S. Melbourne: Sun Books, 1987); ref. 3.11.

12.7:
Because of the larger ratio of ocean to land in the Southern Hemisphere, the thermal inertia (resistance to temperature change) will be greater in the South than in the North. If the war occurs in Northern summer (maximizing acute nuclear winter effects there), then it occurs in Southern winter (minimizing acute effects there); but chronic longer-term effects will be important in both hemispheres. Although such effects have not yet been adequately studied, it is possible that massive amounts of smoke would be carried from North to South. But there could not be as much transported to the South as suspended in the North. Even with some Southern targeting, we

expect nuclear winter effects in the South to be milder than in the North. Note that even in cases where the amount of smoke in the Southern Hemisphere is not directly significant for temperatures or light levels, its effects on rainfall and drought might still be disastrous for agriculture.

Another source of uncertainty is cooling of the oceans. For a summer war the topmost layers of the oceans at midlatitudes are calculated to cool 3 to 5°C (cf. Robock, ref. 11.4) and these cold layers to deepen to 25 meters in the first month after the war (T. R. Mettlach, R. L. Haney, R. W. Garwood, and S. J. Ghan, "The Response of the Upper Ocean to a Large Summertime Injection of Smoke in the Atmosphere," *Journal of Geophysical Research* 92, No. C2, 1987, 1967–1974). No one knows what the longer-term response of the ocean might be, and how the colder ocean might further cool the land in either hemisphere. This is another of the nuclear winter problems crying out for investigation.

12.8:

L. da Silva, "Climatic Consequences of a Nuclear War for South America," *Proceedings, International Symposium on Science, Peace and Disarmament*, G. A. Lemarchand and A. R. Pedace, eds. (Buenos Aires: World Scientific, 1989).

12.9:

A. B. Pittock, "Environmental Impacts on Australia of a Nuclear War," *Ambio 18* (7), 1989, 395–401.

13.1:

Delhi Declaration, January 28, 1985 (United Nations General Assembly, Security Council, 40th Session, February 1, 1985, A/40/114, 1–5. See also, *ibid.*, 39th Session, May 23, 1984, A/39/277, 1–5). Note the joint statement of May 22, 1984, by the same heads of state and government, in which they condemn "the rush toward global suicide," and state that "the people we represent are no less threatened by nuclear war than the citizens of the nuclear weapons states." This is approximately true, except, perhaps, for Argentina. Their view was summarized by Swedish Prime Minister Olof Palme, on receiving the annual Beyond War Award (San Francisco, December 14, 1985) on behalf of the six leaders: After describing nuclear winter as "posing unprecedented peril to all nations," he argued: "[E]ven if we are far removed from the actual nuclear explosions, we will all be affected by the use of nuclear weapons. And therefore we also have a right to have a say about this use. The scientists have laid a foundation for the clear philosophy behind the Five Continent Peace Initiative."

13.2:

"For the Species and the Planet," in *Ending the Deadlock: The Political Challenge of the Nuclear Age* (New York: Parliamentarians Global Action, 1985); also, *Bulletin of the Atomic Scientists 41* (10), November, 1985, 4–5.

13.3:

Javier Perez de Cuellar, address to the United Nations General Assembly, December 12, 1984, Provisional Verbatim Record A/39/PV.97. See also, "Statement by the Secretary-General to the General Assembly on Disarmament Issues," *Disarmament 13* (1), Spring 1985, 3–7.

13.4:

New Zealand Mission to the United Nations, Prime Minister's Statement to the General Assembly, 39th session, September 25, 1984.

Beginning in 1984, and on several subsequent occasions, David Lange, the Prime Minister, stated, "New Zealand does not wish to be defended by nuclear weapons." "New Zealand will never acquire nuclear weapons, and does not ask friendly powers to use them on its behalf. . . . Indeed, it is the Government's view that it is the nuclear weapons themselves that present the real and potentially catastrophic threat." (Government of New Zealand, *The Defence Question* [Wellington, 1985].) The New Zealand diplomat Kennedy Graham writes,

> It is the rationality of nuclear deterrence which New Zealand now queries, and it does so in the name of national security. . . . Taking all of these possibilities into account, the risk of deterrence failing and a nuclear conflict occurring within the next 15 years is assesssed here at one in five. Some will no doubt assign a higher risk factor, others a lower. . . . If the risk is a significant one, then it approaches certainty over a calculable period of time—in

> the assessment made here, within seventy-five years. . . . With the stake so extraordinarily high, is this a tolerable legacy to leave behind? . . .
>
> The 1980s has seen the culmination of all this. The continuing arms build-up, the increasing sophistication of weaponry and the militarization of space have all intensified the apprehensions of the 1970s over the wisdom of strategic trends in recent times. Above all, the world has become one in face of the nuclear threat.

Graham then goes on to describe nuclear winter (Graham, "New Zealand's Non-nuclear Policy: Towards a Global Security," *Alternatives 12* [2], April, 1987, 224, 228, 230).

Views of this sort were unwelcome in Washington, especially after New Zealand refused port visits by U.S. warships if—as is U.S. policy—they would neither confirm nor deny whether they were carrying nuclear weapons. Former Prime Minister Sir Wallace Rowling, the New Zealand Ambassador to the United States in the Reagan years, was nearly *persona non grata*, not permitted even to present his credentials to the President. Secretary of Defense Weinberger, on a trip to Australia, pretended never to have heard of New Zealand. American officials feared that New Zealand's policy on banning U.S. nuclear weapons might prove contagious. Other countries became more likely to catch this virus when it was revealed that the U.S. had neglected to inform closely allied nations and territories (Canada, Iceland, Bermuda, and Puerto Rico) that it had made contingency plans to deploy nuclear weapons on their soil (Leslie H. Gelb, "U.S. Plan for Deploying A-Arms Wasn't Disclosed to Host Nations," *New York Times*, February 13, 1985). The next day the *Times* reported concern about an "unraveling" of support for current U.S. nuclear policy among NATO nations—a "nuclear allergy," it was called (*idem*, "U.S. Tries to Fight Allied Resistance to Nuclear Arms: Says It Seems to Spread," *New York Times*, February 14, 1985).

13.5:

Nigeria: "There will be no hiding place for any of us even though we have no part in, and indeed have continued to warn against, this irrational diversion of . . . resources."

Romania: "The use of merely a small part of existing nuclear arsenals would result in the destruction of all of civilization."

China: "Should they choose to use only a small portion of their nuclear arsenals, not only would the people of these two nuclear powers suffer, but the people of the whole world would be plunged into an unprecedented holocaust. . . . For this reason, the numerous small and medium-sized countries . . . are fully justified in demanding that [the U.S. and U.S.S.R.] immediately halt their nuclear-arms race and take the lead in drastically cutting back their nuclear weaponry."

Canada: "Even for survivors, the world would be virtually uninhabitable after a major nuclear conflict."

Canada's Ambassador to the United Nations Disarmament Conference argued that nuclear winter "should compel the world to rid itself of nuclear weapons" ("Roche Calls for N-arms Cuts," *Winnipeg Free Press*, July 16, 1985).

13.6:

Press release, Office of the Prime Minister, Wellington, New Zealand, August 28, 1986: "Compensation money from the Rainbow Warrior settlement will be spent on New Zealand's first detailed study into the effects a 'nuclear winter' would have on this country."

13.7:

Private communication, Dennis Healey to C.S., December 3, 1986.

13.8:

Henry Kamm, "Greek Chief Flies to Moscow Today: He Tones Down Remarks, Says Squabbles with U.S. Are Just Between Friends," *New York Times*, February 11, 1985.

13.9:

Private communication, Yevgeniy Velikhov to C.S., January 24, 1984.

13.10:

In a meeting on September 29, 1987, at the Institute for the Study of the U.S.A. and Canada of the Soviet Academy of Sciences in Moscow, Major General Boris Trofimovich Surikov of the Soviet General Staff told one of us (C.S.) that nuclear winter is often discussed in the Defense Ministry and has influenced

Soviet strategic policy. The principal policy implication of nuclear winter, he said in response to a question, is the necessity for major reductions in strategic arms. Andrei Kokoshin, one of the Deputy Directors of the Institute, also states that there is widespread knowledge of nuclear winter implications among the Soviet military.

13.11:

This was untrue in the prehistory of nuclear winter. The 1982 Crutzen/Birks paper (see box, "Nuclear Winter: Early History and Prehistory," Chapter 3)—and the entire issue of *Ambio* in which it was published—were classified secret in the Soviet Union, so that even specialists in atmospheric physics and chemistry there were wholly ignorant of it. (G. S. Golitsyn, talk at Cornell University, March 6, 1989.) However, immediately after the October 31–November 2, 1983, conference in Washington, D.C., at which nuclear winter was first publicly discussed—and which included a televised discussion by satellite with Soviet scientists in Moscow—considerable attention was devoted to the new findings in the Soviet media: e.g., G. Vasiliev, "The Scientists' Weighty Word," *Pravda*, November 16, 1983; *Pravda*, December 3, 1983; "Nuclear War Is Impermissible, the Scientists Declare," *Pravda*, December 9, 1983; "Nuclear War: A Threat to Mankind," *Pravda*, December 10, 1983; G. Gontarev, "The Fourth Telebridge U.S.S.R.-U.S.A.: No to Nuclear War," *Vestnik*, Novosti Press Agency (*Soviet Panorama*), No. 236, December 5, 1983; V. Simonov, "The Day After: Looking at the Unthinkable," *Literaturnaya Gazeta*, November 23, 1983; *Izvestia*, November 23, 1983; "Vremia" all-Union TV, December 11, 1983; "TV Conference of Soviet and American Scientists," Moscow Television Service, first broadcast March 24, 1984 (1-hour program; estimated audience: 100 million); *Sovietskaya Rossiya*, December 4, 1983; and N. Paklin, "Preventing a Nuclear Winter," *Izvestia*, January 27, 1984. Since then articles have continued to appear; e.g., *Moscow News Weekly*, No. 13, April 1984, or the article "To Prevent a Catastrophe," by the scientists V. Goldanskii and S. Kapitsa, in *Izvestia*, July 19, 1984, which includes these lines: "All calculations agree that the existing stockpiles of nuclear weapons exceed the threshold beyond which a global geophysical reaction is triggered. This means that our Earth is too small for the nuclear weapons concentrated on it." Subsequent publications include *Sovietskaya Rossiya*, November 30, 1984; *Komsomolskaya Pravda*, March 7, 1985; *Izvestiya*, July 25, 1985; A. Palladin, *Izvestia*, July 28, 1985; *Sotsialisticheskaya Industriya*, March 28, 1985; articles by O. Moroz and Y. P. Velikhov in the *Literaturnaya Gazeta* of, respectively, December 25 and January 22, 1986; the group of articles published in the popular science journal *Priroda*, June 1985; and "Nuclear War: Eternal Darkness," in *Yearbook U.S.S.R. '86* (Moscow: Novosti Press Agency Publishing House, 1986). Among television programs broadcast in this period, those of March 24, 1984, 1440 GMT, and July 16, 1985, 1510 GMT, included discussions on the nature and policy implications of nuclear winter, in Russian translation, by one of us (C.S.). There are also a number of Soviet scientific articles on nuclear winter, some of which are noted elsewhere in these references. The first to be published in Western scientific journals were V. V. Alexandrov, "A Soviet View of Nuclear Winter," *Chemtech 11*, 1985, 658–665; and G. S. Golitsyn and A. S. Ginsburg, "Comparative Estimates of Climatic Consequences of Martian Dust Storms and of Possible Nuclear War," *Tellus 37B*, 1985, 173–181. These papers were preceded by a joint U.S.-Soviet presentation, "Global Climatic Consequences of Nuclear War: Simulations with Three-Dimensional Models," by S. L. Thompson, V. V. Aleksandrov, G. L. Stenchikov, S. H. Schneider, C. Covey, and R. M. Chervin (*Ambio 13* [4], 1984, 236–243), which concluded: "The results are roughly in line with . . . the TTAPS paper." See also K. Ya. Kondratyev and G. A. Nikolsky, "A Survey of Possible Environmental Impacts of a Nuclear Conflict on the Earth's Atmosphere and Climate," Report, U.S.S.R. Committee for the U.N. Environmental Programme (Geneva, 1986); and Yuri Fyodorov, "Nuclear Winter and U.S. Nuclear Policy," *Mirovaya Ekonomika i Mezhdunarodnye Otnosheniya*, June, 1986, 77–82.

The American popular science magazine *Scientific American* has been published in a Russian edition for some years—at first, however, with articles on nuclear war and arms control systematically excised. This is no longer the case. The first such article to be published in the U.S.S.R. was R. P. Turco, O. B. Toon, T. P. Ackerman, J. B. Pollack, and C. Sagan, "The Climatic Effects of Nuclear War," *Scientific American 251* (2), August 1984, 33–43, reprinted in Russian in *V Mire Nauki*, October 1984, 4–16.

A Soviet television drama on the direct effects of nuclear war, as well as nuclear winter, called "Letters from a Dead Man," has been widely watched in the U.S.S.R. beginning in July 1986. And—another indication of Soviet awareness of nuclear winter—Andrei Voznesensky devoted a poem to the dangers of nuclear winter (*Moscow Tass*, May 15, 1984).

13.12:

"Nuclear War and Nuclear Winter," seminar delivered by Carl Sagan at Moscow State University, March 10, 1986. Reported in *NTR*: *Problemi I Rezheniya* 6 (21), March 18, 1986, 1, 7.

13.13:

Vladimir F. Petrovsky, "The Soviet Concept of Comprehensive Security at the Turn of the 21st Century," *Disarmament* 9 (2), Spring, 1986, 77–92.

13.14:

For example, Y. P. Velikhov, ed., *The Night After*: *Scientists' Warning* (Moscow: Mir, 1985); A. Gromyko and V. Lomeiko, *Consequences of Nuclear War* (Moscow: International Relations, 1984); G. S. Golitsyn and A. S. Ginsburg, *Possible Climatic Consequences of Nuclear War and Some Natural Analogues: A Scientific Investigation*, Committee of Soviet Scientists for Peace and Against the Nuclear Threat (Moscow, 1984); N. N. Moiseev, V. V. Aleksandrov, and A. M. Tarko, *Man and the Biosphere*: *A Test of Systems Analysis and Experiments with Computers* (Moscow: Nauka, 1985); A. Khabarov and A. Snegin, *The Consequences of Nuclear War for Asia* (Moscow: Novosti Publishing House, 1985); M. I. Budyko, G. S. Golitsyn, and Y. A. Izrael, *Global Climatic Catastrophes* (New York and Berlin: Springer Verlag, 1988); A. Ginsburg, *Planet Earth in the Post-Nuclear Age* (Moscow: Nauka, 1988); ref. 13.15.

13.15:

Y. P. Velikhov in *Climatic and Biological Consequences of Nuclear War* (Moscow: Nauka, 1986), 183–184. Also, Velikhov in Sadruddin Aga Khan, ed., *Nuclear War, Nuclear Proliferation and Their Consequences* (Oxford: Oxford University Press, 1985).

13.16:

Earlier in his career, Mikhail Gorbachev was in charge of Soviet agriculture, which may have put him in a better position to appreciate the consequences of even a mild nuclear winter than some other leaders.

13.17:

M. S. Gorbachev, address, International Forum for a Nuclear-free World and the Survival of Humanity, Moscow, February 16, 1987. A similar remark was made in a paper delivered in 1984 by Imamullah Khan, Secretary-General of the Karachi-based World Muslim Congress: "The World Muslim Congress rejects the notion that the very survival of mankind should be held hostage to the security interests of a handful of nuclear weapon states." (Khan, "Nuclear War and the Defense of Peace: The Muslim View," paper presented at a conference of scientists and world religious leaders on nuclear winter, Bellagio, Italy, November 19–23, 1984.)

13.18:

For example, C. Sagan, "The Nuclear Winter," *Parade*, October 30, 1983, 4–7; "Nuclear Winter: Effects of Atomic War," *Nightline*, ABC-TV, November 1, 1983; "The Day After: Nuclear Dilemma," *Viewpoint*, ABC-TV special, November 20, 1983; P. Ehrlich, C. Sagan, D. Kennedy, and W. O. Roberts, *The Cold and the Dark: The World After Nuclear War* (New York: Norton, 1984, and many foreign editions); Anne Ehrlich, "Nuclear Winter: A Forecast of the Climatic and Biological Effects of Nuclear War," *Bulletin of the Atomic Scientists 40*, 1984, 3S–14S; C. Meredith, O. Greene, and M. Rentz, *Nuclear Winter: A New Dimension for the Nuclear Debate* (London: SANA, 1984); "Nuclear Winter," *Nightline*, ABC-TV, July 18, 1984; "Nuclear Winter," *Face the Nation*, CBS-TV, December 16, 1984; "The World After Nuclear War," WTBS Superstation, Atlanta, March 1984 (many broadcasts); C. Sagan, "We Can Prevent Nuclear Winter," *Parade*, September 30, 1984, 13–17; M. Harwell, *Nuclear Winter: The Human and Environmental*

*Consequences of Nuclear War* (New York and Berlin: Springer Verlag, 1984); "The Nuclear Winter," televised talk given October 19, 1984, Denver: first broadcast nationally on PBS, April, 1985; "Nuclear Winter: Changing Our Way of Thinking," Marshall Lecture of the Natural Resources Defense Council, Washington, D.C., first televised live in syndication to 150 cities, April 18, 1985; Owen Greene, Ian Percival, and Irene Ridge, *Nuclear Winter: The Evidence and the Risk* (Cambridge: Polity Press, and Oxford: Blackwell, 1985); Michael Rowan-Robinson, *Fire and Ice: The Nuclear Winter* (Harlow, Essex, England: Longman, 1985); A. C. Revkin, "Hard Facts About Nuclear Winter: Everyone Knew That Nuclear War Would Be Hideous, but No One Expected This," *Science Digest*, March 1985, 1, 62–68, 77, 81, 83; S. L. Stephens and J. W. Birks, "After Nuclear War: Perturbations in Atmospheric Chemistry," *BioScience* 35, 1985, 557–562; C. Covey, "Climatic Effects of Nuclear War," *BioScience* 35, 1985, 563–569; Barbara G. Levi and Tony Rothman, "Nuclear Winter: A Matter of Degrees," *Physics Today*, September 1985, 58–65; Lydia Dotto, *Planet Earth in Jeopardy* (New York: Wiley, 1986); P. Crutzen and J. Hahn, *Schwarzer Himmel* (Frankfurt: Fischer Verlag, 1986); Marcus Chown, "Nuclear War: The Spectators Will Starve," *New Scientist* (London), January 2, 1986; *idem*, "Smoking Out the Facts of Nuclear Winter," *ibid.*, December 11, 1986; André Berger, "L'hiver Nucléaire," *La Recherche 17*, 1986, 880–890; G. E. McCuen, ed., *Nuclear Winter* (Hudson, Wis.: Gem, 1987); A. Robock, "New Models Confirm Nuclear Winter," *Bulletin of the Atomic Scientists*, September 1989, 32–35; Christine Harwell and Mark Harwell, *Nuclear Famine* (Burlington, N.C.: Carolina Biology Readers No. 185, 1990); C. Sagan and R. Turco, "Too Many Weapons in the World," *Parade*, February 4, 1990, 10–13; David E. Fisher, *Fire and Ice: The Greenhouse Effect, Ozone Depletion and Nuclear Winter* (New York: Harper & Row, 1990); and many other references (e.g., refs. 2.2–2.7) listed in this book. Nuclear winter has also been depicted in a number of science fiction novels and short stories, and in editorial cartoons (e.g., one in the *Philadelphia Inquirer* captioned "Let There Be Darkness!").

The resistance shown even to punch-pulling portrayals of the consequences of nuclear war was made vivid in the reaction to the November 20, 1983, television dramatization, "The Day

After." The program downplayed fire and radiation sickness, ignored nuclear winter, and managed to end on an optimistic note. Despite that, there was much criticism that showing millions of Americans killed by nuclear war would terrorize the American people into apathy and defeatism. In an ABC television discussion called *Viewpoint* immediately following this dramatic portrayal, one of us briefly presented some of the then very recent nuclear winter findings. Former National Security Adviser and Secretary of State Henry Kissinger expressed his annoyance:

> To engage in an orgy of demonstrating how terrible the casualties of nuclear war are, and translating into pictures the statistics that have been known for three decades, and then to have Mr. Sagan say it's even worse than this, I would say, What are we to do about this? Is it —are we supposed to make policy by scaring ourselves to death . . . ?

A far more realistic dramatization of the aftermath of nuclear war, including at least some of the consequences of nuclear winter, is the BBC drama "Threads," first broadcast in Britain on September 23, 1984, and shown in the United States on the Turner Broadcasting System, beginning January 13, 1985. (It was followed by the BBC nuclear winter documentary, "On the Eighth Day.") The strength of "Threads," reported the *Times* of London, was its demonstration of "how perilously constructed our human society is, and how easily it breaks down—like a spider's web through which a schoolboy pokes his finger" ("The Fragile Web of Human Society," September 24, 1984).

Despite this range and variety of public exposure to nuclear winter, there were some who felt that it did not receive the attention it deserved. For example, here are the comments of the physician and Pulitzer Prize–winning author Lewis Thomas:

> When I first heard the details of this discovery, late in the spring of 1983, I took it to be the greatest piece of good news to emerge from science in the whole twentieth century (a century with less than its fair share of good news up to now). Later in the year, on October 31, an international conference was convened in Washington for the explicit purpose of making the threat of Nuclear Winter as

public as possible. In attendance were several hundred eminent scientists from twenty countries, representing the disciplines and subdisciplines of physics, climatology, biology, and medicine, plus various American and foreign public officials, educators, foreign policy experts, and military and arms control specialists. There were over a hundred newspaper, television, and radio journalists present. On November 1, the Washington assemblage was linked by satellite, direct and live, to a group of Soviet scientists in Moscow for a back-and-forth exchange of views.

And then, in the days and weeks that followed, the strangest of things happened: not much of anything. Some of the major national newspapers carried brief, almost perfunctory, accounts of the conference, most of them on inside pages. I do not recall any mention of the affair on any of the network news programs on the next day, or indeed on any other day, with the single exception of a half-hour discussion on ABC's *Nightline* program.

A month later four Soviet scientists came to testify in Washington on the same topic in the Senate Caucus Room, together with four American counterpart scientists, at the invitation of Senators Kennedy and Hatfield. They expressed their total agreement with the conclusions reached at the end of the October conference. This meeting, also unprecedented, was open to the public and attended by at least a score of media representatives. It received no notice on that evening's television news programs, and virtually none in the next day's newspapers. ["Nuclear Winter, Again," *Discover*, October 1985.]

13.19:

Because, one Sovietologist believes, of military censorship in a running internal factional dispute (ref. 13.27).

13.20:

The authors themselves have given extensive briefings on nuclear winter to the Defense Nuclear Agency and other offices of the Department of Defense, to the Central Intelligence Agency, the National Security Agency, the Arms Control and Disarmament Agency, the National War College, the Senate Armed Services Committee, and many other committees of the Senate and the House.

13.21:

Two fairly representative legislative branch reactions to a joint House/Senate hearing on nuclear winter:

Rep. Parren J. Mitchell of Maryland:

> I'm taking a little time to get myself together. This is an incredible experience, that we're sitting in this crowded hearing room discussing in what appears to be logical and cogent terms the possible extinction of mankind. I don't know how people are able to grasp the dimension of what we are even talking about in our civilized usual smooth language. It's an enormously traumatizing experience for me.

Senator James R. Sasser of Tennessee:

> Thank you very much, Senator, and I want to commend you, Senator Proxmire, for taking the lead in holding these hearings here today. I think they are extremely important and what you have developed today from this panel of very distinguished experts is some most compelling testimony which I think is highly informative and at the same time deeply disturbing, and this is a great service I think to the country . . . a great service to all mankind—to get this information out and get it on the table so that at least some opinion leaders in our society and our Government can see it and hopefully react in a rational way.

[*The Consequences of Nuclear War: Hearings Before the Subcommittee on International Trade, Finance, and Security Economics of the Joint Economic Committee, Congress of the United States*, 98th Congress, 2nd Session, July 11 and 12, 1984 (Washington, D.C.: U.S. Government Printing Office, 1986), 78, 80.]

13.22:

Caspar N. Weinberger, Secretary of Defense, Report to the Congress, "The Potential Effects of Nuclear War on the Climate," 1985 and subsequent fiscal years. These reports were mandated by Congress to be a "comprehensive study on the atmospheric, climatic, biological, health and environmental consequences of nuclear explosions and nuclear exchanges, and the implications that such consequences have for the nuclear weapons, arms control, and civil defense policies of the United States" by the National Defense Authorization Act of Fiscal Year 1987. They are widely considered as unresponsive to the clearly expressed wishes of the Congress. In Fiscal Year 1988 the report was reduced to a single page in length. In reply to a letter from Sen. Timothy Wirth, describing the Department's *pre*-1988 reports as "too thin and lacking in sufficient detail to comply with its implementing legislation," the Secretary of Defense replied that "Nuclear Winter is a hypothesis whose science is not well understood by the scientific community. Assumptions and uncertainty abound in its predictions." (Caspar N. Weinberger to Timothy E. Wirth, October 13, 1987.) Compare this remark with, e.g., refs. 3.9, 3.11, 3.13, 3.14, 3.15, 3.16; see also boxes, "Wildfires, Martian Dust, and Nuclear Winter," Chapter 3, and "Would Billions of People Really Be Killed by Nuclear Winter?," Chapter 5. Even if we were positive that only a 10°C, and not a 20°C, nuclear winter would result from nuclear war, one would think that this deep Ice Age temperature should have a powerful influence on U.S. strategic planning. We are also struck by the reluctance to support further research on nuclear winter by some of those who claim that scientific uncertainties preclude drawing policy implications from nuclear winter.

13.23:

White House Paper, "The President's Strategic Defense Initiative" (Washington, D.C.: U.S. President, 1985), 10 pp.

13.24:

Richard N. Perle, Assistant Secretary of Defense, Testimony, House Committee on Science and Technology, March 14, 1985. Associated Press report, April 2, 1985:

> Washington: The Pentagon admits a nuclear war might wipe out life on the planet but says that's no reason to change nuclear doctrine or stop building atomic weapons. . . . Richard Perle said: "We are persuaded that a nuclear war would be a terrible thing, but we believe that what we are doing with respect to strategic nuclear modernization and arms control is sound and we believe it is made no less sound by the nuclear winter phenomenon."

However, a somewhat different point of view can be found in Perle's testimony to the Senate Armed Services Committee later that same year (*Nuclear Winter and Its Implications*, ref. 2.6) and, e.g., "Perle Says 'Nuclear Winter' Argues for Deep Cuts' (*Defense Daily*, October 10, 1985). Something like both points of view, along with the notion that nuclear winter also lends support to SDI, "strategic modernization," and flexible targeting for escalation control, can be found in a report by the Secretary of Defense, March 1985 (ref. 13.28), which led to the charge by Sen. William Proxmire that "the Pentagon has stolen nuclear winter" (Susan Subak, "Strategists Evade Nuclear Winter," *Nuclear Times*, May–June 1986, 18–19).

That same March 1985 statement by Secretary Weinberger also contains the following memorable sentence: "[E]ven a single-layer [strategic] defense may provide a greater mitigating effect on atmospheric consequences [of nuclear war] than could result from any level of reductions likely to be accepted by the U.S.S.R. in the near term." Ten months later General Secretary Gorbachev called for a three-stage process of arms reduction that, if implemented, would leave the world with no nuclear weapons at all by the year 2000.

Some of those who thought the extinction of the human species by nuclear winter implausible turn out to have their own extinction scenarios. For example, Lowell Wood, after expressing the opinion that at least 80% of the human race, outside of the Northern midlatitude target zone, might survive nuclear war, predicted that "the wave of barbarism" precipitated by the war "may trigger follow-on biological warfare, which could exterminate human life everywhere." He then used this tableau to argue for Star Wars (Wood, "The Evolving Relation Between Defense and Offense in Strategic Nuclear Warfare," in *International Seminar on Nuclear War, 3rd Session: The Technical Basis for Peace* [Frascati, Italy: Servicio Documentazione dei Laboratori Nazionali di Frascati dell'INFN, 1984], 185–186, 298–299.) Wood played a leading role in the ill-starred hydrogen-bomb-driven X-ray laser SDI project, and in the more current "brilliant pebbles."

13.25:

Dyson (ref. 2.8) has expressed concern that, since such future technological developments might "remove the danger of nuclear winter without removing the danger of nuclear war," it may be politically risky to base policy recommendations on nuclear winter alone. However, as we have discussed, the danger of nuclear winter becomes small only when currently configured arsenals fall below a few hundred weapons; a massive reduction of the arsenals is therefore required before this contingency arises. Also, even low yield subterranean nuclear explosions can—through rupture of gas and electrical lines and through static electrical discharges—cause widespread fires and firestorms, as occurred immediately after the 1906 San Francisco earthquake (cf. box, Chapter 2, and Theodore Postol, Testimony, House Science and Technology Committee, September 12, 1984). Further, many (but not all) of the policy implications of nuclear winter follow, although in weaker form, from the prompt effects of nuclear war. This is a measure of the robustness of those implications.

13.26:
U.S. General Accounting Office, *Nuclear Winter: Uncertainties Surround the Long-Term Effects of Nuclear War*, GAO/NS/AD-86-62, March 1986. This report was originally entitled *Nuclear Winter: A Plausible Theory with Many Uncertainties in Science and Policy*, but was changed following a February 12, 1986, letter on White House stationery written by John P. McTague, then acting science adviser to President Reagan. McTague told the GAO, which is supposed to work for the *legislative* branch of the U.S. Government: "I am far less sure of what the results of a nuclear war would be on the climate than is portrayed in the report. . . . Consequently, I recommend that you recast the tenor of the report."

13.27:
S. Shenfield, "Nuclear Winter and the U.S.S.R.," *Millennium: Journal of International Studies 15* (2), 1986, 197–208.

13.28:
Early American concerns about the Soviet response to nuclear winter can be found, for example, in the Memoranda of November 7, 1983, and November 19, 1983, to the Chief of Naval Operations (CNO) from Vice Adm. J. A. Lyons, Jr., Deputy CNO (Plans, Policy and Operations), released through the

Freedom of Information Act, March 14, 1984. In the earlier of the two memos, Admiral Lyons writes,

> In the long term, the [results presented a few days earlier at the first public conference on nuclear winter] deserve serious study to see what, if any, changes in U.S. targeting policy are required. In the short term, however, the conference implications are primarily political. I anticipate that the Soviets will make extensive use of these results, especially in Europe, to demonstrate the dangers of the arms race (by which they mean PERSHING II/GLCM deployment).

However, another official in the same office, the CNO's Deputy Director for Strategic and Theater Nuclear Warfare, stated in another November 1983 memorandum: "Unfortunately, there is very little likelihood of a serious examination of the broad policy implications [of nuclear winter] being done by anyone knowledgeable in the Office of the Secretary of Defense or the Joint Chiefs of Staff." (*Nuclear Winter and Its Implications: Hearings Before the Committee on Armed Services*, United States Senate, 99th Congress, 1st Session, October 2 and 3, 1985 [Washington, D.C.: U.S. Government Printing Office, 1986], 127.)

The Secretary of Defense (Caspar N. Weinberger, "The Potential Effects of Nuclear War on the Climate: A Report to the United States Congress," March 1985), asserted that the Soviets "show no evidence of regarding the whole matter [of nuclear winter] as anything more than an opportunity for propaganda." See also "Soviet Exploitation of the Nuclear Winter Hypothesis," by Leon Gouré, prepared for the Defense Nuclear Agency by SAI, Inc., June 5, 1985, and "An Update of Soviet Research on and Exploitation of Nuclear Winter, 1984–1986," by Leon Gouré, prepared for the Defense Nuclear Agency by SAI, Inc., October 10, 1987.

A different judgment by Sovietologist Stephen Shenfield concludes that Soviet work on nuclear winter "is a genuine, substantive and relatively autonomous product of the Soviet scientific community. This makes it likely that Soviet research findings have had *some* impact on policy makers (while being used at the same time, of course, for propaganda)." But if, he argues, there is "a trend in Soviet thinking towards heightened awareness of the catastrophic consequences of nuclear war and this trend is still far from its end-point, then nuclear winter may well be an issue of central importance" (ref. 13.27). See also Lynch, ref. 2.6, who in 1987 concluded that "Soviet scientists find the nuclear winter hypothesis a compelling one, and . . . have managed to convince the Soviet leadership that the threat of nuclear winter is at least as grave as that posed by any of the other effects of nuclear war."

13.29:

Remarks by K. Ya. Kondratyev, former Rector, Leningrad University, at Plenary Meeting, Soviet Academy of Sciences, Moscow, October 19, 1988:

> Golitsyn is [credited as] one of the authors of the concept of nuclear winter. But nothing of the kind! Foreigners are the authors of the concept. In fact, it's an American label, and it's better not to use it at all. . . . It is the advertisement of American results. I guarantee that this is so. This also is an ethical problem.

Kondratyev was speaking in an unsuccessful attempt to prevent the election of Georgi Golitsyn of the Institute of Atmospheric Physics to the Presidium of the Academy.

13.30:

For many years, Soviet strategists seemed to argue that nuclear weapons did not change the famous judgment of Clausewitz that war is "a continuation of politics by other means." Although other interpretations of former Soviet doctrines are possible (e.g., Robert L. Arnett, "Soviet Attitudes Towards Nuclear War: Do They Really Think They Can Win?," *Journal of Strategic Studies* 2 [2], September 1979, 172–191), the American reaction to such Soviet pronouncements without a doubt propelled the arms race. Richard Pipes, an adviser on Soviet affairs to the White House in the first Reagan term, wrote that the Soviets believe "thermonuclear war is not suicidal, it can be fought and won, and thus resort to war must not be ruled out. . . . As long as the Russians persist in adhering to the Clausewitzian maxim on the function of war, mutual deterrence does not really exist" (Pipes, "Why the Soviet Union Thinks It Could Fight and Win a Nuclear War," *Commentary* 64 [1], July

1977, 21–34). Since then, Gorbachev has explicitly rejected "the Clausewitzian maxim" (*Pravda*, February 17, 1987), and this rejection is tied directly by Soviet Defense Ministry analysts to nuclear winter (Boris Kanersky and Pyotr Shabardin, "The Correlation of Politics, War and Nuclear Catastrophe," *International Affairs*, February 1988, 95–104). This is another demonstration that the prospect of nuclear winter strengthens deterrence.

In his November 1985 arms control speech to the Supreme Soviet, Y. P. Velikhov argued that the United States could not deliver a militarily significant nuclear first strike on the Soviet Union, especially because of the prospect of nuclear winter (reported in *Izvestia*, November 28, 1985, 3). It followed that major reductions in nuclear arsenals were possible. In this period similar arguments were presented at many different high levels of the party, government, and military establishments (cf. also ref. 13.10) and was accepted by Mikhail Gorbachev.

Perhaps the most far-reaching claim for the policy impact of nuclear winter is that made by Tony Hart, Co-Chairman of the World Disarmament Campaign (Nehru Memorial Symposium: Towards a Nuclear Weapon-Free and Non-Violent World, New Delhi, 14–16 November 1988):

> There is certainly a new climate of thinking among the top echelons of the superpowers. It is appealing to believe that it is caused by the realisation during the last year or so that the nuclear terror weapons in their stockpiles cannot be used without wreaking havoc on the whole planet, that the sombre truth of the "Nuclear Winter" scenario has finally penetrated the minds of the war planners.
>
> . . . It was not just the combatants whose societies would perish, but all human life on the planet. [For our view, cf. Chapter 5.] That became apparent in the middle years of this decade. The military machines and their collaborating politicians appeared not to notice or dismissed the new knowledge. Secretly they did notice.

The first words of the INF Treaty (signed December 8, 1987) are "The United States of America and the Union of Soviet Socialist Republics . . . , conscious that nuclear war would have devastating consequences for all mankind . . ." (Treaty Document 100-11, Senate [Washington, D.C.: U.S. Government Printing Office, 1988]), which has been conjectured (Alan Robock, *Technology Review*, September 1988) to refer to nuclear winter. The words quoted are said to indicate "that the threat of nuclear winter has been at least partially responsible for the improved negotiating climate between the superpowers that has resulted in this treaty and progress toward . . . START" (Alan Robock, "Policy Implications of Nuclear Winter and Ideas for Solutions," *Ambio 18* [7], 1989, 360–366).

13.31:

L. W. Alvarez, *Adventures of a Physicist* (New York: Basic Books, 1987).

13.32:

Clearly there were many other factors—including domestic American and Soviet politics and economics, lucid statements by physicians worldwide on the prompt effects of nuclear war, and reasoned exhortations by Roman Catholic and Methodist bishops.

13.33:

Samantha Smith, *Journey to the Soviet Union* (Boston: Little Brown, 1985), 1; E. Chivian, J. P. Robinson, J. Tudge, N. P. Popov, and V. G. Andreyenkov, "American and Soviet Teenagers' Concern About Nuclear War and the Future," *New England Journal of Medicine 319*, 1988, 407–413 (see also J. E. Mack, "American and Soviet Teenagers and Nuclear War," *ibid.*, 437–438); "Summit Hopes High, Expectation Low," *USA Today*, November 12, 1985; Kenneth Callison, Human Survival Foundation, Englewood, Col., private communication, April 4, 1988; W. Green, T. Cairns, and J. Wright, *New Zealand After Nuclear War* (Wellington: New Zealand Planning Council, Ministry of the Environment, 1987), 165.

14.1:

Most American wheat is winter wheat, planted in early fall, lying dormant during the winter, and harvested in June and July. Drought and cold in late fall and winter can destroy this crop. (E.g., "Wheat Crop Faces Threat of Drought," by William Robbins, *New York Times*, November 27, 1989.) Thus fall and early-winter wars can also destroy major cereal crops, even if the chronic effects are short-lived. (Cf. J. Levitt, *Responses of*

*Plants to Environmental Stresses* [New York: Academic Press, 1980].)

14.2:

Only about 1,000 strategic warheads out of the almost 25,000 deployed could destroy all major cities in the Northern Hemisphere. (A somewhat larger number of tactical weapons could play the same role.) Although present trends are against it, if the SALT II treaty were abrogated and new missiles were added at rates planned in the middle 1980s, a substantial nuclear *buildup* could occur over the next ten to twenty years, amounting to as much as a 50% increase in the strategic arsenals (ref. 8.1) and reinforcing the possibility of a severe nuclear winter. Also of concern are direct attacks on military and civilian nuclear power facilities (C. V. Chester and R. O. Chester, "Civil Defense Implications of the U.S. Nuclear Power Industry During a Large Nuclear War in the Year 2000," *Nuclear Technology 31*, 1976, 326–338), which would increase long-term radioactivity burdens by tenfold or so (refs. 3.11, 11.1). Attacks on key coal-fired power plants, with coal stores nearby, would extend the environmental impacts still further. Petroleum stocks are, as we have stressed, extremely vulnerable and extremely dangerous for nuclear winter (ref. 3.14 and Figure 4).

14.3:

This is approximately the case that led to the apocalyptic conclusions of ref. 5.13.

14.4:

Our Classes II, III and IV roughly correspond to cases 1–3, respectively, at the SCOPE Bangkok conference, covering the $\tau_a$ range from 0.3 to 3. Cf. F. Warner et al., *Environment 29*, 1987, 4.

15.1:

A not atypical Third World view is that "the threat of nuclear war is first and foremost a game of political blackmail. . . . The nuclear threat is directed at escalating to the utmost all superpower sanctions against attempts to interfere with their reordering of the globe." (P. T. K. Lin, in S. Mendlovitz, ed., *On the Creation of a Just World Order* [New York: Macmillan, 1975], 285.) An entire volume devoted to the American half of this thesis is Joseph Gerson, ed., *The Deadly Connection: Nuclear War and U.S. Intervention* (Philadelphia: New Society Publishers, 1986). In her essay "Beyond the Facade: Nuclear War and Third World Intervention," Randall Forsberg concludes:

> Suppose that at every level of nuclear warfare . . . the United States has the capability to obliterate Soviet nuclear capability, but that the Soviet Union at best could only destroy part of our capability. Then . . . we might have a monopoly on intervention. The United States could intervene wherever it wanted to, but the Soviet Union would be deterred from intervening, because of the risk that we would dare to challenge Soviet intervention. The Soviet Union would know that at every level of escalation of war, we had a significant and obvious advantage. . . . This is what is driving the nuclear arms race. It has nothing to do with defense, it has little to do with deterrence, except in the sense of deterring their interventions while permitting our own. This is not . . . an extreme interpretation. . . . You can read this yourself if you read the Annual Report of the Secretary of Defense [*ibid.*, 34–35].

Consider how nuclear winter affects the "every level of escalation" part of this argument.

15.2:

Despite the fact that present U.S. policy guidelines on escalation control discourage use of strategic weapons based on American soil in a theater war—e.g., in Europe. There is technically no difficulty with such deterrence. Its principal difficulty is that its credibility is low. Many Western Europeans have been unable to bring themselves to believe that the United States would invite retaliation on its home territory to prevent a conventional Soviet attack on Western Europe.

15.3:

Cf. A. Wohlstetter, "Between an Unfree World and None: Increasing Our Choices," *Foreign Affairs* 63 (5), Summer 1985, 962–994.

15.4:

Noel Gayler, private communication, 1989.

15.5:

E.g., consider the effect of nuclear winter on the following objection to deep cuts in the arsenals: "Deterrence is a psychological phenomenon which is reinforced by large stockpiles. Deep cuts would be perceived as eliminating first-strike weapons, but they could also eliminate second-strike weapons which promote stability" (M. D. Intrilligator, in ref. 9.8).

15.6:

Bernard Brodie, "Implications for Military Policy," in Bernard Brodie, ed., *The Absolute Weapon: Atomic Power and World Order* (New York: Harcourt Brace, 1946), 74.

15.7:

It is evident that political and military alignments in the developing world and between neighboring states are of critical concern to the superpowers, given the enormous resources each devotes annually to influence these alignments.

15.8:

Had Hitler possessed nuclear weapons and their delivery systems in the last two years of World War II, it seems likely that he would have used them—even if massive nuclear retaliation on Germany were a guaranteed consequence. (Carl Sagan, "The Final Solution of the Human Problem: Adolf Hitler and Nuclear War," keynote address, 50th Anniversary Assembly, World Jewish Congress, Jerusalem, January 29, 1986. Excerpted in *Ha'aretz* [Tel Aviv] 68, February 7, 1986, 16.)

16.1:

The quotations from Herman Kahn in this chapter appear on pages 7, 145–150, 297, and 524 of *On Thermonuclear War*, second edition (Westport, Conn.: Greenwood Press, 1961).

16.2:

C. von Clausewitz, *On War*, Anatol Rapoport, ed. (Harmondsworth: Pelican, 1968), 102.

16.3:

Herman Kahn, *Thinking About the Unthinkable in the 1980's* (New York: Simon and Schuster, 1984).

17.1:

Even by national leaders. E.g., then–General Secretary Gorbachev described the quest for parity in overkill as "madness and absurdity." ("Political Report of the Central Committee of the CPSU to the 27th Congress of the Communist Party of the Soviet Union," *Izvestia*, February 26, 1986, 2. This speech and others illuminating the evolution of Gorbachev's thinking in 1986 are collected in Mikhail Gorbatchev, *Pour Un Monde Sans Armes Nucléaires* [Moscow: Novosti, 1987].) The glut of nuclear weapons on both sides has led to a search for new uses, including counterforce and warfighting doctrines, and horizontal proliferation—the spreading of nuclear weapons to other nations.

17.2:

It is also possible that there are other, still undiscovered, adverse consequences of nuclear war beyond nuclear winter. Whatever they are, the fewer nuclear weapons in the world, the less likely it is that these unknown effects will be triggered.

17.3:

"If strategic weapons are effectively invulnerable . . . flexible targeting strategy is readily combinable with a minimum deterrence posture. Stable mutual deterrence by the threat of destroying cities would be possible if each side had a relatively small number of nuclear warheads on submarines and single-warhead mobile missiles. Under these conditions there would be little incentive to preempt, since the other side could devastate many of one's cities in response. But if there were an actual attack, the side attacked would still have the morally preferable option of firing its nuclear weapons only at military targets some distance from cities." (G. S. Kavka, *Moral Paradoxes of Nuclear Deterrence* [Cambridge: Cambridge University Press, 1987], 11.)

17.4:

S. E. Ambrose, *Eisenhower: The President*, Vol. 2 (New York: Simon and Schuster, 1983–84), 553.

17.5:

The technical, strategic, and political deficiencies of SDI are discussed, e.g., in Richard Garwin et al., *The Fallacy of Star Wars* (New York: Vintage, 1984); Hans Bethe et al., "Space-Based Ballistic Missile Defense," *Scientific American* 251, October 1984, 39–49; Ashton Carter, *Ballistic Missile Defense* (Washington, D.C.: Office of Technology Assessment, 1984); Carl Sagan, "The Case Against the Strategic Defense Initiative," *Discover* 6 (9), September 1985, 66–74; "Report to the American Physical Society of the Study Group on Science and Technology of Directed Energy Weapons," *Reviews of Modern Physics* 59, July 1987, S1-S202; John Tirman, ed., *Empty Promise: The Growing Case Against Star Wars* (Boston: Beacon Press, 1986); *SDI: Technology, Survivability, and Software* (Washington, D.C.: Office of Technology Assessment, 1988); S. Lakoff and H. F. York, *A Shield in Space?* (Berkeley: U. California Press, 1989); Crockett L. Grabbe, *Space Weapons and the Strategic Defense Initiative* (Iowa City: Iowa State U. Press, 1990).

Of the several direct connections between SDI and nuclear winter, perhaps the most trenchant relates to permeability. If roughly 100 nuclear explosions over cities and petroleum facilities is enough to generate nuclear winter, and if each side has 5,000–10,000 deployed strategic warheads, then to prevent nuclear winter reliably requires an SDI which is 98–99% impermeable. (Not all warheads are expected to be targeted on cities, certainly, but—to defeat a deployed SDI and thus deter nuclear war—a major fraction of them might be.) Even Star Wars' most enthusiastic (technically competent) advocates never suggested more than 50% impermeability, and many critics believe the most likely figure for a major SDI system, a decade or two hence, after the expenditure of vast national treasure, might, if all goes well, be in the 10–20% range. This is why we disagree with the possibility raised by Colin Gray and some others that "strategic nonnuclear defenses could be the deciding factor in whether or not 'nuclear winter' effects were triggered" (Gray, "Strategic Defense and Peace," in *Nu-*

*clear Deterrence: Ethics and Strategy*, R. Hardin, J. Mearsheimer, G. Dworkin, and R. Goodin, eds. [Chicago: University of Chicago Press, 1985], 297; ref. 13.24). In the real world, all foreseeable SDI systems are far too porous to forestall nuclear winter in a major war. Therefore, START, which would destroy 30 to 50% of Soviet strategic warheads by treaty, would be much more effective from a strictly military, strictly parochial U.S. perspective than an American SDI; it would, compared with SDI, hardly induce a preemptive attack (ref. 19.7); and it would, compared with SDI, cost next to nothing.

17.6:
"Document of the Stockholm Conference on Confidence- and Security-Building Measures and Disarmament in Europe . . . ," September 19, 1986, Appendix E in Wirth (ref. 19.18).

17.7:
*Leo Szilard: His Version of the Facts (Special Recollections and Correspondence)*, edited by Spencer R. Weart and Gertrud Weiss Szilard (Cambridge, Mass.: MIT Press, 1978), 198.

17.8:
Bernard Brodie, "War in the Nuclear Age," in Bernard Brodie, ed., *The Absolute Weapon: Atomic Power and World Order* (New York: Harcourt Brace, 1946), 46, 48.

17.9:
"Merely having more bombs than other countries is not decisive if another country has enough bombs to demolish our cities and stores of weapons." Statement released October 14, 1945, by Robert R. Wilson "for the Association of Los Alamos Scientists, an organization of over 400 scientists working on the atomic bomb."

17.10:
Henry A. Wallace, "American Policy Toward Russia," *New York Times*, September 18, 1946.

17.11:
George F. Kennan to Dean Acheson, "International Control of Atomic Energy," Top Secret, January 20, 1950. *Foreign Re-*

*lations of the United States: 1950*, I, 28–30. Reproduced in T. H. Etzold and J. L. Gaddis, eds., *Containment: Documents on American Policy and Strategy, 1945–1950* (New York: Columbia University Press, 1978).

17.12:

Quoted in Nick Kotz, *Wild Blue Yonder: Money, Politics, and the B-1 Bomber* (New York: Pantheon, 1988), 43.

17.13:

Andrei Gromyko, interview, *Ogonyok 30*, July, 1989, 7. Quoted in ref. 21.7.

17.14:

Alexander Yanov, "An Avoidable 20-Year Race," *New York Times*, October 10, 1984.

17.15:

David Halberstam, *The Best and the Brightest* (New York: Random House, 1972), 72.

17.16:

One notable exception: Harold C. Urey of the University of Chicago. "Atomic Bomb Ban Urged by Dr. Urey," *New York Times*, October 22, 1945, 4.

17.17:

Cf. the very cautious estimate of John D. Steinbruner (in ref. 9.7): "500–2000 warheads delivered in retaliation covers anything that might be considered a reasonable deterrent requirement under any of the prevailing opinions about that requirement." Note that the stockpiles must be larger than the number of weapons delivered. This estimate does not allow for nuclear winter, which will reduce the number of warheads "required."

17.18:

One hundred high-yield warheads, each aboard a separate missile, was the minimum sufficient deterrent imagined by Leo Szilard in 1961. (Szilard, "The Disarmament Agreement of 1988," in *The Voice of the Dolphin* [New York: Simon and Schuster, 1961], 65.)

17.19:

Nick Kotz, *Wild Blue Yonder* (New York: Pantheon, 1988), an instructive account of the history of the B-1 bomber that deserves wide attention.

The strategic role of the B-2, authoritatively described by the Air Force Chief of Staff, Gen. Larry Welch, before the Senate Armed Forces Committee in the week of July 16, 1989, is "cross-targeting" and "routing-out" of mobile missiles. (Cf. Robert R. Ropelewski, "USAF Backpedaling on B-2 Relocatable Target Mission," *Armed Forces Journal International*, July 1989, 14.) Cross-targeting means hitting targets that have been missed by ICBMs, SLBMs, and cruise missiles. In answer to the question "Would it be a fair statement to say . . . that by the time the B-2 is actually used, if it had to be used, it would be after virtual nuclear annihilation of both countries?," General Welch's response was "I would think so." (Rep. Ronald V. Dellums, "A Service in Search of a Bomber," *Washington Post*, July 26, 1989, 25.)

But in a regime of strategic sufficiency, there are not enough warheads for extensive and redundant counterforce targeting by either side, and both sides *want* single-warhead mobile missiles to be invulnerable. There does not seem to be a role for the B-2, or any Soviet counterpart, in an MSD regime. Beyond that, if in a nuclear war the Soviets were saving their mobile missiles—for example, as a reserve force threatening U.S. cities—but discovered (or thought it likely) that B-2s, at their extremely slow speeds, had crossed into Soviet airspace looking for mobile missiles, doesn't this simply provide an inducement for the Soviets to go for broke?

17.20:

G. Dyer, *War* (New York: Crown, 1985), 214.

17.21:

The phrase is W. F. Hanrieder's, in ref. 9.8, p. 56.

17.22:

F. C. Iklé, A. Wohlstetter, A. Armstrong, Z. Brzezinski, W. Clark, W. Clayton, A. Goodpaster, J. Holloway, S. Huntington, H. Kissinger, J. Lederberg, B. Schriever, and J. Vessey, *Dis-*

*criminate Deterrence* (Washington, D.C.: U.S. Government Printing Office, 1988).

17.23:

Final Declaration, NATO Summit Conference, London, July 7, 1990. *New York Times*, July 7, 1990, 5.

17.24:

This possibility was very much on the minds of Soviet officials in January, 1990, during the armed violence between Azerbaijanis, Armenians and Soviet troops in which tanks, armored personnel carriers and helicopters were commandeered by dissident forces. See also Harlan W. Jencks, "As Empires Rot, Nuclear Civil War?" *New York Times*, April 14, 1990.

17.25:

At the June, 1990 Washington summit, Presidents Bush and Gorbachev agreed "upon the basic provisions of the strategic offensive arms treaty," and U.S. Secretary of State James A. Baker, 3rd, announced that "almost all" of the major outstanding issues had been resolved. But it is clear that even on such modest reductions, progress has been slow. Cf. Michael R. Gordon, "Talks Fail to End Disputes on Long-Range Arms," *New York Times*, June 2, 1990, 4.

17.26:

One knowledgeable estimate is that START would entail reducing the U.S. strategic arsenal from roughly 13,000 to roughly 10,000 weapons, and the Soviet strategic arsenal from roughly 11,000 to roughly 9,000 (Desmond Ball, "The Future of the Strategic Balance," in Desmond Ball and Cathy Downes, eds., *Security and Defence: Pacific and Global Perspective* [Sydney: Allen and Unwin, 1990], Chapter 4). This corresponds roughly to a 20% cut by each nation, less than the widely touted 30 to 50%. If, in addition, there are, say, 32,000 tactical warheads on the two sides, START amounts to a reduction in the overall superpower nuclear arsenals of less than 10%.

18.1:

Ted Greenwood, *Making the MIRV: A Study of Defense Decision Making* (Cambridge, Mass.: Ballinger/J. B. Lippincott, 1975), 66.

18.2:

Sagdeev, Kokoshin, and their colleagues (ref. 8.21) argue that an arsenal with only 10 to 100 nuclear weapons would provide an inadequate deterrent because it could deliver a level of devastation "psychologically comparable" merely to that worked by conventional bombing in World War II (a few million people killed). Proponents of existential deterrence, on the other hand, might argue that an invulnerable retaliatory capability of only a few nuclear warheads is sufficient (ref. 9.14; box, Chapter 17).

18.3:

Some of these factors are examined at length in ref. 18.1.

18.4:

The virtues of deMIRVing seem incompletely appreciated in Britain, where the transition (from single-warhead Polaris submarine missiles to multiple-warhead Triton submarine missiles) is swiftly moving toward MIRVing. Britain is opting to risk decreased crisis stability in order to be able to destroy more targets. Something similar is happening in China, where the intermediate-range ballistic missile, the CSS-2, is shortly to be MIRVed—permitting a major escalation in the operational nuclear arsenals without any change in the size of the missile force (Dingli Shen, "The Current Status of Chinese Nuclear Forces and Nuclear Policies," Princeton University Center for Energy and Environmental Studies Report 247, 1990). But Britain and China are only following, with a one- or two-decade delay, the joint example of the U.S. and the U.S.S.R.

18.5:

Glenn Kent, Randall DeValk, and David Thaler, "A Calculus of First-Strike Stability (A Criterion for Evaluating Strategic Forces)," Rand Corporation Note N-2526-AF, June 1988; Glenn A. Kent and David E. Thaler, *First-Strike Stability*, Report R-3765-AF, Project Air Force (Santa Monica, Cal.: The Rand Corporation, August 1989). But see ref. 18.8 and von Hippel and Sagdeev, ref. 19.20.

18.6:

We include this option for completeness, although its costs may be prohibitive and its autonomous, troglodytic mode very worrisome:

ICBMs would be buried so far underground that Soviet missiles couldn't possibly destroy them. They would burrow to the surface sometime after an attack and, despite the already shattered state of the combatants, continue the conflict. A deep-underground basing system would likely achieve its primary objective—invulnerability—but whether it would be effective or affordable is another matter. [Steven J. Marcus, "Doomsday Machines Reconsidered," *Technology Review* 86 (6), August/September 1982, 81.]

18.7:

Such a minisub deterrent—called "Shallow Undersea Mobile," or "SUM"—was first proposed by Richard Garwin and Sidney Drell (see James Fallows, *National Defense* [New York: Vintage Books, 1981], Chapter 6).

Following precedents set in the Reagan years, the Bush Administration has been willing, even eager, to consider treaties that ban land-based MIRVed ICBMs—where the Soviet Union has the advantage—but has remained adamantly opposed to treaties that would ban (or even balance) submarine-based MIRVed SLBMs—where the United States has the advantage. (Cf. Michael R. Gordon, "Soviets Rebuffed by Cheney on Plan Curbing Sea Arms," *New York Times*, April 16, 1990, A1, A8.)

18.8:

H. A. Feiveson, R. H. Ullman and F. von Hippel, "Reducing U.S. and Soviet Nuclear Arsenals," *Bulletin of the Atomic Scientists*, August 1985, 144–150. Also, H. Feiveson and F. von Hippel, *Stability and Verifiability of the Nuclear Balance After Deep Reductions* (Princeton University Center for Energy and Environmental Studies: Report 234, March 1989), 30 pp. The idea, suggested by Soviet analysts, of filling only a few of the submarine's tubes with missiles and verifying that the remaining tubes are empty by satellite inspection just before the sub leaves port, is safe only so long as tubes cannot be filled with missiles at sea, or at clandestine submarine pens.

18.9:

E.g., Richard Garwin, "A Blueprint for Radical Weapons Cuts," *Bulletin of the Atomic Scientists*, March 1988, 10–13;

"Deep Cuts in Strategic Nuclear Weapons: Possible? Desirable?," 38th Pugwash Conference on Science and World Affairs, Dagomys, U.S.S.R., August 29, 1988. Garwin remarks (private communication, 1989), "I don't think there is any necessity to retain bombers, but I *permit* them to be retained, in case someone has an unbreakable attachment to them."

18.10:

The nuclear weapons pioneer (and Manhattan Project Division Leader) Hans Bethe has advocated an MSD in the 200-to-1,000-warhead range (address, Cornell University, March 4, 1989).

18.11:

*Soviet Military Power*, Sixth Edition, United States Department of Defense, April 1987. The U.S. Air Force has proposed eliminating all Minuteman IIs as a way to save money (R. Jeffrey Smith and Molly Moore, "U.S. May Eliminate Minuteman Missiles," *International Herald Tribune*, January 15, 1990).

18.12:

Another difficulty with retrofitting multiple warhead strategic systems is (ref. 18.9) that "each side would feel much more secure if the other had delivery vehicles tailored to a single warhead, so that it would require a much longer and more costly effort to rebuild the force." Similar concerns have been voiced by former National Security Adviser Robert C. McFarlane ("Effective Strategic Policy," *Foreign Affairs* 67 [1], Fall 1988, 33–48). Nevertheless, Garwin has suggested, "Removal of all but one warhead from every ICBM or SLBM, and all but one bomb or air-launched cruise missile from each aircraft would reduce the number of strategic warheads on launchers to about 2000 on each side, and that could be achieved in two years." (Richard Garwin, "Space Defense: The Impossible Dream," *Commonwealth 80*, 1986, 291–293.)

Reducing the number of missiles on U.S. ballistic missile submarines, or reducing the number of warheads per missile (deMIRVing) are among the options considered in face of shortfalls in plutonium triggers because of serious safety problems at the Rocky Flats, Colorado weapons plant (Michael R. Gordon, "Panel Recommends Delay in Restart of Weapons Plant," *New York Times*, June 6, 1990, A18.).

18.13:

The argument in a little more detail: Strategic bombers have a particular relevance to nuclear winter. In a time of crisis they can be widely dispersed among civilian airfields with long runways, as occurred in the U.S. during the Cuban Missile Crisis. But such dispersion, or its mere prospect, provides incentives for nominal "counterforce" attacks on such airfields—which happen to be located preferentially near cities. To the extent that strategic bombers increase the incentive for targeting cities and thereby tend to make a disproportionate contribution to nuclear winter, they are undesirable compared with other weapons systems. While it is true that there are so many such secondary airfields as probably to exceed the number of nuclear weapons in an MSD regime, it is also true that dispersal of bombers to secondary airfields near cities can be detected by satellite reconnaissance and other means, and thereby invite attack on those cities.

18.14:

Whereupon the price of the Exocet promptly quadrupled—from $50,000 to $200,000 (John Stoessinger, "The International Dimension," *Security Management*, November 1984, 67). Even nations with very modest navies longed to buy cruise missiles. Sinking the *Sheffield* created a seller's market. Britain's loss was France's gain.

18.15:

George N. Lewis, Sally K. Ride, and John S. Townsend, "Dispelling Myths About Verification of Sea-Launched Cruise Missiles," *Science 246*, 1989, 765–770; *idem*, "A Proposal for a Ban on Nuclear SLCMs of All Ranges," Center for International Security and Arms Control Special Report, Stanford University, 1989. See also Valerie Thomas, "False Obstacle to Arms Control," *New York Times*, July 13, 1989; Steven Fetter and Frank von Hippel, "Measurements of Radiation from a Soviet Warhead," *Physics Today*, November 1989, 45; von Hippel and Sagdeev, ref. 19.20.

18.16:

Unresolved in this discussion is to what extent a bomber force is perceived to be required for "force projection" with conventional arms in the developing world—as in Vietnam, Libya, Afghanistan, Chad, Iraq, and the Falklands/Malvinas (to take, respectively, some examples from recent American, Soviet, French, Israeli, and British history). Strategic bombers are of course dual-capable, as readily fitted with conventional as with nuclear bombs. The U.S. use of the Okinawa-based B-52, its premier strategic bomber, in Vietnam is the clearest example. (It was judged necessary to drop 8 megatons of *conventional* explosives on Southeast Asia, and no other delivery system was anywhere near so suitable.)

But shorter-range bombers or fighter-bombers could serve coercive or retaliatory functions while posing less of a strategic nuclear threat in the U.S./Soviet confrontation, although no such distinction can be drawn for less widely separated rivals —e.g., in the Middle East. In addition, aerial refueling can effectively convert a medium-range conventional bomber into a long-range strategic bomber. Aircraft carriers, too, are mainly intended for "force projection" but can also be employed, near the adversary's shores, for strategic purposes; and they are increasingly vulnerable to increasingly widespread conventionally armed cruise and other missiles. Perhaps it is time to reassess "force projection."

18.17:

Any nuclear or nuclear/conventional agreement possible now might be undermined by future technologies, especially long-range, high-accuracy conventional weapons. This implies the need for something like the present ABM Treaty for emerging conventional or quasi-conventional technologies, but with beefed-up inspection provisions. For a late-Reagan-era assessment of the intersection of nuclear winter, SDI, and the enhanced capabilities of conventional weapons, see Peter de Leon, *The Altered Strategic Environment* (Lexington, MA: Heath, 1987).

19.1:

Defense Policy Panel, Committee on Armed Services, U.S. House of Representatives, *Breakout, Verification and Force*

*Structure: Dealing with the Full Implications of START* (Washington, D.C.: U.S. Government Printing Office, 1988).

19.2:

An additional reason that the development of antisatellite (ASAT) weaponry is counterproductive for the United States is the long-standing fact that a larger fraction of American military communications traffic than Soviet traffic is routed by satellite. This is a matter of those living in glass houses throwing stones. (The preceding argument is now being challenged: "War Games Imply ASATs Will Give Edge over Soviets," by Vincent Kiernan, *Space News*, March 26–April 1, 1990, 24).

On the other hand, Soviet satellites that detect the launch of ballistic missiles—and so provide early warning of a nuclear war—spend time in each orbit at low altitudes, where they are vulnerable to American ASATs; their American counterparts, in geosynchronous orbit, are well out of range of existing Soviet ASAT technology. Why would the U.S. seek the ability to blind Soviet eyes to the launch of an American first strike? It is an interesting question and worth working through the possible answers. What would be the likely Soviet response to such blinding? That also is worth thinking about. The Department of Defense in recent years has been eager to test U.S. ASAT capabilities, but Congress has banned such tests as long as the Soviet voluntary moratorium on its own ASAT testing remains in effect.

19.3:

Theodore B. Taylor, "Verified Elimination of Nuclear Warheads," *Science and Global Security 1* (1–2), 1989, 1–26. The author is a longtime designer of American nuclear warheads. See also Robert L. Park and Peter D. Zimmerman, "Megawaste: Junking Nuclear Bombs," *Washington Post*, June 5, 1988; von Hippel and Sagdeev, ref. 19.20.

19.4:

Noel Gayler, "How to Break the Momentum of the Nuclear Arms Race," *New York Times Magazine*, April 25, 1982.

19.5:

See, e.g., Frank von Hippel and Barbara Levi, "Controlling Nuclear Weapons at the Source: Verification of a Cutoff in the

Production of Plutonium and High-Enriched Uranium for Nuclear Weapons," in K. Tsipis, D. A. Hafmeister, and P. Janeway, eds., *Verification of Arms Control: The Technologies That Make It Possible* (London: Pergamon-Brasseys, 1986), 338–388; F. von Hippel, Prepared Statement for presentation, Hearings, Armed Services Committee, U.S. House of Representatives, June 6, 1989.

19.6:

Cf. Michael R. Gordon, "Agreeing How to Spy on Each Other," *New York Times*, June 28, 1987.

Here is an account by the Defense Policy Panel of the House Armed Services Committee (ref. 19.1):

> The answer often given for the hidden missile problem is anytime, anywhere, short-notice, no-refusal, on-site inspection. The INF verification regime is often faulted because it limited challenge inspections to declared sites and did not provide for short-notice inspections at suspect sites. The latter was the United States' original position but after it was accepted in principle by the Soviets, the United States backtracked, largely to protect the security of American installations. . . . [For future treaties] both sides are unlikely to accept anything close to no-right-of-refusal, which would leave each exposed to "fishing expeditions" aimed at collecting intelligence. The United States is also bound by legal constraints concerning proprietary information of private manufacturers and the Fourth Amendment ["unreasonable search and seizure"] rights of citizens.

The reader is invited to weigh the benefits and liabilities of intrusive inspection in this light. We note that the Fourth Amendment protections do not seem to have much force with respect, e.g., to drug issues; mandatory urinalysis and random, warrantless inspection of autos for contraband have become aspects of everyday life in America. The possible Fourth Amendment ramifications of intrusive inspection seem tenuous; is it likely that many Americans will be storing nuclear weapons or missiles in their homes, or converting their garages into clandestine factories? Homes do not seem to be the issue.

19.7:

With anything like the present arsenals, SDI is destabilizing: An American SDI, say—while wholly ineffective against a massive Soviet first strike—is well adapted to mopping up the residual Soviet retaliatory force after a massive American first strike (and vice versa). It can therefore be understood as a means of forcibly removing the Soviet deterrent. Such a prospect understandably would make Soviet policymakers nervous. Thus SDI deployment is an inducement to a Soviet preemptive strategic attack. Better to destroy as much of the American forces as we can while we still have a chance, the argument goes, than to be completely at their mercy once their shield goes up. The counter to this argument is that any "shield" will be stupefyingly porous. But what if Soviet officials, believing the assertions of American officials, think the shield will be anything like impermeable?

19.8:

E.g., "Interagency Report on Ballistic Missile Defense Systems," *Aviation Week and Space Technology*, October 17, 1983, 16.

19.9:

There have been several reports on how space-based laser weaponry could be used to ignite flammable materials in cities and bring about a so-called "laser winter" (A. Latter and E. Martinelli, "S.D.I.: Defense or Retaliation?", R & D Associates, Report, May 28, 1985; Carolyn Herzenberg, "Nuclear Winter and the Strategic Defense Initiative," *Physics and Society 15* [1], 1986; T. S. Trowbridge, "Laser Winter: Ground Incendiary Threat from the SDI," Red Mesa Research Corp., Los Angeles, 1988). The concept has been criticized as a cost-ineffective way to ignite cities in an age of nuclear weapons (even with optimistic SDI configurations, the laser forces might be exhausted in such an attack on cities, leaving them useless for strategic defense); at the same time, certain "advantages" are noted—an attack on $C^3I$ facilities and a limited number of strategic targets would be instantaneous, without warning, and, from the target's point of view, nonnuclear. (H. Lynch, "Technical Evaluation of Offensive Uses of S.D.I.," working paper, Center for

International Security and Arms Control, Stanford University, 1987.)

Soviet sources are said to be concerned that SDI (and antisatellite weapons), if used, may lead to a belt of fine missile and satellite fragments in Earth orbit that will have a significant optical depth, cooling and darkening the Earth: "In principle, 'space winter' is possible. It will resemble 'nuclear winter,' but will last for a longer time." (*Defense Daily*, June 21, 1988.)

19.10:

Solly Zuckerman, *Proceedings*, American Philosophical Society, August 1980. Similar views have been expressed in Lord Zuckerman's books—e.g., *Nuclear Illusion and Reality* (New York: Viking Penguin, 1983).

19.11:

There are many other proposed tripwires. Until they were removed in accordance with the INF treaty, Pershing IIs and ground-launched cruise missiles were tripwires. Hundreds of thousands of young American soldiers in Europe are also tripwires; it is hard to imagine many of them dying in a Soviet conventional attack without some sort of U.S. retaliation. The essence of Western tripwire thinking is that neither Western Europeans nor Americans believe that the U.S. really would retaliate with strategic weapons against WTO conventional aggression, for fear of a Soviet rejoinder on American soil; the function of the tripwire is to enhance the plausibility that the U.S. will, under such circumstances, begin a nuclear war. The primary target of tripwires is the human mind.

19.12:

Gwynne Dyer, *War* (New York: Crown, 1985), 190.

19.13:

For example, P. Lewis, "Soviet Offers to Adjust Imbalance of Conventional Forces in Europe," *New York Times*, June 24, 1988, A1. Cf. "Proposal of WTO Political Consultative Committee," Budapest, June 1986; and Warsaw Treaty States, Session of the Political Consultative Committee, Berlin, May 28–29, 1987, which called on NATO to join in reducing "the armed force and conventional armaments in Europe to a level where neither side, maintaining its defense capacity, would have the means to stage a surprise attack against the other side or offensive operations in general"; and "Memorandum of the Government of the Polish People's Republic on Decreasing Armaments and Increasing Confidence in Central Europe" ("Jaruzelski Plan"), July 17, 1987, Appendix D in Wirth, ref. 19.18. In the absence of a substantive NATO response for several years, the Soviet Union began withdrawal of troops and armor unilaterally (see text below). But by 1990 "wholesale reductions" in NATO tactical nuclear weapons in Germany were said to be imminent (Apple, ref. 21.4), and a fundamental review of tactical "forward defense" (e.g., Airland Battle 2000) was underway [David White, "Warsaw Pact 'No Longer a Threat' " *Financial Times* (London), May 23, 1990, 1.]

19.14:

Jonathan Dean, "The NATO-Warsaw Pact Confrontation in the Twenty-first Century: Rough Model for an Optimum Force Posture," in *Alternative, Defensive Postures for NATO and the Warsaw Pact: Possibilities and Prospects for Conventional Arms* (Washington, D.C.: The American Committee on U.S.-Soviet Relations, 1988), 17–36. Dean is the former U.S. Ambassador to the Mutual Balanced Force Reduction Negotiations (1973–1981). See also Jonathan Dean, *Meeting Gorbachev's Challenge: How to Build Down the NATO-Warsaw Pact Confrontation* (New York: St. Martin's, 1990).

19.15:

Secretary of Defense Richard Cheney, statement issued November 10, 1989 (see, e.g., "Pentagon Says Risk of War Is at Postwar Low, but Warns Against Euphoria," by Michael R. Gordon, *New York Times*, November 11, 1989); and National Intelligence Estimate, September 1989 (discussed in "Soviet Changes Mean Earlier Word of Attack," by Michael R. Gordon and Stephen Engelberg, *New York Times*, November 26, 1989). Joint Chiefs of Staff/Central Intelligence Agency/Defense Intelligence Agency joint assessment, *Using Earlier Warning to Improve Crisis Deterrence and Warfighting Capabilities*, reported in "Study Finds NATO War Plans Outdated: Report Concludes Alliance Overestimates Soviet Attack Capability," by Patrick E. Tyler and R. Jeffrey Smith, *Washington Post*, November 29, 1989. See also Bernard E. Trainor, "With Reform, Tough Times for the Warsaw Pact," *New York Times*, December 20, 1989, A16; David White, Ref. 19.13.

An illuminating dispute arose in early March 1990, after William H. Webster, the Director of Central Intelligence, testified before the House Armed Services Committee. The Soviet Union was unlikely to pose a major conventional warfare threat to NATO or the United States, the CIA had estimated, and this was true even if Mikhail Gorbachev were replaced by a more bellicose Soviet leadership. Secretary Cheney objected to these conclusions, chiefly, it seemed, on the grounds that they made more difficult his task of convincing Congress to allocate money to the Department of Defense. (Cf. Michael Wines, "Webster and Cheney at Odds over Soviet Military Threat," *New York Times*, March 7, 1990, A1, A13.)

Events were moving so swiftly that one week later an analysis by the DoD's own Policy Planning Staff judged that WTO had become defunct as an effective military organization, and a Military Net Assessment by the Joint Chiefs of Staff implied that—in part because of the deterioration of the armed forces of Eastern Europe—NATO could effectively defend Western Europe against a WTO conventional attack without resort to nuclear weapons (Michael R. Gordon, "Aide Differs with Cheney on the Soviet Threat," *New York Times*, March 13, 1990; Michael R. Gordon, "U.S. Shift Seen on Defense of Europe," *ibid.*, March 14, 1990). While U.S. officials were clearly reluctant to say so flatly, traditional justifications for tactical nuclear weapons in Europe and extended deterrence seemed to be unraveling.

Even conservative members of Congress began to conclude that major cuts in the defense budget were desirable and prudent—in part because much of Eastern Europe had become a kind of *de facto* buffer zone insulating the West against a putative Soviet invasion. Proposals for a 50% cut in the defense budget in the decade of the 1990s—which would necessarily embrace many of the steps we advocate in Chapters 18–20—were denounced by Secretary Cheney as implying "a radical change in our global status" and a fall from superpower ranking (Michael R. Gordon, "Cheney Calls 50% Cut a Risk to Superpower Status," *New York Times*, March 17, 1990; R. W. Apple, Jr., "Bush Is Reported Willing to Accept Big Military Cuts," *ibid.*, March 18, 1990). This mode of thinking imagines that military force is the only measure of national security (ref. 20.1).

19.16:

R. W. Apple, Jr., "Bush Calls on Soviets to Join in Deep Troop Cuts for Europe as Germans See Path to Unity," *New York Times*, February 1, 1990, A1, A12.

19.17:

"Surely an Alliance with the wealth, talent, and experience that we possess can find a better way than extreme reliance on nuclear weapons to meet our common threat. We do not believe that if the formula $e = mc^2$ had not been discovered, we should all be communist slaves." (Robert S. McNamara, "The United States and Western Europe: Concrete Problems of Maintaining a Free Community," speech, Ann Arbor, Michigan, August 1, 1962. Reported in *Vital Speeches of the Day* 28 [20], 1962, 627–629.)

19.18:

The evidence for this has been available for some time. See, for example, G. Perkovich, *Defending Europe Without Nuclear Weapons* (Boston: Council for a Livable World Education Fund, 1987); the annual *Military Balance* (London: International Institute for Strategic Studies, all recent vintages); and studies by the Rand Corporation, the Brookings Institution, the Congressional Budget Office and, in its annual reports to Congress, the Defense Department itself—as cited by Jane M. O. Sharp (*New York Times*, November 6, 1986). See also Sen. Timothy E. Wirth, "Intermediate Nuclear Forces Treaty and the Conventional Balance in Europe: Report to the Committee on Armed Services, United States Senate," February 3, 1988 (Washington, D.C.: U.S. Government Printing Office); K. Silversteen, "Conventional Arms Myths in Europe," *The Nation*, June 11, 1988; T. K. Longstreth, "The Future of Conventional Arms Control in Europe," *Journal of the Federation of American Scientists 41* (2), February 1988, 1; "NATO and Warsaw Pact Forces: Conventional War in Europe," *The Defense Monitor* (Center for Defense Information) *17* (3), 1988, 1; Michael R. Gordon, "Cutting Arms in Europe: It's Down to the Details," *New York Times*, March 9, 1989, A12; Robert D. Blackwill and F. Stephen Larrabee, eds., *Conventional Arms Control and East-West Security* (Durham, N.C.: Duke University Press, 1989). For a discussion of the inefficiency of Soviet conventional forces in recent combat, see A. Alexiev, *Inside the Soviet Army in Afghanistan* (Santa Monica, Cal.: The Rand Corporation, Publication R-3627-A, 1988).

Here is one reason that the deficiencies of the notion of massive Soviet conventional superiority were so rarely heard in the West: " 'The "soft on communism" argument is always a disincentive to question the "overwhelming superiority" argument. No two ways about it,' says an aide to a ranking Senate Armed Services Committee member . . ." (Perkovich, *Defending Europe*, 52.)

A Soviet perspective, from Central Committee member Georgi Arbatov:

> [There] is a good deal of hypocrisy on your [the U.S.] side. Your authorities complain that we have superiority in conventional weapons. Perhaps we do in some categories and we are prepared to build down in these areas. But you've been complaining about this for forty years, despite the fact that the West's GNP is two and a half times bigger than ours. If you really thought we had such superiority, why didn't you catch up? Your automobile and tractor industries are much stronger than ours. Why didn't they build tanks? No, I think you've used this scare about alleged Soviet superiority to hold your NATO alliance together and to justify building an absolutely irrational number of nuclear weapons. [Interview with Arbatov, "America Also Needs Perestroika," in Stephen F. Cohen and Katrina vanden Heuvel, eds., *Voices of Glasnost: Interviews with Gorbachev's Reformers* (New York: W. W. Norton, 1989), 317.]

But if American factories produced tanks rather than cars, trucks, and tractors, the civilian economy would have suffered greatly. This is just what happened to the U.S.S.R. Nuclear weapons buy "deterrence" much more cheaply: "More bang for the buck," as the slogan went in the 1950s.

19.19:

Additional reassurance, if it is needed (we very much doubt that it is), could be provided by comparatively inexpensive, purely defensive measures, such as fortifying the 780-km-long NATO/WTO boundary (ref. 19.18). But this option became exceedingly remote as German unification proceeded. It is sometimes argued (e.g., Catherine M. Kelleher, in ref. 9.8) that for political, economic, and demographic reasons no additional NATO conventional defenses can be accommodated in Western Europe. This remains to be demonstrated, but if it is true it also argues for conventional force reductions.

19.20:

Some of these studies appear in 1989 and 1990 issues of the journal *Science and Global Security: The Technical Basis for Arms Control and Environmental Policy Initiatives* (New York: Gordon and Breach). See also Frank von Hippel and Roald Sagdeev, eds., *Reversing the Arms Race: How to Achieve and Verify Deep Reductions in the Nuclear Arsenals* (New York: Gordon and Breach, 1990). Other high-level U.S./Soviet discussions on how to get to minimum sufficiency have informally taken place (e.g., "Focus on Further Arms Cuts Urged: Anticipating Agreement on a 50% Cut in Nuclear Warheads, U.S. and Soviet Conferees Set Their Sights on 'a Minimum Nuclear Deterrent,'" by Michael Parks, *Los Angeles Times*, October 19, 1989); and official discussions seem likely (Michael R. Gordon, "U.S. Invites Ideas from the Soviets on Strategic Cuts: Moscow's Earlier Suggestion on Reducing Nuclear Arms Have Been Spurned," *New York Times*, February 12, 1990, A1, A11). While acknowledging divisions in the Administration on whether to discuss deep cuts with the Soviets, one anonymous high official quoted in the *New York Times* article said, "The willingness to hear the Soviet ideas on START-2 does not mean that there might not be a pause after START. But the general direction is not in favor of a pause." A program of major cuts in U.S. nuclear and conventional forces during the decade of the 1990s—although not on the scale that we advocate here—has been proposed by the influential defense analyst William W. Kaufmann (*Glasnost, Perestroika, and U.S. Defense Spending* (Washington, D.C.: The Brookings Institute, 1990). Soviet Prime Minister, Nikolai I. Ryzhkov proposed cutting the U.S.S.R.'s military budget by one-third, or one-half, over a shorter period of time (*ibid.*, 23–24).

19.21:

E.g., U.S. Arms Control and Disarmament Agency (ACDA) advertisement in *Scientific American*, January 1990, 151, inviting applicants for a competition among faculty members in American universities to become visiting scholars at ACDA. Their "perspective and expertise" is solicited.

19.22:

Roald Z. Sagdeev, private communication, 1989.

20.1:

Cf. Carl Sagan and Ann Druyan, "Give Us Hope: An Open Letter to the Next President," *Parade*, November 27, 1988, 1, 4–9. Reprinted in pamphlet form by the Council for a Livable World (Boston, 1989).

20.2:

E.g., the August 15, 1978, letter to President Jimmy Carter from the nuclear weapon designers Norris Bradbury, Richard Garwin, and J. Carson Mark:

> The assurance of continued operability of stockpiled nuclear weapons has in the past been achieved almost exclusively by non-nuclear testing—by meticulous inspection and disassembly of the components of the nuclear weapons, including their firing and fusing equipment. . . . It has also been rare to the point of non-existence for a problem revealed by the sampling and inspection program to *require* a nuclear test for its resolution.

Also, the May 14, 1985, letter to Dante Fascell, Chairman, Committee on Foreign Affairs, U.S. House of Representatives by Hans Bethe et al.: "Continued nuclear testing is not necessary in order to insure the reliability of the nuclear weapons in our stockpile." See also Hugh E. DeWitt and Gerald E. Marsh, "Stockpile Reliability and Nuclear Testing," *Bulletin of the Atomic Scientists*, April 1984, 40–41.

20.3:

Radioactive decay of tritium occurs at a rate of about 5.5% per year. The greatest uses of tritium are in fission-fusion-fission and "enhanced radiation" weapons. The present inventory of tritium in all U.S. (or Soviet) nuclear weapons is a state secret, but probably amounts to about 100 kilograms (T. Cochran, W. Arkin, R. Norris, and M. Hoenig, *Nuclear Weapons Databook*, Volume II: *U.S. Warhead Production* [Cambridge, Mass.: Ballinger, 1987], 223 pp.; also, ref. 19.3)—the mass (but considerably less than the bulk) of a typical American football player. With reductions in the existing nuclear arsenals, curtailment of new warhead production and testing, and recycling of the existing tritium in warheads, the present tritium inventory should be sufficient for the next two decades in almost any nuclear arms reduction regime. We endorse the proposal for a verified reduction of nuclear forces on both sides paced by (but, we hope, faster than) the reduction in thermonuclear weapons driven by the natural radioactive decay of tritium. (J. Carson Mark, Thomas D. Davies, Milton M. Hoenig, and Paul L. Leventhal, "The Tritium Factor as a Forcing Function in Nuclear Arms Reduction Talks," *Science 241*, 1988, 1166–1168. See also David Albright and Theodore B. Taylor, "A Little Tritium Goes a Long Way," *Bulletin of the Atomic Scientists*, January/February 1988, 39–42.)

20.4:

Article VI of the Treaty on the Non-Proliferation of Nuclear Weapons (signed, 1968; ratification completed, 1970) commits the United States, Britain, and the Soviet Union

> to pursue negotiations in good faith on effective measures relating to cessation of the nuclear arms race at an early date and to nuclear disarmament, and on a treaty on general and complete disarmament under strict and effective international control.

This is the *quid pro quo* because of which other signatory nations promise not to develop nuclear weapons. The U.S., U.K., and U.S.S.R. are still in flagrant noncompliance with their obligations under this treaty—a fact to which other nations often point when queried about their own real or purported violations.

The preamble to the 1963 Limited Test Ban Treaty proclaims as the "principal aim" of these three nations

> the speediest possible achievement of an agreement on general and complete disarmament under strict interna-

> tional control . . . which would put an end to the armaments race and eliminate the incentive to the production and testing of all kinds of weapons, including nuclear weapons.

Protocol I of the Geneva Convention of 1949, Article 85, outlaws "launching an indiscriminate attack affecting the civilian population . . . in the knowledge that such attack will cause excessive loss of life." This Convention, ratified by the United States in 1977, is, like all treaties, through Article VI of the U.S. Constitution, "the supreme law of the land." No loss of life in nuclear war is more "excessive" than that potentially caused by nuclear winter, and no sophistries about populations not being targeted "*per se*" (ref. 8.15) can circumvent U.S. and Soviet obligations under the terms of the Geneva Convention. Only reducing nuclear forces below the level capable of generating a nuclear winter constitutes serious intent to comply with U.S. law.

20.5:

Among Soviet weapons systems that would be cancelled under such an arrangement are the 10-warhead SS-18 (Mod-5), the Typhoon strategic submarine, the "Blackjack" strategic bomber, and all nuclear-armed cruise missiles.

20.6:

Plus a greatly expanded international research program, funded by the governments, on nuclear winter. This would include studies of organisms and ecosystems as well as environment and climate, cover the full range of nuclear winter cases, utilize new computer systems to improve the spatial resolution of the general circulation models, include the influence of ocean cooling and ocean circulation, and emphasize long-term global consequences. We think it would be helpful for policymakers and the public to see computer-derived motion pictures of the time-evolution of surface temperature, drought index, etc., on a world map, for each of a wide variety of nuclear winter cases.

21.1:

Lal Kishanchand Advani, President of the Bharatiya Janata Party in India, commented:

> Ever since the Chinese invasion of 1962, and since China and Pakistan went nuclear, we feel that *Realpolitik* demands that we also become nuclear. I would be happy if the whole world becomes non-nuclear, but the situation being what it is, even though nuclear weapons may not be used, they do give political leverage to a country, particularly in the limited relationships and communications we have with Pakistan and China. [Barbara Crossette, "Militant Hindu Leader Wins King-Maker's Role," *New York Times*, November 28, 1989.]

Pakistani politicians might argue that they need nuclear weapons because India has "gone nuclear."

In fact, only one nuclear explosion has occurred so far on the subcontinent, that set off by an Indian "device." Under Prime Minister Indira Gandhi, India's nuclear weapons capability increased from an estimated 1 bomb per year in 1974 to about 30 bombs per year a decade later (Leonard S. Spector, *Going Nuclear* [Cambridge, Mass.: Ballinger, 1987], 91). An entire enrichment plant for converting uranium powder to uranium hexafluoride was smuggled into Pakistan from West Germany between 1977 and 1980. In March 1985 a West German court convicted one Albrecht Migule of the deed; he was given a $10,000 fine and a six-month suspended sentence (*ibid.*, 103–104, 282). This occurred under the "safeguards" imposed by the Vienna-based International Atomic Energy Agency. There seems to be a great deal of winking, nodding, and dozing going on in national legal systems and international organizations that are supposedly dedicated to preventing nuclear proliferation.

21.2:

Possible examples include Iraq, whose Osiraq nuclear reactor was demolished by an Israeli air strike in 1981 (although "it seems almost impossible that by the time of the Israeli raid . . . Osiraq could have secretly produced plutonium in the quantities needed for a clandestine nuclear-weapons program" [Spector, ref. 21.1, p. 162]). Iraqi nuclear capability is again growing. Also in 1981, Libya entered into negotiations to acquire nuclear weapons with an American former (or perhaps then still current) CIA officer named Edwin Wilson. Wilson had previously supplied Libya with timers, detonators, and 20 tons of C-4 plastique explosive, as well as former Green Berets to teach Libyans how to use this technology (especially useful for blowing up airliners). With a Belgian associate, Wilson offered Col. Muammar Khadaffi, the Libyan leader, a facility for producing nuclear weapons. It would have, they said, an annual productivity of 8 bombs of 1 megaton yield each and 25 of 500 kiloton yield—suitable for delivery by aircraft—as well as larger numbers of lower-yield weapons. The offer was a scam to which Libya soon tumbled (*ibid.*, 150–153). The importance of the incident lies in Libya's interest in acquiring nuclear weapons and in the existence of arms dealers, of ambiguous official or quasi-official status, who are willing to supply any weapon to any party—for a price. Their existence was brought forcibly to public attention during the Iran-Contra fiasco.

21.3:

For example, the remark of Nigerian Foreign Minister Bolaji Akinyemi: "Nigeria has a sacred responsibility to challenge the racial monopoly of nuclear weapons." (*New York Times*, November 23, 1987, A12.) Ali Mazrui, in *The Africans: A Triple Heritage* (Boston: Little, Brown, 1986), 315, and elsewhere, suggests that only when African nations, with what he describes as their "underdevelopment and instability," seem about to acquire them will the major powers understand the need to abolish nuclear weapons.

21.4:

This was a major element in the 1990 debate about whether a united Germany would have ties to both NATO and WTO, NATO only, or to some new European treaty organization. Continuing NATO nuclear weapons on German soil is supposed to obviate the need of some future German leader for a German nuclear arsenal. France and the USSR, in particular, are imagined to be relieved by such a prospect. But the Soviet view, in rapid flux as this book goes to press, may be that a unified Germany in NATO is acceptable so long as there are *no* nuclear weapons stationed on its soil. [R.W. Apple, Jr. "Arms and Germany," *New York Times*, June 29, 1990; David Goodhart, "Bonn Ponders Nuclear Trade-off with Moscow," *Financial Times* (London) June 6, 1990, 3.]

21.5:

Sanjoy Hazarika, "India Is Reportedly Ready to Test Missile with Range of 1,500 Miles," *New York Times*, April 3, 1989; *Missile Proliferation*, Congressional Research Service Report 88–642 F, revised February 9, 1989; James T. Hackett, "The Ballistic Missile Epidemic," *Global Affairs* 5 (1), Winter 1990, 38–57; Janne E. Nolan and Albert D. Wheelon, "Third World Ballistic Missiles," *Scientific American*, August, 1990, 34–40.

However, with fewer armed conflicts in the world and global tensions declining, stress on the manufacturers of such missiles is increasing. A clear-cut case is Avibrás Aerospacial S.A., the leading Brazilian manufacturer of missiles and in 1987 the leading exporter of any private corporation in Brazil. Its profits derived mainly from the Iran-Iraq war. When the war ended, the company fell on hard times. In January 1990 it filed for bankruptcy (James Brooks, "Peace Unhealthy for Brazilian Arms Industry," *New York Times*, February 25, 1990).

21.6:

Paul Lewis, "Nonaligned Nations Seek Total Nuclear Test Ban," *New York Times*, November 15, 1989.

21.7:

Stephen Shenfield, "Minimum Nuclear Deterrence: The Debate Among Soviet Civilian Analysts," Center for Foreign Policy Development, Brown University, November 1989.

While Chinese officials have more or less consistently cited a 50% reduction in superpower strategic arsenals as a precondition for China entering talks on nuclear disarmament (ref. 9.12), some hint of the force levels required for China to reduce its arsenals has been given by Di Hua, a Director of the China International Trade and Investment Corporation. He envisions ratios in the 5:1 to 3:1 range ("China and the Bomb," *Science* *239*, 1988, 972–973), consistent with what the Soviet Foreign Ministry analysts are discussing. Chinese nuclear weapons are thought to be targeted almost exclusively on the Soviet Union.

21.8:

D. Albright, "Israel's Nuclear Arsenal," *Public Interest Report* (Federation of American Scientists) *41* (5), May 1988, 4–6.

21.9:

Book of Joshua, 6:1–27. Also, when it was announced on March 18, 1988, that Chinese intermediate-range missiles would be purchased by Saudi Arabia, the U.S. government declared that it had been assured by the highest levels of the Saudi government that these missiles would never be used to carry nuclear warheads. But no guarantees were, or could be, offered.

21.10:

E.g., Israeli then–Foreign Minister Yitzhak Shamir, in an address to the U.N. General Assembly, October 1, 1981, in which he also called for a Middle East nuclear-weapon-free zone.

21.11:

In 1966, the Israel Atomic Energy Commission was reorganized, and the Prime Minister appointed himself chairman; by 1969 the Dimona plutonium facility was activated; in 1974 the CIA reported operational Israeli nuclear weapons; and in 1979 Israel and South Africa, working together, are suspected of having tested a low-yield nuclear weapon (B. Beit-Hallahmi, *The Israeli Connection: Whom Israel Arms and Why* [London: Tauris, 1987]). See also ref. 21.8.

21.12:

Joseph Stalin, *Pravda*, September 25, 1946. Teng Ch'ao, "Piercing Through the Myth of Atomic War," Beijing, 1950 [cited in Henry Kissinger, *Nuclear Weapons and Foreign Policy* (New York: Harper, 1957), 367]; Mao Zedong, January 1955, remarks to the Finnish envoy to China, "The Chinese People Cannot Be Cowed by the Atom Bomb" in *Selected Works of Mao Tsetung*, Volume 5 (Beijing, 1977), 152–153; "A Magnificent Victory for Mao Tsetung's Thought," *Jiefangjun Bao* [Liberation Army Daily], June 18, 1967, quoted in John Wilson Lewis and Xue Litai, *China Builds the Bomb* (Stanford, Cal.: Stanford University Press, 1988), 210.

22.1:

There would still remain the possibility of conventional (nonnuclear) explosives, launched on strategic rockets, causing the burning of distant cities or petroleum repositories. But the scale of rocketry needed to make enough smoke to generate a "high explosives winter" is out of reach for all nations except the superpowers. These considerations underscore the importance of arranging the world so that—nuclear weapons or no nuclear weapons—no nation has large stocks of strategic missiles for any purpose, real or ostensible, spaceflight included. This prohibition is also relevant to capabilities for chemical and biological warfare, which surely would be also under divestment as an MSD regime were approached.

22.2:

But this danger exists today as well, and it is possible to imagine a nuclear-armed terrorist group that is immune to the threat of nuclear retaliation, or even a national leader so captured by ideology and fanaticism as to be unmoved by the prospect of destruction of his homeland (cf. ref. 15.8). Moreover, there are potent nonnuclear means of retribution available to nations fully divested of nuclear weapons.

22.3:

Frederick S. Dunn, "The Common Problem," in Bernard Brodie, ed., *The Absolute Weapon: Atomic Power and World Order* (New York: Harcourt Brace, 1946), 15. The general idea has a long history: "In small numbers," Thomas Hobbes wrote in *Leviathan*, "small additions to the one side or the other make the advantage of strength so great as is sufficient to carry the victory."

22.4:

Nuclear winter provides, however, a possible counterargument to this leisurely pace. Consider the possibility that even the smallest credible sufficient deterrent could produce a climatic catastrophe. We have based our estimates of the size of a minimum sufficient nuclear force on the smallest strategic arsenal that might generate a nominal (Class III) nuclear winter. We found that a global inventory of perhaps a few hundred weapons would suffice—with the U.S. and U.S.S.R. having the largest share: say, 100 weapons each. If the weapons systems are highly reliable and invulnerable to attack, this would still provide, we argued, a robust deterrent. But in discussing Figure 6 (near the end of Chapter 16), we noted that under the most unfavorable conditions of targeting and the remaining uncertainties in the governing physics, as few as 50 weapons explosions might be enough. (For a "mild" [Class II] nuclear winter, still much worse than the events following the 1815 Tambora volcanic explosion, even fewer weapons might suffice.) If so small an inventory were all that was permitted to the nuclear-armed nations, the U.S. and U.S.S.R. might expect only 20 weapons each, or fewer. Because of intrinsic uncertainties in the reliability of nuclear weapons and their delivery systems, so sparse an arsenal might not be considered a secure deterrent. (There are those who have the same reservations about a 100 weapons limit.) Were further research to support such low estimates of the nuclear winter weapons threshold, nuclear winter would then suggest that we move quickly from minimum sufficiency to abolition—in line with Kahn's tradeoff between the reliability of the nuclear deterrent and the likelihood that it constitutes a Doomsday Machine. Deterrence of conventional attack would then require some other, nonnuclear means. But so low a nuclear weapons threshold is, on present evidence, unlikely and we have ignored the foregoing argument heretofore, and in the remaining pages of this book.

22.5:

Although there is no good argument that nuclear weapons discourage conventional war between a superpower and a nuclear-unarmed nation such as Vietnam or Afghanistan; nor among the client states of the superpowers.

22.6:

For example, "In the face of such evidence [on nuclear winter], it is clear that the institution of war is running on empty. It is simply no longer possible for the major powers to achieve anything against each other by means of war. Indeed, even to try is to risk obliterating not only their own futures, but everyone else's too." (Gwynne Dyer, *War* [New York: Crown, 1985].)

22.7:

Among its advocates is President Mikhail Gorbachev:

> If we start orienting ourselves to a "minimal" nuclear deterrence now, I assure you that nuclear weapons will start

> spreading around the world, rendering worthless and undermining even what we can achieve at Soviet-American talks and at talks among the existing nuclear weapons states. [Quoted in M. M. Kaplan, "Towards a Nuclear Weapons–Free World" (paper DEL-023), 6, from the Nehru Memorial Conference "Towards a Nuclear Weapon–Free and Non-Violent World," New Delhi, November 14–16, 1988.]

Kaplan, the former Secretary-General of the Pugwash international conferences of scientists working to end the nuclear arms race, adds,

> Quite apart from the moral objections to reliance on retaliation with weapons of mass destruction, nuclear deterrence is not a reliable means to ensure security, if based on current arsenals and the inherent danger of an accidental war. Nuclear disarmament down to a "minimum deterrence" will still not bring stability; in the long run a new arms race is bound to occur, and more nations are likely to acquire nuclear weapons for their security.

However, he continues,

> the known means of verification can ensure compliance with undertakings to reduce nuclear arsenals down to a very low level, but cannot be made 100 percent effective, a necessary condition for the complete elimination of nuclear weapons [*ibid.*, 5–6].

Ingvar Carlsson, Prime Minister of Sweden, writes to us (private communication, 1989):

> The concept of a minimum nuclear deterrent is reprehensible both morally and in terms of national security. The findings you and Professor Turco have documented so thoroughly show us that even a limited use of nuclear weapons would have catastrophic consequences on a global level. It is a cynical fallacy for any state to base its national security on the premise of a world-wide holocaust. To completely eliminate the threat posed to us all by nuclear weapons, all such weapons must be abolished.

Freeman Dyson (private communication, 1989) argues:

> I think it will be easier and quicker, as well as more desirable, to get a worldwide agreement to scrap nuclear

weapons altogether than to get a "minimum deterrent" agreement. Of course, this is a matter on which reasonable people differ. The important thing is that we seize rapidly the present opportunity for deep reductions, without waiting for the experts to side-track us into arguments about what a minimum deterrent requires.

22.8:

Official Soviet opinion has lately been shifting away from abolition and toward an intermediate period of minimum sufficiency of unspecified duration. E.g., the necessity of "without delay working out a definition of the specific parameters of minimum nuclear deterrence, including tactical nuclear weapons." ("Joint Soviet-Finnish Declaration: New Political Thinking in Action," *Pravda*, October 27, 1989.)

22.9:

E.g., Jonathan Schell, *The Abolition* (New York: Knopf, 1986). For a brief summary of Soviet opinion, see ref. 21.7.

22.10:

Einstein, in an unfinished address drafted after a meeting on April 11, 1955 with Israeli Ambassador Abba Eban and Consul Reuven Dafni in Princeton, New Jersey; in *Einstein on Peace*, edited by Otto Nathan and Heinz Norden (New York: Simon and Schuster, 1960), 641.

22.11:

These themes are developed in greater detail in a forthcoming book by Carl Sagan and Ann Druyan (New York: Random House).

# APPENDIX A

# CLIMATE: THE GLOBAL ENERGY MACHINE

The light and heat of the [Earth] come from the Sun, and its cold and darkness from the withdrawal of the Sun.

—*The Notebooks of Leonardo da Vinci* (New York: Random House, Modern Library edition), p. 21

This night methinks is but the daylight sick.

—Portia, in William Shakespeare, *The Merchant of Venice*, Act 5, Scene 1

**What Is Climate?**

Predicting the weather—especially more than a few days into the future—is, as everyone knows, difficult. Accurate long-term weather forecasts may be beyond the abilities of modern science for a considerable time to come. There are even scientists who think it will be beyond our abilities forever. But climate is not the same as weather, and predicting future climate —while hardly easy—may be within our grasp even now.

Weather is the local state of the atmosphere at any given moment; it varies continuously. Climate can be defined, very loosely, as the average weather over long periods and large regions. So some sort of averaging is always necessary to connect weather with climate. Climate may be expressed as a set of average meteorological parameters, such as the average surface temperature* and average precipitation for a particular

---

* Temperatures in science, as in everyday life almost everywhere in the world, are commonly measured in degrees Centigrade, or °C. This is sometimes also called degrees Celsius and still abbreviated °C. Another measure of temperature is degrees Fahrenheit, or °F—used in everyday life in the United States and about nowhere else. The

month—July, say. The parameters may be given as daily, weekly, monthly, seasonal, annual, decennial, centennial, millennial, etc., time averages. The climate is usually described for regions that are geographically, ecologically, economically, or politically continuous; for example, the Great Plains of North America, the Tibetan Plateau, the Rhine Valley, or New Jersey. Even more broadly, climate parameters averaged over a continent, a hemisphere, or the entire globe may be used in studies of "global change" and climate history. It is mainly in this broadest sense that in the present book we discuss climate and climatic perturbations caused by nuclear war-generated smoke and dust.

### Vulnerability of Life to Climate Change

The climate of our little world has been in a state of continuous change since the earliest development—some four billion years ago—of a stable environment in which life could arise and evolve. Changes in climate have occurred at enormously different rates, from the slow warming (by several degrees Centigrade above previous global average temperatures) in the Cretaceous epoch, occupying 70 million years of geological time, to the abrupt global cooling just at the end of the Cretaceous—now identified with a cometary or asteroidal impact and an ensuing "impact winter" (Chapter 5). The extinction of the dinosaurs and most of the other species then alive may have been caused by this devastating event. On a different time scale, ice ages represent periods of cooling (by about 5°C below the global average) extending over thousands of years—inter-

---

conversion between these two temperature scales is given by the simple relationship:

$$°C = 5/9\ (°F - 32).$$

Equivalently,

$$°F = 9/5\ °C + 32.$$

On the Centigrade scale, the freezing point of pure water is 0° C and the boiling point, 100° C. On the Fahrenheit scale, the corresponding temperatures are 32° F and 212° F. It's easy to see that the Centigrade scale is easier to master. In physics there is also an absolute or Kelvin temperature scale, which starts not from the freezing point of water, but from absolute zero—colder than which nothing can be. Absolute zero is at just about −273°C, so

$$°K = °C + 273.$$

rupted by interglacial warmings of perhaps ten thousand years' duration. On still shorter timescales, there is a 0.2°C variation of inland temperatures with an eleven-year cycle—corresponding to small changes in the Sun's brightness caused by the periodic presence and absence of sunspots. (So, to be readily detectable, any nuclear winter or other climatic change would have to be not much less than about 0.2°C.)

The long-term global climate, as measured by the average surface temperature over the planet, has not varied by more than 10°C from present values during the entire climatic history of the Earth accessible to modern science. Extreme events such as the terminal Cretaceous impact winter have on occasion punctuated the gentle background trend, with disastrous ecological consequences. The prodigious scale of mass extinctions—including, at times, the extinction of up to 90 percent of all existing species—is due in part to a kind of evolutionary complacency: Adaptations to extreme environments tend to be lost by mutation if, under stable long-term climatic conditions, there is no advantage to such adaptations. As immense durations of time pass, life, comfortably adjusted to prevailing conditions, becomes increasingly at risk from sudden environmental change. It is worthwhile noting that human civilization has evolved entirely in an epoch of benign climate. At least some essential building blocks of our civilization—it is natural to think first of agriculture—may therefore be precariously vulnerable, even to comparatively small changes in the climate.

The SCOPE study (ref. 3.11) calls attention to

> the extreme sensitivity of human life on Earth to disruption in the agricultural, economic, and societal bases that maintain populations far above the carrying capacity of natural ecosystems; i.e., the levels possible without any agricultural production. . . . The Earth's human population has a much greater vulnerability to the indirect effects of nuclear war, especially through impacts on food productivity and food availability, than to the direct effects of nuclear war itself.

### Predicting the Future: What Is an Atmospheric "Model"?

Scientists interested in the workings of the natural world (as well as the rest of us) often wish to predict the future. By studying the present and the past, scientists develop theories that purport to describe how the phenomena of interest (the average temperature, say) are forged by the forces of nature. Such theories are initially expressed as general concepts, which are then rigorously formulated as a set of mathematical equations quantifying relationships among various physical parameters—the so-called Navier-Stokes equations, for example, that describe the motions of a viscous fluid such as air. Of course, the defining mathematical expressions are themselves based upon fundamental laws of physics and chemistry (e.g., the perfect gas law, connecting pressure, temperature, and density at any point in the atmosphere). The equations, along with information on the initial state of the system and the values of the key physical parameters, constitute a "model" of the phenomena. The model can be solved analytically (by direct manipulation of the governing equations) or numerically; in the latter case, the equations are rewritten in a form suitable for a digital computer. Solutions of the model in a sense provide predictions of possible future events.

To make sure that the model is giving a realistic picture of the future, its predictions are checked against events that have already occurred. If the model forecasts are not in accord with what has actually happened, then the theory upon which the model is based must be modified, the equations reformulated, and new solutions sought. If, after reasonable alteration and diligent attention to detail, the theory cannot be brought into accord with observations, then it is discarded. The failure of a model when confronted with relevant data usually suggests alternative ways to explain the observed phenomena. Often, the theory initially proposed can account for a large part of the observations, but not every detail. Then it may be refined through small modifications that improve the overall accuracy of the predictions without contradicting the underlying principles. The development of a physical model—particularly for the complex phenomena that characterize the global environment—is an evolutionary and interactive process in which data are continually sought to improve, or disprove, the basic theory.

Imagine, for example, that you—who've slept until noon for the last few months—wish to predict the exact time of sunrise tomorrow. You could guess, based on your memory of the change with the seasons of the time of sunrise last year. Unlike the weather, the time of sunrise on Earth is regular as clockwork, and this would be a reasonable approach even to pretty high accuracy, depending on how good your memory is. Alternatively, you could construct a more precise "model" of sunrise. You would take into account Newton's law of gravity, the motions of the planets around the Sun, the masses of the Sun, the Moon, and the planets and their distances, the rotation rate of the Earth on its axis and the tilt of this axis with regard to the Earth's orbit around the Sun. Based on these concepts, you would then write down the equations that describe the motion of the Earth through space and the position of a point on the Earth's surface in relation to the Sun. You would solve these equations in a convenient way, perhaps on your personal computer, to obtain a precise solution for the times of sunrise at your geographical locale. You would then check your answers against what you knew of sunrise in years gone by. Perhaps you would want to allow for any mountains on your Eastern horizon. If all has gone well you would correctly predict tomorrow's sunrise. If not, you've made a mistake, or you've left something out. Back to the drawing board.

Now suppose that we wish to predict the weather over the next few weeks, or the climate over the next few years. In each case we can develop a theoretical concept, write down appropriate equations, formulate a numerical solution, make predictions, and check their accuracy. If the predictions are inaccurate, we go through the process again with some changes in the assumptions. Atmospheric scientists have been developing models for weather and climate prediction along these lines for the past two decades. These models are extraordinarily complex, requiring all of the number-crunching power of the largest supercomputers to obtain answers of practical utility. But models of very complex environmental phenomena are always approximate; not every physical detail or effect can be included for a place as intricate as the whole Earth.

Sometimes we aren't certain how a particular aspect of the environment works, and we are forced to approximate it or leave it out. But just because a model doesn't include every possible detail doesn't mean that it's inaccurate, much less useless. The accuracy can be checked by comparison of predictions with observations—past, present, and future. Moreover, the basic conceptual (or mathematical) model can be employed to predict the effects of analogous phenomena and checked against observations of those phenomena. (For example, radiative climate models developed for nuclear winter aerosols can be used to calculate the cooling effects of forest fire smoke or windblown dust. Then, when there's a forest fire or a dust storm, we can check the models out.)

There is far greater elegance and utility in an accurate, simple conceptual model than in an accurate, complex one. But to devise good simple models, our insight into the physical world must be deep, and nature must be generous to us. There is no guarantee that simple models will be accurate or even useful.

## One-Dimensional Versus Three-Dimensional Models

Models of the atmosphere can be formulated in three spatial dimensions: up and down (vertically); north and south (meridionally—i.e., along a meridian of longitude); and east and west (i.e., along a parallel of latitude). Typically, the most dramatic variations in parameters such as temperature, pressure, heating rates, particle concentrations, etc., occur in the vertical direction. (The air is much thinner only 20 kilometers up than it is anywhere on the Earth's surface.) Accordingly, many analyses of phenomena that are geographically widespread (e.g., stratospheric ozone depletion, volcanic-aerosol effects on global radiation) are carried out using vertically oriented one-dimensional (1-D) models. On the other hand, for describing winds, precipitation, and weather in general, horizontal (place-to-place) variations are obviously also important, and three-dimensional (3-D) models, taking into account all three spatial directions, are commonly used. One-dimensional models are usually much easier to apply, and their results are easier to interpret. Thus, they have gained widespread acceptance in defining basic concepts and theories, and in exploring the sensitivity of predicted effects to uncertain physical parameters. Ideally, if it were practical, all models would include the effects of three-dimensional motions, and 3-D models someday —perhaps in a decade or two—may well displace 1-D models as the tool of routine exploratory scientific analysis.

It's sometimes thought that if a model used to describe a phenomenon is depicted in more spatial dimensions, then it's necessarily more sophisticated than its lower-dimensional sis-ter models. However, sophistication more often lies in the details of the underlying physics treated in the model, than in dimensionality alone. For example, in describing the radiative properties of atmospheric gases and particles, you need to calculate accurately a number of parameters related to the transfer of radiation through the atmosphere, including the sizes and composition of the aerosols. For this latter problem to be handled adequately you must treat the physics of the aerosol particles in detail, including the effects of settling under gravity, coagulation of the particles by Brownian (random thermal) motions, growth and evaporation of the aerosols, and removal by clouds and precipitation. In the original one-dimensional TTAPS work, and in many subsequent 1-D studies, the details of the aerosol physical and radiative properties were calculated in excruciating detail, to explore the sensitivities and uncertainties inherent in the smoke modeling. In the 3-D work that was later carried out to improve on the TTAPS study, many simplifying assumptions were made about the particles. From time to time, 3-D modelers have assumed that the particles are all the same size; that they don't absorb infrared radiation; that they are uniform in composition; that dust and smoke exhibit similar radiative properties; that particle sizes don't change with time; that the particles have no mass and do not fall out; or that they are completely flushed out of the air by clouds. For many problems such assumptions might be more profoundly limiting than treating aerosols as if they vary only in the (essential) vertical direction. The 3-D modelers have sought to correct as many of these approximations as their computer codes will allow, but even today they have not resolved all the ambiguities of such simplifying approximations.

Nature cannot be perfectly replicated in a computer. Models are used to explore basic phenomena, to better understand how the world works, and to predict how the environment may be altered in the future. One-dimensional models are well suited to treat certain aspects of the physics for which three-dimensional models are ill suited; thus, 1-D models are ideal for exploring the effects of aerosols on the radiation budget and climate, while 3-D models are ideal for defining the geographical and seasonal impacts of radiation changes. The 1-D models account only for vertical movement and vertical variation of aerosols, and hence must employ additional assumptions about horizontal transport. The 3-D models do not fully resolve aerosol properties and thus must employ simplifying assumptions about particle microphysics.

Which is the greater deficiency? In fact, the two approaches are complementary. Each learns from the other. Both types of models predict essentially the same surface temperatures, when care is taken to use the same starting conditions and the same physical parameters in the models. And both kinds of models reach essentially the same conclusion regarding nuclear winter: With the massive smoke injections of a nuclear war, the land surfaces of Earth will cool at an average rate, and to an extent, unprecedented in human history.

We now describe some of the key elements of "models" of the Earth's environment used to understand weather and climate—and nuclear winter.

## Visible and Infrared Light

To understand the climate we must first recognize that there are different kinds of light. All light can be considered as waves, consisting of troughs and crests, like waves in the ocean. But light doesn't need some material, or medium, like water to travel in. Light can propagate through a vacuum. The distance from crest to crest (or trough to trough) is called the wavelength. The kind of light our eyes can detect is called, reasonably enough, visible light—which also turns out to be the kind of sunlight for which the Sun's intensity is greatest. We and our eyes have evolved to utilize sunlight. Now the human eye-brain combination perceives the wavelength of visible light as color. That's what color is. Wavelength is often measured in microns; 1 micron (short for micrometer) is a millionth of a meter. A microbe that measures a micron across is much too small to see with the unaided eye. The wavelengths of visible light are a little smaller than a micron, but most of us have no difficulty detecting color. Red light has a wavelength of almost 0.7 microns, and violet light a little more than 0.4 microns. All the other colors of the rainbow fit in between, at intermediate wavelengths.

But the Sun emits much more than visible light. It puts out light at wavelengths shorter than violet (less than 0.4 microns wavelength): This is called ultraviolet. It also puts out light at wavelengths longer than red (more than 0.7 microns wavelength): This is called infrared. The human eye cannot detect either ultraviolet or infrared light (also called ultraviolet or infrared radiation)—but that's a deficiency in the design of humans. These are as legitimate kinds of light as blue or yellow.

Every object in the universe gives off radiation; the hotter it is, the shorter the wavelength of the radiation. The surface of the Sun shines mainly in the visible, and in the near infrared (about 1 to 5 microns). The Earth, much cooler than the Sun, mainly gives off longer infrared wavelengths (between about 5 and 20 microns). This is sometimes called heat or thermal radiation, because these are the wavelengths emitted by warm objects in our everyday lives. (But really all wavelengths of light can be thermal radiation, corresponding to the temperature of the emitting body.) So the Earth is mainly heated by visible light from the Sun, and mainly cools to space by emitting long-wave, or thermal infrared radiation.

## How Does the Climate System Work?

The Earth's climate is a *system*, a vast engine, its output depending on the interaction among sunlight, the atmosphere, the land, the oceans, and even the life forms of our planet. The principal determining factors are:

(1) the balance between the visible (and near infrared) sunlight that warms the Earth on its daytime hemisphere and the far infrared or thermal radiation emitted to space from the entire planet, which works to cool the Earth;

(2) the reservoirs of heat on the Earth, mainly the oceans; and

(3) the additional warming caused by the greenhouse effect of the Earth's atmosphere.

The Sun is the source of energy that drives our climate system. Our world uninterruptedly receives about 100 quadrillion (100,000 trillion) watts of power from the Sun,* which sur-

* The unit of power, the watt, represents the rate of production or use of energy; that is, energy per unit time. A typical household incandescent light bulb in the U.S. uses roughly 100 watts of power. It radiates

passes by 10,000 times all of the power generated by our global civilization. Of all the Sun's energy intercepted by the Earth, about 33 percent is reflected back to space by clouds, air molecules, and the ground (especially bright desert and snowy or icy surfaces). The rest of the energy (about 67 percent) is absorbed either by the atmosphere (about 22 percent) or by the ground (about 45 percent). (Slightly different percentages are shown in Figure 8, reflecting the time variation and uncertainty in our knowledge of these numbers.) The visible sunlight is converted into heat as soon as it is absorbed.

When you turn toward the Sun and feel its warmth on your face, you are converting sunlight into heat; you are radiating energy in the infrared (invisible to you) back toward space; you are participating in the climate system. You've become a little cog in the great climate engine.

Ultimately, the heat must either be stored somewhere on the Earth or it must escape to space. If heat were not lost, the temperature would rise to intolerable levels in only weeks or months. Fortunately, the atmosphere and surface can get rid of the excess heat by radiating it to space. Then, very quickly, the incoming sunlight and the outgoing thermal radiation come into balance—at some temperature dictated by relatively simple laws of physics:

Every object, even the air around you, radiates energy as heat. *You* certainly do. (This is the principle on which some "night vision" systems are based. The infrared thermal emission of an intruder—cloaked by darkness from our ordinary vision—is detected.) The hotter the object, the more energy it emits. (An intruder with a fever is a little easier to make out.) The flame of a fire is very hot (about 1500°C) and its thermal emission is easily detectable without scientific instruments. The rate at which an object emits energy is proportional to the absolute temperature, T, of the object raised to the fourth power (i.e., $T^4$). Thus, if the absolute temperature of the object is doubled, its thermal emission is not twice as much, but 2 × 2 × 2 × 2, or 16 times as much.

---

visible light (and thermal infrared radiation that we perceive as heat) because of the high temperature to which its tungsten filament is brought by the electricity running through it. A red-hot poker glows because it is hot. So does the filament in the bulb. So does the Sun.

Now we can see how the Earth's climate comes into equilibrium with the Sun. As sunlight is absorbed, the Earth (and its atmosphere) warms. But as the temperature increases, the rate of heat emission to space increases quickly, until the rate of loss of thermal energy equals the rate of gain of energy from sunlight. At this point the temperature remains stable and unchanging; an equilibrium between input and output has been achieved. Once this energy balance is established, it is closely maintained over time even as the heating and cooling rates fluctuate about their average values. However, a sustained shift, up or down, in either the energy input or the energy output leads to climatic drift. If the drift proceeds too far in one direction, a new climatic state may be switched on; for example, sustained cooling over hundreds of years can trigger an ice age lasting thousands of years. In the opposite direction, sustained warming could, under certain circumstances unlikely for the planet Earth, lead to runaway greenhouse heating to broiling surface temperatures—as occurred for Venus early in its history.

The response of the climate system may at the same time be both fast and slow. Consider how temperatures can vary on both sides of a long-term average temperature. The best example of short-term temperature change is the daily (or diurnal) cycle in surface temperatures. The diurnal temperature variation is particularly pronounced in dry desert regions where, during the night, temperatures can fall by 30°C or more from daytime highs, only to increase again with the reappearance of the morning sun. Extremes of seasonal temperature change—from stifling summers to frigid winters—are caused almost entirely by the tilt of the Earth's axis of rotation (measured with respect to the plane in which it orbits around the Sun). The seasons have nothing to do with when the Earth is closer to or farther from the Sun in its annual orbit. Our hemisphere of the Earth—whichever one, Northern or Southern, we happen to live in—is pointed toward the Sun in summer and away from the Sun in winter, because the Earth's axis of rotation stays fixed in space as the planet makes its annual voyage around the Sun. The axis, very nearly, points always to the same region of the sky—which is why the North Star is always overhead at the North Pole. The seasons illustrate how the global climate can change dramatically over periods as short as weeks to months,

FIGURE 8

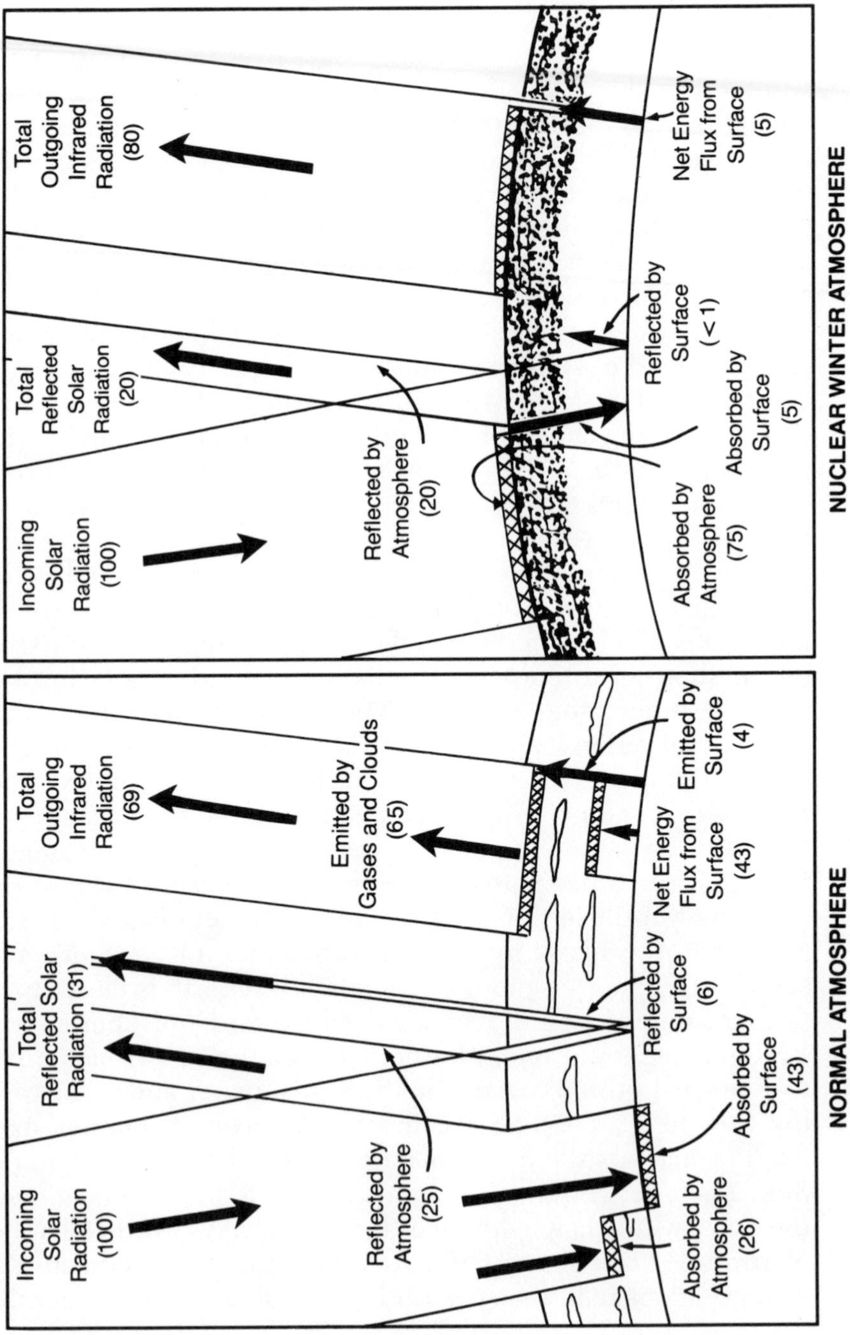

**NORMAL ATMOSPHERE**

Incoming Solar Radiation (100)

Reflected by Atmosphere (25)

Absorbed by Atmosphere (26)

Total Reflected Solar Radiation (31)

Absorbed by Surface (43)

Reflected by Surface (6)

Total Outgoing Infrared Radiation (69)

Emitted by Gases and Clouds (65)

Net Energy Flux from Surface (43)

Emitted by Surface (4)

**NUCLEAR WINTER ATMOSPHERE**

Incoming Solar Radiation (100)

Reflected by Atmosphere (20)

Absorbed by Atmosphere (75)

Total Reflected Solar Radiation (20)

Absorbed by Surface (5)

Reflected by Surface (<1)

Total Outgoing Infrared Radiation (80)

Net Energy Flux from Surface (5)

The energy budget of the normal atmosphere, at left, contrasted with the energy budget of the atmosphere after a nuclear war, at right. The thicknesses of the beams of radiation in which the arrows are embedded are roughly proportional to the energy fluxes—that is, to the amount of energy absorbed or emitted by a given area in a given period of time. You might measure these fluxes in, for example, the number of watts received or emitted per square centimeter of the Earth's surface. For simplicity, we here measure energy fluxes in arbitrary units, so that 100 of these units describes how much sunlight is falling on a given area at the top of the Earth's atmosphere (say, a square centimeter) in a given period of time (say, a second). In the normal atmosphere some of the incoming solar radiation is reflected back to space by clouds and the atmosphere; some of it is absorbed by the atmosphere; and a goodly percentage of it reaches the surface of the Earth, where it is mainly absorbed and heats the ground. The atmosphere and the surface then achieve an energy equilibrium by radiating infrared radiation back to space. But in the aftermath of a nuclear war, a nearly opaque pall of smoke and dust can be generated that prevents most of the sunlight from reaching the surface. Instead, it is mainly absorbed by the smoke and reflected back to space by the dust. The top of the smoke cloud will be much warmer than that region of the atmosphere ordinarily gets; the surface of the Earth will be much colder. In both sketches, "reflected by atmosphere" accounts for the effect of clouds as well as the molecules and particles in the air.

driven by relatively minor changes in the amount of sunlight received. In tropical regions, the received solar flux is less variable with season than at other latitudes, and the variations in seasonal climate are correspondingly small. In polar latitudes, where seasonal variations in sunlight are most extreme—from midnight sunshine in summer to a winter's night lasting for months—the climatic response is likewise extreme.

On longer timescales, it appears that fluctuations of only a few percent in the amount of sunlight received—associated with small variations in the Earth's orbit around the Sun and in the angle of tilt of its rotation axis—may be sufficient to initiate ice ages. (On 10,000 year timescales the Earth's axis does not always point to the North Pole.) Although the periods between ice ages correlate very well with the periods of these changes in the Earth's orbit and axis, the amount of climate change that results is more than we would estimate. The specific mechanisms that amplify these small changes in orbits and tilts into large fluctuations in global climate are still unknown. There is some essential machinery built into the world climate system that still eludes us. Its existence is a warning to us—it suggests that the climatic response to an imposed change in the intensity of sunlight received at the Earth's surface may, in some circumstances, be much more than we calculate.

## Heat Reservoirs: The Climate Flywheel

Not all the sunlight absorbed by the Earth is instantaneously converted to thermal energy and reradiated back to space. Some of it may be stored for a time in the Earth's heat reservoirs —the air, the continents, and particularly the oceans. The atmosphere is not a substantial reservoir of heat because it is so tenuous and, compared with the land or the sea, has a trivial mass (a mere 5 quadrillion tons). The land, which is much more massive, is also an inefficient heat reservoir, because soil and rock are very poor conductors of heat.

But water is an excellent conductor of heat. (Recall how rapidly people can freeze to death in Arctic or Antarctic seas). The oceans of the Earth are about 300 times more massive than the atmosphere. Their surface layers are well-mixed by winds to a depth of 100 meters (300 feet, the length of a football field), which greatly accelerates the transfer of heat through these layers. The surface waters of the world ocean can store an amount of heat roughly equivalent to twenty years' worth of solar energy input. So the Earth's oceans act as a secondary heat *source* driving the climate engine during periods when the primary sources (especially direct sunlight) are, for whatever reason, reduced. The oceans can also provide a heat *sink* when the primary heat sources become, for whatever reason, more intense. These influences of the oceans are felt on time scales of months to decades. The transfer of energy from the oceans to the atmosphere occurs efficiently through the evaporation and recondensation of water.

The role of the oceans in the climate system is similar to that of a flywheel in an automobile engine. The momentum of a massive flywheel smooths out fluctuations in the mechanical forces imposed on the crankshaft and so provides a uniform and steady torque to the drive shaft. Likewise, the thermal inertia of the oceans—hard to warm, hard to cool—smooths out variations in the Earth's radiative energy balance (sunlight and thermal emission to space) to provide a uniform and steady global climate. Up to a point.

## The Greenhouse and Antigreenhouse Effects

The greenhouse effect makes our planet habitable. The basic cause of the greenhouse effect is the fact that the atmosphere is nearly transparent to visible sunlight (except in smog-shrouded cities like Los Angeles or Mexico City), but at the same time is partially opaque to the long-wavelength infrared thermal (or heat) radiation. Solar radiation, as we have said, falls mainly into two spectral intervals—the visible and the near-infrared. (We ignore a smaller contribution in the ultraviolet, which is environmentally important for ozone depletion but not for greenhouse warming.) The total amount of energy radiated to space by the Sun is almost equally divided between visible and near-infrared radiation. Gases in the atmosphere are almost completely transparent to visible radiation, which is why on a clear day we can see mountains that are 100 or 200 kilometers away. But there's more in the air than gases. Clouds can reflect 50% or more of the incident sunlight back into space. The atmosphere and clouds together can absorb 50% or more of the incident near-infrared solar radiation. Nevertheless, *most* of the energy in the solar spectrum that is not reflected back to space reaches the surface of the Earth and is converted there into heat.

The atmosphere contains minor amounts (about 1% and about 0.03%, respectively) of water vapor ($H_2O$) and carbon dioxide ($CO_2$), which are strong absorbers of infrared thermal radiation. As the warm surface of the Earth radiates infrared radiation upward, the overlying atmosphere absorbs part of that radiation, retains it, and prevents the heat from escaping directly to space. Some of the heat is reradiated by the atmosphere in the direction it was going—to space. Some is reradiated back to the surface. The result is that, for the same input of solar energy, the surface becomes warmer than it would had the Earth been an airless world. It is fairly easy to calculate that, if there were no $H_2O$ or $CO_2$ in the atmosphere (but with the same major constituents, nitrogen [$N_2$] and oxygen [$O_2$]), the oceans would be solid ice, and the Earth would be a frozen and lifeless planet.

An analogy to the greenhouse effect is the way that a blanket can keep you warm in an unheated room on a cold winter's night. In this case, the source of energy is your own body heat, generated metabolically. The blanket simply prevents your body heat from escaping into the room; instead, it allows the heat to accumulate in the space between you and the blanket. The Earth's atmosphere is also a blanket, but made of gas. Both your body heat and the warmth of the Earth's surface are powered by sunlight. We use food as a convenient intermediary to store the energy of sunlight, while the Earth is heated by sunlight more directly, but stores most of the heat in the oceans.

The greenhouse effect does not alter the overall energy balance of the Earth. The absorbed solar energy flux is still precisely balanced by the infrared or thermal radiation emitted to space, greenhouse effect or no greenhouse effect. In the presence of a greenhouse effect, though, the emission to space is partly from a warmer atmosphere/surface system, but now also, in part, from cooler layers of infrared-opaque air fairly high in the atmosphere. The climate system knows how to adjust itself rapidly so that the overall balance of energy is maintained.

The greenhouse effect is of fundamental importance to our well-being. Without the trapping of heat by the $H_2O$ and $CO_2$ blanket, the average surface temperature of the Earth would be about 35°C colder—as we have said, well below the freezing point of seawater. On the other hand, too much greenhouse effect and the Earth would be a miserable sweltering hothouse. There is very little carbon dioxide in the Earth's atmosphere; doubling or trebling it (so the $CO_2$ abundance would still be less than 0.1%) is probably enough to produce extremely dire consequences through the disruption of agriculture and ecosystems, and the rising of sea level.

We are beneficiaries of the greenhouse effect. We owe our lives and well-being to a delicate balance of invisible gases established through no effort and no understanding on our part. Now that we do understand, this fragile and providential equilibrium should elicit prudence and humility—prudence about tinkering with the finely tuned engine, and humility in the face of the incompleteness of our knowledge.

The greenhouse effect can be enhanced by increases in the $CO_2$ or $H_2O$ abundance, by additions of other gases such as methane or the chlorofluorocarbons (that, rounding out their malevolent influence, also attack the ozone layer), and by changes in the Earth's cloud cover. If the abundance of greenhouse gases increases, the atmosphere and surface will tend to warm. But the role of clouds in the greenhouse effect is complex and as yet incompletely understood. Depending on the density and height of clouds, they may act either to warm or to cool the underlying surface—although satellite studies suggest that the net influence of clouds and the water vapor that sustains them is to warm the Earth further.

The greenhouse effect can be short-circuited if sunlight is blocked or attenuated in the upper atmosphere while infrared thermal emission to space from the lower atmosphere and surface remains unaffected. We call this the "antigreenhouse" effect. It was first explicitly defined and calculated by us and our TTAPS colleagues in the development of nuclear winter theory [although Fred Hoyle and Chandra Wickramasinghe seem to have been the first to describe a similar effect (due to cometary dust, not urban smoke); see box, "Impact Winter," Chapter 5]. Smoke absorbs sunlight before it can reach the ground, so there is little surface illumination and the surface is poorly warmed; hence, there is less heat to trap beneath the atmospheric greenhouse blanket. If, in addition, the smoke (opaque to visible light) is transparent to the Earth's infrared thermal emission, the surface can still cool itself off by radiating to space just as when there's no smoke. Under such circumstances, surface temperatures could drop sharply.

The actual situation in the atmosphere is more complicated, and careful calculations must be carried out to study the greenhouse and the antigreenhouse effects. The results of such calculations are in substantial accord with the implications of the conceptual models just discussed. However, in the more detailed calculations we can take explicit account of land versus ocean surfaces, the chemical composition of air, the spectral properties of the various atmospheric gases, the reflection and absorption of radiation by clouds, the uneven distribution of solar heating over the globe, and so on. The fact that numerous, mutually consistent detailed calculations have been made of the greenhouse and antigreenhouse effects greatly increases our confidence in the conceptual models. The same concepts, applied to the atmospheres of the Earth and the other planets, yield predictions in excellent accord with what we find there.

### The Influence of Smoke and Dust: Nuclear Winter

The global climatic effects of cometary or asteroidal collisions with the Earth, volcanic eruptions, wildfires, and nuclear war are all expressed through the action of small atmospheric particles—called aerosols—on the radiative energy balance. Particles in the atmosphere can affect the Earth's radiation balance in several ways: by reflecting sunlight back to space, by absorbing sunlight and heating the air, and by absorbing and emitting infrared thermal radiation. In general, a cloud of fine particles tends to warm the surrounding air by intercepting and absorbing sunlight, but it can either warm or cool the underlying surface—depending on whether the particles absorb infrared radiation more readily than they absorb and scatter solar radiation. Generally speaking, aerosols that are strong absorbers in the infrared can produce an additional greenhouse effect, provided they are not highly reflective as well.

The antigreenhouse effect of an aerosol is maximized for particles that are highly absorbing at visible wavelengths and highly transparent at infrared thermal wavelengths. Then, the sunlight is ineffective in heating the surface, but the surface can still cool itself rapidly by emitting energy directly to space.

Much less sunlight reaches the surface when an aerosol consists of black particles such as soot, which efficiently *absorb* visible light, than when it consists of bright particles such as soil dust (mainly silicates), which primarily *scatter* visible light. Soot, which is copiously produced in large fires, particularly of common fuels such as petroleum, is not only a strong absorber of sunlight; it is also a relatively poor absorber of infrared thermal radiation. This is why soot is an ideal aerosol for creating a powerful antigreenhouse effect. Surprisingly small thicknesses of soot (a layer, say, a few microns, or one ten-thousandth of an inch thick) can, if distributed in the air over large areas, turn off most of the greenhouse effect—through the grace of which our planet is comfortably above the freezing point, and to which we owe our lives.

The degree to which an aerosol will cool the surface (by blocking sunlight) or warm the surface (by enhancing the greenhouse effect) also depends on the size of the particles. It turns out that very fine particles—with sizes smaller than about one micron—are best at cooling the surface. One of the most common examples of particles this small is cigarette smoke, which sometimes appears blue because the very small smoke particles (like the molecules of air that make the sky blue) are much more efficient at scattering blue light than any other color (and visible light much more than infrared radiation). Soot particles from raging fires can be as small as the particles in cigarette smoke, but are much more absorbing, which is why soot can go so far in turning off the greenhouse effect.

The influence of aerosol layers on the energy balance also depends on the thickness and density of the layer. The combined impact of all these factors can be expressed, for a given type of aerosol, in terms of a single number, the optical depth of the aerosol layer. As we've mentioned, for materials such as dust or cigarette smoke that absorb very little visible light, the optical depth of interest is for scattering; for materials such as soot that absorb strongly, we use the absorption optical depth (ref. 7.9). This distinction is further developed in the following discussion, and in Figure 9.

If a beam of light is shining down on a layer of aerosols, then the more fine particles in the way, the less light that manages to emerge from the bottom of the layer. Also, the more *absorbing* the particles are, the less light passes through them. Scat-

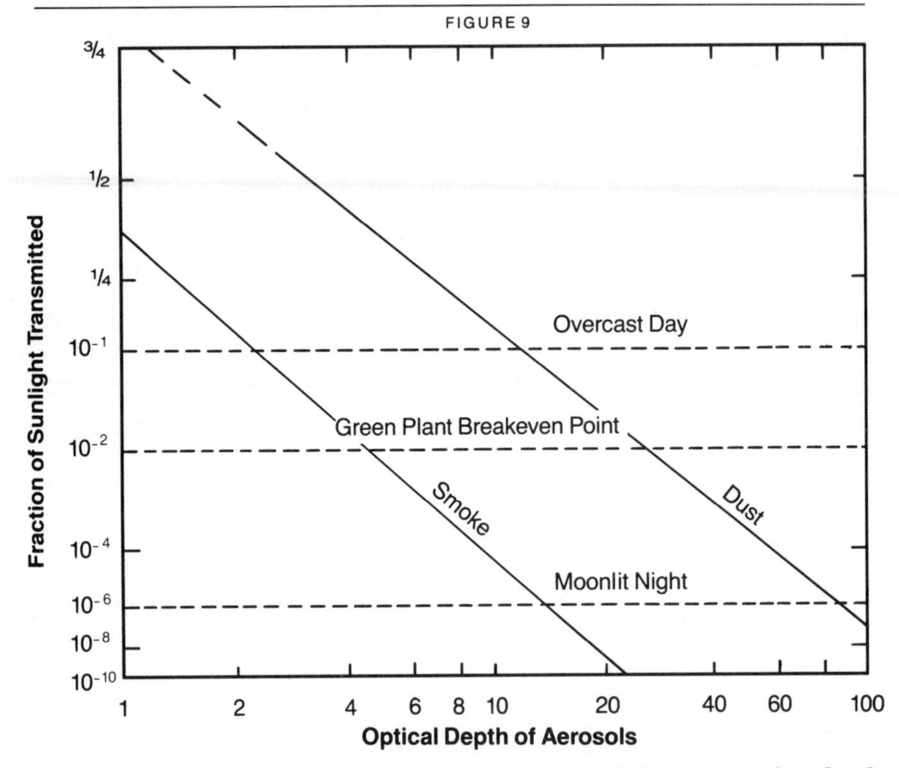


How much light gets through an aerosol haze, or cloud of smoke or dust, depends on the optical depth of the layer of fine particles (ref. 7.9). For optical depth 1, most of the light penetrates through the scattering dust cloud. But less than half penetrates through the darker absorbing smoke cloud (which here is not as dark as pure soot). At optical depth 2, the light that gets through a cloud of soot is only as much as on a gloomy, overcast day in the normal atmosphere. At optical depth 5 for smoke, so little light gets through that green plants can just barely harvest enough sunlight for photosynthesis to keep pace with plant metabolism. At an absorption optical depth greater than 5, photosynthesis grinds to a halt. At optical depth around 12 for smoke clouds, it is as dark at midday as it would be at midnight on a clear night with a full moon in the normal atmosphere. To achieve the same darkness at noon with dust takes an optical depth as large as 80. Clearly, smoke optical depths much larger than 1 can be very dangerous if distributed over large areas of the Earth. In these illustrative calculations, the absorption optical depth (for smoke) is given for the Sun overhead, but the scattering optical depth (for dust) is for the Sun about 35° from the horizon. A light transmission of $10^{-2}$ is 0.01 or 1% the sunlight falling on the top of the atmosphere; $10^{-4}$ is 0.0001 or one ten-thousandth; $10^{-6}$ is one-millionth; etc.

tering doesn't reduce the total amount of energy in the beam (as absorption does); it merely redirects the energy—some of it continuing forward, some of it reflected back. This is why, for the same optical depth, scattering is so much less effective than absorption in reducing the intensity of the beam. How much light will be gone by the time it passes through the layer also depends on how big the particles are. The optical depth measures the overall effectiveness of an aerosol in reducing—through absorption or scattering or both—the intensity of radiation traversing the aerosol (see ref. 7.9). The optical depth also varies with the wavelength of the radiation, and is usually specified at a reference wavelength corresponding to the mid-visible color green. The absorption optical depth of smoke is a convenient indicator of the potential for the smoke to affect the climate, and we so use it in this book.

The influence of absorbing smoke—or scattering soil dust particles—on the transmission of sunlight for a range of optical depths is illustrated in Figure 9. With absorption or scattering optical depths much less than 1, the resulting radiative perturbations are small and only minor climatic effects are expected. For absorption optical depths close to or greater than 1, the resulting radiative energy balance is highly disturbed, since most of the solar radiation would be absorbed in the atmosphere. For aerosols that mainly scatter rather than absorb sunlight, optical depths greater than about 5 are required to cause a similar disruption. In either case—strong absorption or strong scattering—the climate would rapidly shift into unprecedented modes. You can see in Figure 9 how quickly it gets dark as the smoke optical depth increases by a little. Like the Richter Scale for earthquakes, optical depth is a logarithmic scale; as the optical depth increases arithmetically, the diminution of sunlight increases geometrically. If you double the absorption optical depth when it is already more than 1, you do much worse than halve the amount of sunlight that gets through the Earth's atmosphere.

The onset and duration of climatic effects depends on how long the aerosols stay up in the air. The more quickly the particles are removed, the shorter and less extreme are the climatic effects; the more persistent the aerosols, the longer and more intense the climate anomalies. Experience with large-scale forest fires, volcanic eruptions, and dust storms—and simulations with sophisticated atmospheric aerosol models—suggest that, following a nuclear war, large quantities of soot aerosols would be suspended in the upper atmosphere for several months to several years. There is a "bootstrap" effect here that we call self-lofting. The tiny, high-altitude, dark soot particles absorb sunlight and are heated; their high temperature in turn warms the air around them, which rises and carries the suspended particles to higher altitudes, tending to prevent them from falling out or being carried by the air down to the surface. This extends the atmospheric lifetime of the soot and helps create a stable soot layer on a worldwide scale. The sunlight spreads the soot, and the soot blocks the sunlight.

Since ocean surface temperatures require several years to respond to changes in the radiation balance, only relatively small variations in ocean temperatures would be expected. But the temperatures of the land and the immediately overlying air can change much more quickly than the temperatures of the ocean—indeed on a timescale of hours, as everyday experience indicates. When the atmosphere and land cool, stable layers of air develop near the surface that suppress convection and allow even stronger surface cooling. Counteracting this effect is the tendency for water to condense from the air in the form of fogs, dew, and frost, which can for a time inhibit sustained cooling. The heat stored deep in the land also slowly diffuses to the surface, initially slowing the rate of temperature decline, but later slowing the rate of temperature recovery as the soil's heat deficit is recharged.

The most severe nuclear winter physically possible occurs when the entire greenhouse effect is turned off. If the average temperature of the Earth (land and ocean) is normally about 13°C, complete cancellation of the greenhouse effect means an eventual global temperature decline to about −22°C (22 Centigrade degrees below freezing). Such a cooling must last for decades to fully influence the oceans, but only weeks to fully influence large land masses. To the extent that sunlight reaches the lower atmosphere where most of the principal greenhouse gases are, the nuclear winter cooling will be less. A complete undoing of the greenhouse effect has as a necessary, but not sufficient, condition that no sunlight reaches the surface of the Earth at midday. By the foregoing standards, typical nuclear winter global temperature declines of 10 to 20°C correspond to roughly 30 to 60% efficiency (100% efficiency being the maximum cooling permitted by the physics of the atmosphere). But the greenhouse effect can be very far from being totally turned off for a global climatic catastrophe to be worked.

A comparison of various temperature regimes on Earth is given (Chapter 2) in Figure 1. The middle range of nuclear winter effects moves continental climates to a region intermediate between the Earth with its present greenhouse effect and the Earth with no greenhouse effect at all. Were the more severe end of nuclear winter effects to prevail for extended periods of time, most forms of life on Earth would become extinct. There are no calculations we know of that suggest the duration of the acute phase of nuclear winter to last more than a few months—or the chronic phase more than a few years—in which case the thermal reservoir represented by the Earth's oceans would not have time to cool, or the oceans to freeze.

## Summary

The present climate of the Earth is wonderfully optimal for most life on Earth. This is no coincidence—most of those organisms maladapted to the present climate are dead. But our fortunate distance from the Sun—not too close, nor too far away—the extensive, stabilizing oceans, and the modest greenhouse effect, are jointly responsible for an overall climate that is clement and hospitable for life and for our global civilization. Because this worldwide society is so new—agriculture was invented only some ten thousand years ago—and because we never before have had to think of such contingencies, we are unprepared for major, and especially major and rapid, climatic change.

The climate system is disquietingly sensitive to small changes in the energy balance: Over long time periods, the onslaught of the ice ages seems to have been triggered by subtle variations (of a few percent at most) in the Earth's orbital parameters and in the tilt of its axis of rotation.* Such climatic

* Recent research has found that in the interglacials the carbon dioxide abundance in the Earth's atmosphere was high, while in the ice ages it was low. But the changes in greenhouse effect from the varying amounts of $CO_2$ do not account for the extent of these changes in

disturbances, if they came again, would not be by the hand of man. But nuclear winter reminds us that we ourselves can contrive still more catastrophic climatic changes.

A single major volcanic explosion can produce a "year without a summer," induce crop failures, and lead to widespread famine. The impact of a large asteroid or comet with the Earth would, through a variety of direct and indirect environmental effects, threaten the continuing existence of our species and many others. The most likely nuclear winter climate effects fall between these two cases. Our global civilization is precariously propped up by technologies and infrastructures that would not survive a nuclear war. We are sensitively dependent on a benign climate that has not varied significantly over the last hundred years, and only a little over the last few thousand years—but that, in a nuclear winter, would become brutally severe and unpredictable. We are vulnerable.

[For further reading on climate and climate change, we recommend H. H. Lamb, *Climate: Present, Past, and Future* (London: Methuen, 1972–1977); R. Londer and S. H. Schneider, *Coevolution of Climate and Life* (San Francisco: Sierra Club Books, 1984); John Imbrie and Katherine Imbrie, *Ice Ages: Solving the Mystery* (Hillside, N.J.: Enslow Publishers, 1979); G. Genthon, J.M. Barnola, D. Raynaud, C. Lorius, J. Jouzel, N. I. Barkov, Y. S. Korotkevich and V. M. Kotlyakov, "Vostok Ice Core: Climatic Response to $CO_2$ and Orbital Forcing Changes Over the Last Climatic Cycle," *Nature 329* (1987), 414–418; Wallace S. Broecker and George H. Denton, "What Drives Glacial Cycles?," *Scientific American*, January 1990, 49–56; and James F. Kasting, "Long-term Stabil-

---

ancient temperatures. Somehow the small astronomical variations—through some feedback effect we do not yet understand—are driving the temperature changes. Wallace Broecker of Columbia University proposes that there is a fundamental instability in the Earth's atmosphere-ocean system so that, if pushed, it can flip quickly into an ice age mode, taking millennia to recover. Also, between 900 million and 600 million years ago, glaciers seem to have been abundant on all the continents, even those at very low latitudes. James Kasting of Pennsylvania State University remarks, "If . . . there were glaciers in the tropics, then the [world] climate system must have been operating in a wholly different mode than it has throughout most of the Earth's history." How this might come about we simply do not know.

ity of the Earth's Climate," *Palaeogeography, Palaeoclimatology, Palaeoecology 1* (1989), 83–95. The 0.18°C inland eleven-year temperature cycle was discovered by R. G. Currie (*Journal of the Atmospheric Sciences 38*, 1981, 808–818); its explanation in terms of the solar cycle was proposed by J. A. Eddy, R. L. Gilliland, and D. V. Hoyt, *Nature 300*, 1982, 689–693.]

# APPENDIX B

# NUCLEAR WINTER THEORY:

## EARLY PREDICTIONS COMPARED WITH LATEST FINDINGS

. . . hot, cold, moist, and dry, four champions fierce, strive here for mast'ry.

—John Milton, *Paradise Lost*

Here is a compilation and comparison of the early TTAPS conclusions on nuclear winter (numbered 1 through 14) with the most recent results (refs. 3.11, 3.13, 3.14, and references given there), followed by a graphical summary of the work to 1990 of all leading researchers on the problem:

1. Fire ignition by nuclear detonations would be efficient: burnout of flammable materials in cities would be extensive (more than about 50%).

   Latest Findings:

   *Urban Fires*: Nearly instantaneous ignition and flashover occurs in buildings exposed to intense thermal pulse. Frequent ignitions to distances of up to 15 km for a 1-megaton airburst (see frontispiece). Fire burnout models indicate fire spread well beyond the ignition zone in many cases, and nearly complete burnout in the fire zone. Hiroshima and Nagasaki provide direct evidence for such fire effects.

   *Rural Fires*: Ignition of wildlands and particularly croplands is limited (more limited than we first estimated), with a strong seasonal dependence. However, multiburst effects can cause widespread carbonization of vegetation, particularly on missile fields. Delayed wildland fires in tracts of dead vegetation might be important.

2. Urban damage in a nuclear exchange would be extensive: urban fires would contribute most of the sooty smoke.

   Latest Findings:

   Analyses of urban collateral damage in an extended counterforce nuclear exchange range up to 25 to 50% of the total urbanized area in the nations involved. Countervalue targeting would cause at least 50% urban destruction. Less extensive but specific high-priority targeting, such as oil refineries, assures adequate soot production for global climatic effects. Urban fires in general are much sootier than rural fires.

3. Urban smoke plumes would rise into the middle and upper troposphere, some into the lower stratosphere; wildfire plumes would rise up to 5 km.

   Latest Findings:

   *Urban Fires*: State-of-the-art fire plume models predict smoke injections reaching the middle and upper troposphere, with substantial smoke deposition in the lower stratosphere in firestorms.

   *Wildfires*: Observations of modest-sized (100-hectare) forest burns in Canada, and wildfires in California, show plumes typically reaching 5 to 6 km.

4. Prompt rainout scavenging would reduce smoke emissions by approximately 50% in large urban fires.

   Latest Findings:

   Field measurements and laboratory studies show that fresh soot particles (the critical light-absorbing component of smoke) are very poor cloud condensation nuclei (a few percent are active at typical cloud supersaturations). Urban plume/microphysics models predict less than about 10% prompt scavenging in this case. In large-scale fire experiments to date, precipitation removal has been observed on only one or two occasions, with maximum removal of the oily, not the sooty, smoke of about 30%. "Black rain" is occasionally seen in massive wildfires. Hiroshima and Nagasaki "black rains" occurred in humid summer maritime environments; the soot removal efficiency is unknown for these events.

5. Soot production in a nuclear exchange could create absorption optical depths around 1 on a global scale.

   Latest Findings:

   Current estimates for fire ignition and smoke emission indicate most likely hemispherical-average absorption optical depths of 1 to 3 (with possible values ranging down to 0.3 and up to 10) in a full-scale nuclear exchange. While estimates of quantities of flammable materials have decreased, estimates of the light absorption by a given amount of smoke have increased, and the two changes effectively offset one another.

6. Mesoscale (100 to 1000 km) dispersion of smoke and dust clouds would be rapid and does not limit global effects.

   Latest Findings:

   Satellite and aircraft observations of extensive forest fire plumes, volcanic debris, and Saharan dust show rapid dispersal over large geographical regions. Recent mesoscale computer simulations of large smoke plumes show early solar heating and stabilization. Heating and self-lofting of a soot plume from an oil fire has been observed in a field experiment. Localization of individual smoke sources seems to have little effect on large-scale dispersion and subsequent climatic effects.

7. Deep land surface temperature declines would occur beneath massive smoke clouds. TTAPS originally estimated maximum land temperature drops of about 25°C (after correction for ocean moderation and for seasonally averaged sunlight) and mean land temperature drops beneath extended smoke clouds of roughly 15°C for their baseline nuclear war scenario.

   Latest Findings:

   All existing climate model calculations (1-, 2-, and 3-dimensional) show deep land cooling for plausible smoke injections. The smallest average temperature drops over smoke-covered land masses (for the baseline NRC or SCOPE smoke emissions) are approximately 10°C in summer, with drops at some locations to 35°C, averaged over day and night. Significant continental land freezing is evident in summer in many baseline forecasts. The cooling caused by smoke from forest fires is measured to be as large as 20°C after a few days, but only a few °C below thin, mainly scattering haze. These results are consistent with nuclear winter calculations.

8. Nuclear war perturbations could spread rapidly to tropical regions and the Southern Hemisphere.

   Latest Findings:

   The most recent global dispersion computer models show hemispheric-scale dispersion of smoke clouds within two weeks, with significant transport over the tropics on this time scale. For large smoke injections, substantial quantities of smoke reach the Southern Hemisphere within several weeks. Unprecedented transequatorial atmospheric circulation is predicted. Anomalous circulation patterns of this sort are also seen in Martian dust storms.

9. Heating and stabilization of the upper atmosphere would be caused by large smoke injections, leading to prolonged soot lifetimes.

   Latest Findings:

   Global circulation models predict the formation of widespread stabilized air layers in the upper troposphere and lower stratosphere that effectively lower the boundary of the stratosphere (also called the tropopause) by 5 kilometers or more. After several weeks, most of the residual smoke in the atmosphere is trapped in the stabilized region, with an effective residence time of approximately one year. Significant self-lofting of smoke as a result of solar heating is also computed. Thus, smoke need not be injected directly into the stratosphere to reside there eventually. Indirect evidence for stabilization and lofting is found in large historical fires, such as the Alberta, Canada, forest fires of 1950, and direct evidence has been provided by recent large-scale fire experiments.

10. Soot aerosol does not significantly chemically react in the atmosphere.

   Latest Findings:

   Laboratory studies of soot/ozone reactions indicate no significant reduction in the atmospheric lifetime of the most important light-absorbing component of smoke.

11. The global biological and ecological effects of the environmental changes associated with nuclear winter would be disastrous, particularly for humans.

   Latest Findings:

   The SCOPE report, volume II (ref. 3.11), clearly defines the potential impacts of climatic and other anomalies on agricultural production and on human survival.

Agricultural crops are found to be particularly sensitive to decreases in average temperatures, minimum temperatures (frosts), light levels, and rainfall, and can be severely damaged by ultraviolet radiation, toxins, and radioactive fallout. Up to several billion human deaths from starvation have been predicted.

12. Nuclear winter could be triggered by relatively small numbers of warheads and/or megatonnage.

    Latest Findings:

    Certain flammable materials such as petroleum are concentrated at relatively few sites, and are sufficient in quantity, if burned, to cause major global environmental effects. Major urban centers are limited in number and are extremely vulnerable to nuclear attack; enough flammable material is present there to cause nuclear winter.

13. Counterforce nuclear strikes could trigger significant climatic effects.

    Latest Findings:

    The combined quantities of smoke and dust generated by a full-scale "pure" counterforce nuclear exchange could substantially reduce solar energy at the surface over long timespans; anomalous and possibly agriculturally destructive climatic variations might follow.

14. Intermediate timescale radioactive fallout would deliver integrated radiation doses about ten times greater than earlier predictions, amounting to tens to hundreds of rem (i.e., up to 100 times background).

    Latest Findings:

    New calculations confirm the higher dose estimates; the increase in long-term fallout dose is related to changes in the yields of nuclear weapons. Radioactive hot spots could deliver doses ranging up to several hundred rem. Additional local contributions from attacks on nuclear reactors could augment the long-term radioactivity doses from weapons fallout by an additional factor of ten.

FIGURE 10

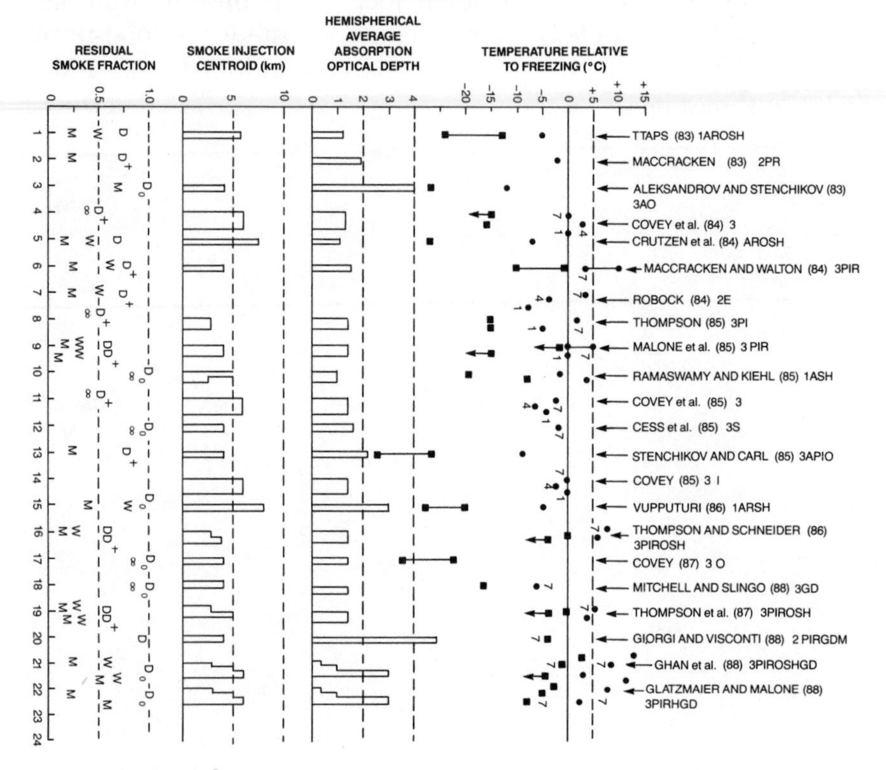


TEMPERATURE:
- ● AVG. LAND UNDER SMOKE (MAX)
- ■ INLAND (MAX)
- # = MONTH OF THE YEAR

TEMPERATURE OFFSET USED =
- 0°C; WINTER
- 13°C; ANNUAL, FALL, SPRING
- 25°C; SUMMER
- 35°C; SOME LLNL CASES

MODEL TREATMENTS:
- 1, 2, 3 = DIMENSIONS
- A = ANNUAL AVERAGE SOLAR INSOLATION
- P = PATCHY SMOKE INJECTION
- I = INTERACTIVE TRANSPORT
- R = REMOVAL BY PRECIPITATION
- O = OPTICAL PROPERTIES EVOLVE
- S = SCATTERING INCLUDED
- H = INFRARED-ACTIVE SMOKE
- E = ENERGY BALANCE
- G = GROUND HEAT CAPACITY
- D = DIURNAL VARIATION
- M = MESOSCALE, 48-HOURS

SMOKE REMOVAL:
- D = DAY (PROMPT REMOVAL)
- $D_o$ = ARBITRARY INITIAL INJECTION
- W = WEEK
- M = MONTH
- + = ASSUMPTION IMPLICIT IN SMOKE SCENARIO ADOPTED
- ∞ = NO SMOKE REMOVAL AFTER INJECTION

Summary of nuclear winter climate model calculations made by many researchers. Data are shown for:

(i) ● —average land temperatures (coastal plus inland) in regions beneath widespread smoke layers for the coldest one- to two-week period in the simulation. (In some reports, only temperature *changes* are given; for them, the absolute temperatures have been deduced by subtracting the computed average temperature decrease from the temperature offset for each season listed in the legend.) The month of the simulation is indicated by a numeral near the filled circle. For the one-dimensional radiative/convective models, the average land temperature decreases are taken as one-half of the "all-land" temperature decreases to account for the effect of ocean moderation. Annual average insolation also applies in these cases. High-altitude dust (as opposed to smoke) is accounted for in the TTAPS models, but in no others.

(ii) ■ —minimum land temperatures beneath smoke (again, where necessary, absolute temperatures were obtained as above). The freezing point of fresh water is 0°C, and the average temperature of the Earth is about 13°C. Most nuclear winter calculations indicate temperatures dropping below freezing over large areas at some time.

(iii) Hemispheric-average absorption optical depth, $\tau_a$, of the total smoke injection assumed in these calculations. Values of $\tau_a$ as high as 5–10 seem possible in a global thermonuclear war (ref. 3.14), but no modern three-dimensional general circulation models have calculated cases of such severity.

(iv) Height centroid of the smoke mass injection (cf. Fig. 2).

(v) Residual smoke fractions at several times (e.g., after a day, a week, a month) in each simulation. A value of 1 means no smoke removed; 0 means all smoke removed.

The legend defines the symbols used. Data have been obtained from the references noted; some values were estimated using related published information. Certain data, missing in this figure, were unavailable. The calculations shown roughly correspond to recommended "baseline" smoke injection scenarios; less severe and more severe cases have been investigated, but not as frequently as baseline cases. The studies have been ordered roughly in chronological sequence. For a given study, several cases may be illustrated. For more detail, see reference 3.14.

# APPENDIX C

# A SHORT HISTORY OF THE TTAPS NUCLEAR WINTER STUDY

> Man's mind cannot grasp the causes of events in their completeness, but the desire to find those causes is implanted in man's soul.
>
> —Leo Tolstoy, *War and Peace* (1868), XIII, 1

As is true of many discoveries in science, our findings on nuclear winter were the result of a long preparatory effort—during which the environmental consequences of nuclear war were far from our thoughts. We were absorbed in other matters, including the exploration of nearby worlds. Carl Sagan's 1960 doctoral thesis at the University of Chicago was mostly on the Venus greenhouse effect. Radiotelescopes had shown that the planet Venus was an unexpectedly bright emitter of radio waves. After examining a range of alternative possibilities, Sagan argued that the only explanation that made sense was that the surface of Venus was very hot, and calculated that a greenhouse effect involving massive amounts of carbon dioxide and small amounts of water vapor might explain the high temperatures. Mainly at Harvard University in the middle to late sixties, Sagan and his first graduate student, James B. Pollack, extended and refined these results. Pollack's Harvard doctoral thesis also was devoted to the planet Venus, and he later performed the first Venus greenhouse calculations involving detailed synthetic spectra. The series of Soviet *Venera* spacecraft directly confirmed that the surface of Venus was indeed very hot, and the American *Pioneer Venus* probes in 1978 confirmed that an atmospheric greenhouse effect in which carbon dioxide and water vapor played major roles was indeed responsible. Both scientists have studied the greenhouse effect on a number of other worlds. Since a main mechanism by which nuclear winter works is the turning down of the greenhouse effect, this research was a kind of unconscious preparation for nuclear winter.

In 1971 the United States launched *Mariner 9*, a robotic spacecraft that became the first artifact of the human species to orbit another planet. Sagan and Pollack were members of the NASA imaging team, with responsibilities in mission design and in the interpretation of television pictures radioed back to Earth. But when *Mariner 9* arrived at Mars in mid-December, 1971, it found a planet about as interesting—at least to the cameras—as a tennis ball (but without the seams). There was no detail anywhere, just a featureless disk. Mars was enveloped in a great global dust storm that lasted for months. In Pasadena, California, at NASA's Jet Propulsion Laboratory, imaging team members were cooling their heels. In this period, Sagan, Pollack, and their coworkers used *Mariner 9* to take and interpret the first close-up pictures of Phobos and Deimos, the two moons of Mars. But until the dust storm cleared there was very little else to do.

Other scientific instruments aboard the spacecraft were getting useful data, however—among them the IRIS (*I*nfra*r*ed *I*nterferometric *S*pectrometer) instrument. By measuring the intensity of infrared radiation at various wavelengths received from Mars, IRIS was able to determine temperatures at various levels in the Martian atmosphere. It found that the atmosphere of Mars—especially where the dust was intercepting sunlight—was warmer than expected, and that the very surface was colder than expected. Sagan and Pollack attempted some elementary calculations to see if they could understand this result—because the bigger the temperature anomalies, the more dust is required to explain them, and the more dust in the atmosphere, the longer it would be before the cameras could see the surface. When the dust storm finally cleared in March 1972 and the wonders of the Martian landscape were revealed, these calculations were put aside. But they played an important role in the evolution of our thinking.

In 1975, Owen B. Toon received his doctorate in physics from Cornell University, with Sagan as his thesis adviser (the dissertation was on climatic change on Mars and Earth); he then went to NASA's Ames Research Center to work with Pollack. Toon, Pollack, and Sagan collaborated on a number of studies on the influence of fine particles on the atmosphere and climate of the Earth. They successfully calculated the roughly 1°C hemispheric temperature drop resulting from large volcanic explosions. Sophisticated optical and climate (energy balance) models were developed in this work and were soon used to analyze the sulfuric acid clouds of Venus and windblown Saharan dust. The three also collaborated in a study, published in *Science*, of some influences of human technology on the Earth's climate—including the burning of forests in slash-and-burn agriculture.

By this time, Richard Turco had joined the Ames team. He spent eight months at Ames in 1971 as a National Research Council postdoctoral fellow, after graduating from the University of Illinois with a doctorate in electrical engineering and physics. One of Turco's first projects at Ames was to develop, with Pollack, a model for the evolution of the Venus atmosphere and its greenhouse over the past 4.5 billion years. Turning toward Earth, Turco constructed one of the earliest detailed models of the stratospheric ozone layer. In the fall of 1971, Turco left Ames to join R & D Associates, then in Santa Monica, California, as a Defense Department consultant. RDA had been spun off the Rand Corporation, a Defense Department "think tank" at which much of the original formulation of U.S. strategic policy and weapons systems had been performed. But Turco continued to collaborate with the NASA scientists at Ames on ozone and aerosol problems.

Beginning in 1975, Turco joined Toon, Pollack, and other Ames-based scientists in constructing a detailed aerosol microphysical computer model, unique in the scientific community at that time, to study stratospheric particles, meteoric debris, volcanic eruption clouds, and clouds so high up that they are still in daylight well after night has fallen on the ground below. These optical/climate, aerosol, and ozone models were all later applied to the nuclear winter problem. Turco, Toon, Pollack, and others, using their new aerosol model, showed that the effluents from even a very large fleet of supersonic aircraft would have little effect on the Earth's climate, contrary to some then-current opinion.

In the late 1970s Sagan was making the *Cosmos* television series and writing the accompanying book. The last episode/

chapter, called "Who Speaks for Earth?," was centrally devoted to the nuclear arms race. In this period Sagan proposed to Pollack and Toon that they join in a study, along the lines of their previous collaborative work, of the effect of nuclear war–generated dust on the Earth's climate.

In 1980, Luis Alvarez and co-workers at the University of California at Berkeley hypothesized that the extinction of the dinosaurs 65 million years ago, at the end of the Cretaceous Epoch, was caused by an asteroid or cometary impact that generated a massive global dust cloud. To discuss the general problem of asteroid impacts with the Earth, and the physical and biological consequences of such impacts, a meeting was convened in Snowbird, Utah, on October 19–22, 1981, under the auspices of the Geological Society of America. A presentation was delivered by Toon on the response of the atmosphere to the dust lofted by a large impact, and conclusions were drawn regarding possible effects on climate and life. Two papers dealing with the microphysical evolution and climatic effects of the Alvarez dust cloud were subsequently written by a scientific team that included Toon, Pollack, and Turco.

Attending the Snowbird meeting were two National Research Council (NRC) staff members—Lee Hunt and Adm. William Moran (U.S. Navy, ret.). On the basis of Toon's presentation of new evidence on the environmental significance of massive dust clouds, they decided that a more careful look at the problem of dust raised by nuclear explosions was in order. They arranged an ad hoc meeting at the National Academy of Sciences (NAS) on April 6, 1982. The NRC asked Toon and Turco if they could make some preliminary estimates of nuclear dust climate effects.

Turco's work at RDA had led him to study the physical effects of nuclear weapons explosions, and such issues as nuclear and conventional command, control, and communications; systems design and deployment; countermeasures; and war-fighting policy. The company always had close ties to the Defense Nuclear Agency, for which it had developed a range of analytic models on nuclear explosions. Turco's experience on nuclear weapons effects, acquired in more than a decade's work at RDA, was helpful when the TTAPS team began to construct the original nuclear winter model.

By 1982 Turco was working on the military implications of

the dust injected into the air in a nuclear war. He had access to a variety of unclassified information, some of it unique, on the quantity and size distribution of the soil particles raised by nuclear explosions. The nuclear dust data base was used to estimate the optical properties of a hemispheric-scale cloud produced by 10,000 megatons of high-yield surface detonations (a scenario originally adopted by the NRC in its 1975 Report, *Long-term Worldwide Effects of Multiple Nuclear-Weapon Detonations*). Toon and Thomas Ackerman—a young expert on radiation and climate who had recently joined NASA Ames—then made rough estimates of surface temperature drops for the nuclear war case by scaling against the earlier "dinosaur" predictions on obscuring dust. The results were surprising. Major optical and climate perturbations were found to be possible for such a war, in contradiction to earlier findings of the NRC (*ibid.*). The Toon/Ackerman/Turco estimates were reported by Turco at the April 6 NAS meeting.

On March 23–25, 1982, two weeks before the April 6 NAS meeting, a workshop on atmospheric infrared radiation had been held at the Kaman/TEMPO Company, a defense contractor in Santa Barbara, California, under sponsorship of the Defense Nuclear Agency (DNA). In attendance were Turco and Eric Jones (a Los Alamos National Laboratory scientist who had also been invited to join the ad hoc NAS panel). Turco and Jones met and talked with Fred Fehsenfeld (a National Oceanic and Atmospheric Administration scientist who is a colleague and friend of Paul Crutzen's). Fehsenfeld gave Turco and Jones preprints of the paper by Crutzen and John Birks (*Ambio 11* [1982], 114) that contained startling estimates of the smoke emissions in a nuclear war; it proposed that major optical perturbations could result.

Turco and Jones brought the Crutzen and Birks work to the attention of the April 6 NAS panel meeting. Because the panel believed it had identified several potentially serious implications of nuclear dust and smoke, a letter was drafted to Frank Press, President of the Academy, urging further action on the problem, possibly in cooperation with a Defense Department agency such as the Defense Nuclear Agency, and possibly involving classified data. But it was nearly a year before an Academy Committee was convened to begin an in-depth assessment of the questions raised in the letter to Press. (The first official panel meeting occurred on March 7–8, 1983, by which time the main TTAPS results were already in hand.)

Meanwhile Sagan—having emerged from his *Cosmos* responsibilities and the Saturn flybys of the *Voyager 1* and *2* spacecraft—contacted his former graduate students Pollack and Toon early in 1982 to ask them again to join him in studying the climatic consequences of nuclear war. At a meeting on the origin of life held at Ames Research Center, Pollack, Toon, and Sagan met to discuss the question. Because Turco, Toon, and Ackerman were by now already involved with the NAS ad hoc committee, it was decided that all five scientists would join forces in an informal research effort. In this same period Sagan had also heard—from Joseph Rotblat of the University of London—about the forthcoming Crutzen/Birks study, and it was decided to investigate the effects of smoke as well as dust.

Following the April 6, 1982, NAS meeting, the TTAPS workers expanded their effort to define more precisely the smoke and dust problem. The Crutzen and Birks calculations of smoke emissions were preliminary, and what the resulting climatic and environmental effects would be had been left unresolved. By the fall of 1982, the newly instituted TTAPS team had developed a methodology for carrying out extensive calculations for various nuclear war scenarios and for testing how the results depended on other choices of imperfectly known parameters. A large data base had also been assembled. Specific topics considered during this period were:

properties of urban fires (Crutzen and Birks had based their analysis on forest fires, which proved to be climatically much less important);

types and amounts of flammable materials in cities;

composition and optical properties of smoke;

toxic compounds released by urban fires;

heights of smoke plumes and extent of "black rain";

physical and optical properties of nuclear-war-generated dust;

nuclear-war/volcanic-eruption analogs;

oxides of nitrogen and water vapor content of nuclear clouds;

intermediate timescale radioactive fallout;

nuclear-war-related ozone depletion and resulting ultraviolet exposure;

dust and smoke particle microphysical processes;

perturbations of atmospheric visible and infrared radiation fields;

changes in air temperatures and light levels;

meteorological implications of smoke and dust injection;

interhemispheric transport of nuclear-war-generated aerosols;

sensitivity of effects to the size of nuclear arsenals and targeting;

sensitivity of climatic perturbations to uncertainties in physical parameters; and

analogs of nuclear winter in phenomena on other planets, especially Martian dust storms.

Because of our long (although inadvertent) preparation for this study over many years, and because of the TTAPS team's access to the Cray computer at NASA's Ames Research Center, we were able to make very rapid progress. A paper was prepared for presentation of our initial findings at the 1982 fall meeting of the American Geophysical Union (AGU) in San Francisco, and a summary was published in *EOS*, the *Transactions* of the American Geophysical Union. However, at the last moment senior managers at the Ames Research Center insisted that there should be no verbal presentation of the new results at the AGU meeting. Although the proposed work had been laid out to Ames management before we proceeded and subjected to internal review afterward, the claim was made by the

Ames Director, Clarence Cyvertson, and his assistant, Angelo Gustafero, that the paper had not received adequate internal review. They also admitted to being concerned about the political implications of the results. As was explained to one of us, "Two weeks ago some nut tried to blow up the Washington Monument; last week the Senate killed the MX missile [a short-lived death, as it turned out]; and this week you want me to be responsible for telling the President that his whole nuclear strategy is wrong?"

As nearly as we can tell, this was not a case of a government official in Washington calling up Ames officials and telling them to prevent the paper from being presented at a scientific meeting; it was voluntary censorship—out of concern for what might happen to Ames in the climate of the early Reagan years, *if* the paper were presented. This was shortsighted, of course, because it would have been politically much worse for NASA if it were known that it was trying to keep from the American people a discovery on the dangers of nuclear war. Press reports on the withdrawal (for example, "NASA Withdraws Presentation," *Aviation Week and Space Technology*, December 20, 1982, 67) caused some consternation. James Beggs, who was then NASA Administrator, in discussions with one of us, understood the issue very well and promised that continuing research on this problem would be permitted and access to the Cray computer maintained.

The Ames management then requested an independent scientific review of the work, and also established an internal review committee of three senior scientists. They independently reviewed the TTAPS study on two separate occasions, in March and August 1983, and what was subsequently published benefited from these reviews.

During the winter of 1982/1983, however, middle-level managers at NASA headquarters in Washington had become concerned about funding work that they felt was outside the mandate of NASA. Accordingly, they reduced the research budget of the Ames TTAPS workers by $40,000 to inhibit further research on the climatic effects of nuclear war. At this point, however, most of the work had been completed. But the Ames management, acknowledging a request by the National Academy of Sciences for completion of the research, allowed *internal* Ames funds to be spent on the nuclear winter research.

There was no monolithic position at NASA on the TTAPS research but, instead, many different voices on how best to serve the nation—and even the planet.

In 1983 Turco participated in a lengthy seminar at RDA, attended by the company's leading scientists, that discussed key aspects of the nuclear winter theory. The review session provided another early indication that no important errors had been made in the formulation of the theory.

Meanwhile, beginning in June 1982 a group of environmentalists and foundation executives had concluded that inadequate attention was being given to the potential environmental consequences of nuclear war. Environmental organizations had raised public consciousness on many local, regional, and global hazards, but had somehow neglected the most serious hazard by far, nuclear war. The group asked Carl Sagan to join them, and only then discovered the ongoing TTAPS research. This led to the creation of a Steering Committee—chaired by George M. Woodwell of the Marine Biological Laboratory, Woods Hole, Massachusetts—to consider the possibility of holding

> a major public Conference so that the TTAPS study and the biological findings on the consequences of nuclear war could be made available to educators, scientists, business executives, public officials, and other citizen leaders and representatives of other nations, as well as environmentalists. . . . At the suggestion of Dr. Sagan, arrangements were made to have the TTAPS paper undergo peer review at a meeting of eminent physical scientists. The data would then be shown to a large number of expert biological/ecological scientists so that they could consider how extensive the long-term worldwide impacts would be on humankind, as well as on the planet's life-support systems. It was understood that only if the data held up after peer review would the proposed public Conference be scheduled. . . .
>
> In late April, 1983, approximately one hundred scientists from the United States and other countries met for the peer review process at the American Academy of Arts and Sciences in Cambridge, Massachusetts. The invited scientists represented a broad variety of fields. At the first meeting, organized and chaired by Dr. Sagan (who was still recovering from the near-fatal aftermath of an appendectomy performed the previous month), about forty phys-

ical scientists and ten biological scientists considered and evaluated the preliminary draft of the TTAPS study. The group generally agreed with the conclusions of the report as to the potential for substantial reductions in the amount of solar light reaching the Earth's surface and for severe climatological changes, although suggesting minor adjustments. . . . [This was then followed by a preliminary meeting on the biological consequences.]

With the assurance from the assembled scientists that the analysis was valid, and that the conclusions had to be taken very seriously, the Steering Committee decided to go ahead with plans for the Conference, and thirty-one national and international scientific, environmental, and population organizations or institutes agreed to help sponsor it. [From Preface, *The Cold and the Dark: The World After Nuclear War*, by Paul R. Ehrlich, Carl Sagan, David Kennedy, and Walter Orr Roberts (New York: W. W. Norton, 1984).]

In preparation for the closed April 22–23 and 25–26, 1983, meetings, TTAPS prepared a detailed description of the findings, which, because of its unmarked blue cover, became known as "The Blue Book." It was distributed to about 150 scientists for review and comment, including those who were to attend the intensive critical review meetings in Cambridge. (It was at these meetings that the acronym TTAPS was first coined by Dr. Newell Mack of Harvard University.) The letter of invitation to these meetings read in part,

From the physical sciences community, we are especially concerned to receive critiques on errors of omission or commission; evaluations of whether the full range of significant parameterization has been employed; and suggestions for rough order-of-magnitude calculations and simple physical insights which might help to clarify the analysis. . . . We are, of course, keenly aware that the public has a significant right to know about this issue, but are concerned that the premature discussion of these results before they are critically reviewed may lead to misinterpretations and misuse. We therefore ask you to exercise all reasonable precautions against general release of the contents of this paper. . . . This is a difficult multidisciplinary problem of pressing worldwide importance. We very much appreciate your help.

Following the review meeting in Cambridge, the TTAPS report was condensed and on August 4, 1983, submitted as an article to the journal *Science*. As is typical for the refereed scientific literature, the editors of *Science* submitted the paper for critical review to three experts whose identities were not revealed to the authors. After the referees' comments were received, the paper was revised and accepted for publication. It appeared in the December 23, 1983, issue of *Science*.

The term "nuclear winter" had been coined by Turco in the original TTAPS Blue Book report. Our attachment to it grew when we discovered, after an official but last-minute NASA review of the paper in press in *Science*, that it was in NASA's view impermissible to include such phrases as "nuclear war" or "nuclear weapons" in the title. These prohibitions applied to our co-authors and not to ourselves, but obviously we needed a paper acceptable to all its authors. NASA seems to have worried that some bureaucrat in the White House or the Office of Management and Budget, idly flipping the pages of *Science* magazine, might become enraged on discovering nuclear war being thought about in an unapproved agency. The phrase "nuclear winter," we gather, was expected to pass through this filter.

On October 31 and November 1, 1983, the public Conference was held—called "The World After Nuclear War: The Conference on the Long-Term Biological Consequences of Nuclear War." The proceedings of this Conference were published in the book *The Cold and the Dark* (*ibid.*), which includes a transcript of the discussion between American scientists in Washington and Soviet scientists in Moscow. The authors of this book participated in the Washington Conference, but our colleagues Brian Toon, Tom Ackerman, and Jim Pollack—strongly discouraged by NASA management even from attending—did not.

Members of the TTAPS team continued to play a role in the National Academy of Sciences study (ref. 3.10) and in the massive international SCOPE study (ref. 3.11). Sagan initiated a major effort, beginning with his *Foreign Affairs* article (ref. 2.3), to explore the policy implications of nuclear winter and to make the nuclear winter findings known to government leaders and the public worldwide. After Congress mandated that the Department of Defense (DoD) investigate the nuclear winter problem, support for continuing TTAPS research in the years following 1983/1984 flowed mainly from the DoD's Defense Nuclear Agency (which strangely has the same acronym, DNA, as the central molecule of life on Earth). Toon and Ackerman were funded to carry out more sophisticated climate studies. Turco, in part because of his past ties to DNA, became a key technical adviser for the development and overview of DNA's "Global Effects Program" (a DoD euphemism for nuclear winter)—which ranged from predicting global climate change to simulating thunderstorms induced by large fires, and which encompassed a variety of numerical, laboratory, and field experiments. As we have stressed elsewhere in this book, though, the research program was never adequate to the seriousness of the problem. It is now essentially defunct.

But the DNA program dominated nuclear winter research. It has, in essence, been largely responsible for supporting the confirmation of the basic TTAPS theory—this despite the fact that high officials in the DoD had perceived the nuclear winter thesis as a threat to existing policy, and had gone to great lengths in attempting to discredit it. Again, we discern many different voices in the federal bureaucracy.

In subsequent years, the TTAPS team has remained active in the science and policy of nuclear winter. Because of the background and planetary perspective of the researchers and the way in which study of other planets has fed back into nuclear winter, we believe this is a good example of the practical benefits for life on Earth of scientific exploration of the planets.

Nuclear winter ideas are now also cycling back into planetary science: Could major impacts of asteroids or comets with other planets have temporarily turned off their greenhouse effects? Was there a time in the history of the Earth when the impact flux was so high that a permanent pall of fine dust enveloped the Earth, erasing the greenhouse effect for hundreds of millions of years? (Carl Sagan and David Grinspoon, "Was the Early Earth Shrouded in Impact-Generated Dust?," *Bulletin of the American Astronomical Society 19* [3], 1987, 892.) Could the environments of other planets be made more Earth-like by artificially generating a planet-enshrouding cloud layer, modulating the local greenhouse effect? Might the global warming on Earth produced by the increasing greenhouse effect be controlled by an artificially generated dust pall? (Carl Sagan,

"Planetary-Scale Ecotechnology: The Honda Prize Address for 1985" [Tokyo: The Honda Foundation]; James Pollack and Carl Sagan, "Planetary Engineering," in preparation, 1990.)

Of course, it is not necessary to have a nuclear war in order to put fine particles into the Earth's atmosphere, and no one is proposing nuclear winter as the answer to greenhouse warming. But might we maintain a carefully controlled amount of fine atmospheric aerosols worldwide so we do not have to find an alternative to the global fossil fuel economy? Our tentative answer—independent of how the fine particles are put up there—is no. Within the limits of our present knowledge, such a technological "fix" seems too uncertain and too dangerous. We will simply have to make our hard choices down here on Earth. But this debate—on another possible practical application of nuclear winter theory—is likely to continue.

# PERMISSIONS ACKNOWLEDGMENTS

Grateful acknowledgment is made to the following for permission to reprint previously published materials:

*American Broadcasting Company:* Excerpts from "Viewpoint," broadcast on November 20, 1983, and "Nightline," broadcast on July 18, 1984. Copyright © 1983, 1984. Reprinted by permission of ABC News.

*Michael J. Altfield and Stephen J. Cimbala:* Excerpt from "Targeting Nuclear Winter" by Michael F. Altfield and Stephen J. Cimbala, originally published in *Parameters*. Reprinted by permission of the authors.

*American Association for the Advancement of Science:* Excerpts from "Long-Term Biological Consequences of Nuclear War" by Paul R. Ehrlich, "Mutual Deterrence or Nuclear Suicide," and "A Run Worth Making" from *Science* Magazine. Copyright © 1983, 1984 by the American Association for the Advancement of Science. Reprinted by permission.

*Christopher Anvil and The Scott Meredith Literary Agency, Inc.:* Excerpt from *Torch* by Christopher Anvil. Copyright © 1957 by Street & Smith. Renewed by Davis Publications, Inc. Reprinted by permission of the author and the author's agent, Scott Meredith Literary Agency, Inc.

*The Atlantic Monthly:* Excerpt from "Einstein on the Atomic Bomb" from the November 1945 issue of *The Atlantic Monthly*. Copyright 1945. Reprinted by permission.

*Bilingual Press/Editorial Bilingue and Anvil Press Poetry Limited:* Excerpts from "Songs of the Fallen" from *Flower and Song, Poems of the Aztec Peoples,* translated by Edward Kissam and Michael Schmidt. Rights throughout Canada and the Open Market are controlled by Anvil Press Poetry Limited. Reprinted by permission of Bilingual Press/Editorial Bilingue, Hispanic Research Center, Arizona State University, Tempe, Arizona 85287 and Anvil Press Poetry Limited.

*Bucknell University Press:* Excerpts from "Odes" by George Santayana. Reprinted by permission of Bucknell University Press.

*Bulletin of the Atomic Scientists:* Excerpts from "How Dangerous Are Atomic Weapons" by Edward Teller (February 1947) and "Science and Our Times" by J. Robert Oppenheimer (September 1956). Copyright 1947, © 1956 by The Educational Foundation for Nuclear Science, 6042 South Kimbark Avenue, Chicago, IL 60637, USA. Reprinted by permission of the *Bulletin of the Atomic Scientists*, a magazine of Science and World Affairs.

*Columbia Pictures, Inc., and Stanley Kubrick:* Excerpts from the screenplay of "Dr. Strangelove" by Stanley Kubrick, Peter George, and Terry Southern. Copyright © 1963 by Columbia Pictures, Inc. Reprinted by permission of Columbia Pictures, Inc., and Stanley Kubrick.

*Doubleday, a division of the Bantam, Doubleday, Dell Publishing Group:* Excerpts from *Mandate for Change: The White House Years* by Dwight Eisenhower and "Spring Prospect" by Du Fu, translated by Greg Whincup, from *The Heart of Chinese Poetry*. Copyright © 1987 by Greg Whincup. Reprinted by permission of Doubleday, a division of the Bantam, Doubleday, Dell Publishing Group.

*Foreign Affairs:* Excerpt from "To Cap the Volcano" by McGeorge Bundy (October 1969) and "The Danger of Thermonuclear War" by Andrei Sakharov (Summer, 1983). Copyright © 1969, 1983 by The Council on Foreign Relations, Inc. Reprinted by permission.

*Harcourt Brace Jovanovich, Inc.:* Excerpts from *The Absolute Weapon: Atomic Power and World Order* by F. S. Dunn, B. Brodie, A. Walfer, and W. T. R. Fox. Copyright 1946 by Yale Institute of International Studies. Copyright renewed 1974 by Yale Concillium on International and Area Studies. Reprinted by permission of Harcourt Brace Jovanovich, Inc.

*Harcourt Brace Jovanovich, Inc., and Faber and Faber, Ltd.:* Two lines from "Little Gidding" from *Four Quartets* by T. S. Eliot. Copyright 1943 by T. S. Eliot. Copyright renewed 1971 by Esmé Valerie Eliot. Rights throughout the world excluding the United States are controlled by Faber and Faber, Ltd. Reprinted by permission of Harcourt Brace Jovanovich, Inc., and Faber and Faber, Ltd.

*Harper and Row, Publishers, Inc.:* Excerpt from *Famous Fires* by Hugh Clevely. Copyright © 1957 by Hugh Clevely. Copyright renewed. Reprinted by permission of Harper and Row, Publishers, Inc.

*Hill and Wang, a division of Farrar, Straus, Giroux, Inc.:* Excerpts from *On the Brink: Americans and Soviets Reexamine the Cuban Missile Crisis* by James G. Blight and David A. Welch. Copyright © 1989 by James G. Blight and David A. Welch. Reprinted by permission of Hill and Wang, a division of Farrar, Straus and Giroux.

*The International Institute for Strategic Studies:* Excerpt from "Targeting for Strategic Deterrence" by Desmond Ball (Adelphi Paper 185). Reprinted by permission of The International Institute for Strategic Studies.

*Joint Publishing (H.K.) Co. Ltd.:* Excerpt from *Reverberations: A New Translation of the Complete Poems of Mao Tse-tung* by Nancy T. Lin. Copyright © 1980 by Joint Publishing Company (H.K.) Co. Ltd. Reprinted by permission.

*Alfred A. Knopf, Inc.:* Excerpts from *The Fate of the Earth* by Jonathan Schell. Copyright © 1982 by Jonathan Schell. Reprinted by permission of Alfred A. Knopf, Inc.

*William Morrow and Company and The Longman Group UK:* Excerpts from *The Longman History of the US* by Hugh Brogan. Copyright © 1986 by Hugh Brogan. Rights throughout the world excluding the United States are controlled by The Longman Group UK. Reprinted by permission of William Morrow and Company and The Longman Group UK.

*The National Review:* Excerpts from "The Scandal of Nuclear Winter" by Brad Sparks (November 15, 1985), "The Death of Truth" by Jeffrey Hart (November 7, 1986), and editorial entitled "Reichstag Fire II" (December 9, 1986). Copyright © 1985, 1986 by National Review, Inc., 150 East 35th Street, NY, NY 10016. Reprinted by permission.

*The New York Times:* Excerpts from "Administration Policy Towards Russia" (September 8, 1946); excerpts from "An Avoidable 20-Year Race" (October 10, 1984); excerpts from "Interview with Ronald Reagan" (February 12, 1985); excerpts from "Rethinking Nuclear War" (September 9, 1985); excerpts from "Militant Hindu Leader Wins King-Maker's Role" (November 28, 1989); excerpts from "Toxic Smoke Killed Some in Seconds" (March 27, 1990). Copyright 1946, © 1984, 1985, 1989, 1990 by The New York Times Company. Reprinted by permission.

*W. W. Norton and Company, Inc.:* Excerpts from *The Cold and the Dark* by Carl Sagan. Copyright © 1984 by Carl Sagan et al.; excerpts from *Voices of Glasnost* by Stephen F. Cohen and Katrina anden Heuvel. Copyright © 1989 by Stephen F. Cohen and Katrina anden Heuvel. Reprinted by permission of W. W. Norton and Company.

*Penguin Books, Limited:* Excerpts from *The Koran,* translated by N. J. Dawood (Penguin Classics, 5th Edition). Copyright © 1956, 1959, 1966, 1968, 1974, 1990 by N. J. Dawood. Reprinted by permission of Penguin Books, Limited.

*Pocket Books, a division of Simon and Schuster, Inc.:* Excerpts from *The Atomic Age Opens* by Donald Peter Geddes. Copyright 1945, © 1972 by Simon and Schuster, Inc. Reprinted by permission of Pocket Books, a division of Simon and Schuster, Inc.

*Center for International Studies at Princeton University:* Excerpts from *On Thermonuclear War* by Herman Kahn. Reprinted by permission of the Center for International Studies at Princeton University.

*Society of Authors on behalf of the Estate of George Bernard Shaw:* Excerpt from *Back to Methuselah* by George Bernard Shaw. Reprinted by the kind permission of the Society of Authors on behalf of the author's Estate.

*The Estate of Leo Szilard:* Excerpts from *The Disarmament Agreement of 1988* and *His Version of the Facts* by Leo Szilard. Reprinted by permission of The Estate of Leo Szilard.

*Lewis Thomas:* Excerpt from "Nuclear Winter, Again" by Lewis Thomas. Copyright © 1984 by *Discover.* Reprinted by permission of the author.

*Unwin Hyman, Ltd.:* Excerpts from *Common Sense and Nuclear Warfare* and *Has Man a Future,* both by Bertrand Russell; excerpts from Euripedes *Medea,* translated by Gilbert Murray. Copyright 1906 by Gilbert Murray. All excerpts reproduced by the kind permission of Unwin Hyman, Ltd.

*Winnipeg Free Press:* "Roche Calls for N-Arms Cuts" from *The Winnipeg Free Press.* Reprinted by permission of The Canadian Press.

# INDEX

ABC, 375–76
aboveground nuclear testing, radioactive fallout from, 337
absorption optical depths, 102–3, 113, 193–96, 385, 453
  classes of nuclear winter in terms of, 193–96
  in extreme nuclear winter, 196
  in marginal nuclear winter, 194
  in nominal nuclear winter, 194
  scattering optical depths compared with, 437–39
  in severe nuclear winter, 196
  of smoke, 96–99, 102–9, 117, 125–26, 341–44, 352, 439
  of soot, 102–3, 109–10, 341, 447
  in substantial nuclear winter, 195
  TTAPS on thresholds for, 108–9
Acheson, Dean, 231, 390–91
Ackerman, Diane, 32
Ackerman, Thomas P., 20, 64, 67, 104, 303, 311
  *see also* TTAPS
acrolein ($C_3H_4O$), 48
actuarial risk, 81
Advani, Lal Kishanchand, 410–11
aerosol particles, physics of, 425–26
  *see also* dust; soot
Aeschylus, 10, 13
Afghanistan:
  Soviet invasion of, 225
  Soviet withdrawal from, 153
*Agamemnon* (Aeschylus), 10, 13
Agent Orange, 48
agriculture, 160–63, 174, 441
  absence of studies on effects of pyrotoxins, radioactive fallout, and ozone layer depletion on, 59, 328–29
  absorption optical depths related to, 103
  climatic vulnerability of, 43, 76–78, 99–102, 149–50, 172–73, 194–96, 341, 384–85, 421, 449–50
  greenhouse effect and, 89, 435
  impact of temperature declines on, 22, 43, 99–102, 341
  in marginal nuclear winter, 194
  after Mt. Etna eruption, 112
  after Mt. Rabaul eruption, 111
  natural temperature extremes and, 98, 340
  in nominal nuclear winter, 194–95
  in noncombatant nations, 172–73
  research on nuclear winter effects on, 338
  SCOPE on, 76–78, 172–73
  in severe nuclear winter, 196
  in Soviet Union, 149–50
  in substantial nuclear winter, 195
  Tallent Report on, 77–78
  thresholds for disruption of, 98–99, 101, 103, 340–41
  TTAPS vs. latest findings on human survival and, 449–50
  volcanic eruptions and, 99–101, 111–12, 442
Ahmadu Bello University, 39
Air Force, U.S., 253
Airland Battle 2000, 268
airlines, liability settlements offered by, 336–37
Akhromeyev, S. F., 334
Akinyemi, Bolaji, 412
Akizuki, T., 106–7
Alberta, forest fires in, 38
Alexandrov, Vladimir V., 135–42, 357

Altfeld, Michael F., 245, 307
Alvarez, Luis, 64, 66, 188, 458
Alvarez, Walter, 64, 66
"America Also Needs Perestroika" (Arbatov), 406
American Academy of Arts and Sciences, 463
American Astronomical Society, 75
American Civil Defense Association, 316, 364
*American Epic, An* (Hoover), 161–62
American Geophysical Union (AGU), 75, 461
*Americans Talk Security*, 153
analysis, methods of, 15
Andronov, Iona, 141
Angier, Natalie, 326
Antiballistic Missile (ABM) Treaty, 225, 265–66, 398
antigreenhouse effect, 66, 435–37
antisatellite (ASAT) weaponry, 399
Anvil, Christopher, 44
Arakawa, A., 125–26
Arbatov, Georgi, 406
Argentina, 286
arguments from authority, 15
Arkin, William M., 350
*Armed Forces of the Soviet State, The* (Grechko), 154
arms control:
  and abolition of nuclear weapons, 299, 417
  Alexandrov on, 139
  bilateral, 225
  as central to U.S. policy, 278–79
  China on, 134, 354–55
  and confirmation of nuclear winter hypothesis, 309–10
  and controversy over nuclear winter, 316
  for conventional forces, 153, 267–69, 271, 277, 282, 407
  cruise missiles and, 254
  deterence and, 151, 223–24, 304
  to dismantle Doomsday Machines, 218–19
  first strikes and, 145
  France on, 134, 355–56
  and freeze in nuclear weapons arsenals, 313, 315
  Gorbachev on, 268, 353, 356, 379, 393
  Great Britain on, 134, 355–56
  Gulf crisis and, 282
  in light of nuclear winter, xix, 131, 133–35, 227, 229, 248, 305, 309–12, 316, 353–56, 383–84
  MIRVing and, 228, 242, 395
  MSD and, 227–29, 232–35, 242, 259
  noncombatant nations and, 172, 179–80, 369
  in preventing proliferation of nuclear weapons, 290–93
  role of public in, 262–64
  SDI and, 390
  and severity of nuclear winter, 204
  Soviet Union and, 133–35, 204, 223–25, 229–30, 232–35, 260–64, 277–81, 311, 354–56, 383, 390, 393, 400, 409–10
  timing for, 280
  treaties on, *see specific treaties*
  unilateral, 225, 248
  verification and, 229–30, 259–64, 280, 400, 417
Arms Control and Disarmament Agency (ACDA), 318, 408
arms race, xviii-xix, 285, 299
  and abolition of nuclear weapons, 297
  and asymetry of perception about nuclear winter, 149
  forces driving, 386
  MSD and, 249–50, 255, 398
  New Zealand's criticisms of, 368
  noncombatant nations and, 173
  public support for, 20
  suppression of reports on dangers of, 20, 303
  and winnability of nuclear war, 382
*Art of War, The* (Sun Tzu), 144
ashes, 28
Associated Press, 378–79
asteroids, impact of, 64–66, 420, 442, 458, 466
atmosphere, thickness of, 15–16
"Atmosphere after a Nuclear War: Twilight at Noon, The" (Crutzen and Birks), 327

atmospheric models:
one-dimensional vs. two-dimensional, 424–26
physics of aerosol particles in, 425–26
in predicting future climates, 421–24
visible and infrared light in, 426–27
atmospheric sciences, nuclear winter insights related to problems in, 34, 320
Atomic Energy Commission (AEC), 303
Australia, nuclear winter effects in, 175
Auton, David L., 124
Avibrás Aerospacial, S.A., 413
Ayres, Robert U., 317
Aztec civilization, 18

*Back to Methuselah* (Shaw), 80
Baker, James A., III, 393
Ball, Desmond, 351, 393
BBC, 71, 375
Bee, R. J., 305–6
Beer's law, 342
Beggs, James, 462
Beit-Hallahmi, B., 414
Bethe, Hans, 396
Bhopal disaster, 48–49, 86, 336
Bing, G. F., 351
Birks, John, 41, 327, 370, 459–60
black rains, 447
Blight, James G., 87–88, 289
Bovin, Alexander, 248
Bradbury, Norris, 408
Brazil, medium- and/or long-range ballistic missiles of, 286, 413
Brezhnev, Leonid, 70, 261, 353
Broad, William J., 317, 359
Brodie, Bernard, 83, 145, 230, 357–58, 392
Broecker, Wallace, 42*n*
Brogan, Hugh, 260
Brooks, James, 413
Brown, Harold, 353
Brownian (random thermal) motions, 425
Buckley, William F., 315
Buencuchillo, Father, 109
*Bulletin of the Atomic Scientists*, 68–69
Bundy, McGeorge, 233, 357
Bunyan, John, 130
Burke, Arleigh, 232
Burrows, William E., 165
Bush, George, 395
arms control and, 269, 393
on conventional forces, 269
on winning nuclear war, 183–84
Butterfield, Daniel, 20*n*
Byron, George Gordon, Lord, 94–95

Caidin, Martin, 124
Canada, 321, 369
Canonical Deterrent Force (CDF), 226
Cao Hongxing, 174
carbon dioxide ($CO_2$) concentrations, 24
carbon monoxide (CO), 48
Carey, William D., 308
Carlsson, Ingvar, 181, 417
Carter, Jimmy, 408
on MSD, 232–33
on targeting, 122
Cassandra, Princess of Troy, fall of Troy prophecied by, 13–15
cataracts, 58
Central Europe, nuclear weapons-free corridor in, 268
Central Intelligence Agency (CIA):
Alexandrov and, 141
on Israeli nuclear weapons, 414
on risk of conventional war in Europe, 269, 404
*Challenger* disaster, xviii, 84–85
Chapleau experiment, 51
*Charioteer*, 119
chemical plants, 48–49, 86, 336
Cheney, Richard, 404
Chernobyl accident, xviii, 53–54, 84–85, 166
Chester, C. V., 385
Chester, R. O., 385
Chicago, Ill., 1871 fire in, 29
China, People's Republic of, 126
on arms control, 134, 354–55
conventional forces of, 152
on global implications of nuclear winter, 369
internal ethnic conflicts of, 233

China, People's Republic of *cont'd*
medium- and/or long-range ballistic missiles of, 286, 413–14
on MIRVing, 394
nuclear forces of, 134, 201, 285, 354–55, 411
on nuclear weapons, 287–89
prospects of causing nuclear winter, 201
projected nuclear winter casualties in, 288–89
and proliferation of nuclear weapons, 290–91, 413
targeting and, 73, 117, 332, 346–47
wildfires in, 38–39
chlorinated benzenes, 48
chlorine, 55–57
chlorofluorocarbons (CFCs), 51, 55–57, 88–89, 303
*Christian Science Monitor*, 356
Cicero, 4
*Le Cid* (Corneille), 116
Cimbala, Stephen J., 245, 307
cities:
absorption coefficient for smoke from, 343
collateral damage in, 446
fires in, 29, 34–35, 48, 343, 445–46
leadership targets in, 351
smoke injection profiles for, 37, 321–23
targeting of, 73, 117–26, 197–203, 210, 240, 270–71, 330–31, 346–47, 349–52, 364, 397
threatened release of pyrotoxins in nuclear ignition of, 48–49
TTAPS vs. latest findings on fires in, 445–46
*see also specific cities*
civil defense:
at Chernobyl, 166
on survival after nuclear war, 163
civilization, global, 442
destruction of, 63, 69–70, 73–74, 90, 334–35
extinction of humans vs. destruction of, 73–74
peril presented by nuclear winter for, 26
replacement cost of, 86–88
in substantial nuclear winter, 195–96
Clarke, Arthur C., 44
Clausewitz, Carl von, 216*n*, 382–83
Clevely, Hugh, 29
climate:
carrying out nuclear exchange without detriment to, 131–33
changes due to smoke injected by post-nuclear fires, 324
definition of, 419–21
determinants of, 6
deterrence due to anomalies in, 149
general confirmation of impact of nuclear winter on, 34–41, 324
impact of dust from volcanic vs. nuclear explosions on, 42
impact of dust on, 23, 42, 317–18, 436–41
impact of firebombing on, 124, 126
impact of major strategic attacks on, 304–5
impact of nuclear testing on, 124–26
impact of nuclear war on, 5–6, 20–21, 23, 26, 98, 131–33, 304–5, 323, 339–40
impact of smoke from nuclear explosions on, 23–24, 98–99, 317–18, 324, 339–40, 436–41
impact of sub-threshold wars on, 304
impact of volcanic eruptions on, 42, 125–26
in late Cretaceous catastrophe, 64–65
life on earth dependent on, 22, 419–43
nuclear autumn vs. nuclear winter simulations on, 35–40, 321–22
PAN-EARTH project on, 320
petroleum facility targeting and, 201, 203
relationship between number of warheads detonated and, 199–202
in severe nuclear winter, 196
Soviet agriculture and, 150
in substantial nuclear winter, 195

and triggering nuclear winter, *see* thresholds, climatic effects
TTAPS on, 20–21, 42
vulnerability of life to changes in, 420–21
*see also* atmospheric models; temperatures
climate feedbacks, 362–63
Climate Research Laboratory, 136
climate system:
greenhouse effect in, 427, 429, 433–36
heat reservoirs in, 427, 432–33
how it works, 427–32
infrared light in, 426–29
as sensitive to small energy balance changes, 441
visible light in, 427–32
*Climatic Change,* 323
Coates, Ken, 354
Cohen, Sam, 348, 352
Colbeck, I., 324
Colby, William, 165
*Cold and the Dark: The World After Nuclear War, The* (Ehrlich, Sagan, Kennedy, and Roberts), 362, 464–65
Cold War, xvii, 6, 269
comets:
in impacts with Earth, 64–66, 420, 442, 458, 466
in near collisions with Earth, 66
Command-destruct systems, 279
Committee of Soviet Scientists Against Nuclear War, 273
Conference on the Long-Term Biological Consequences of Nuclear War, 137
"Consequences of Execution," 159
conventional forces, 152, 186, 207, 229
and abolition of nuclear weapons, 298, 414–16
arms control for, 153, 267–69, 269, 271, 277, 282, 407
creating high explosives winter with, 414–15
DoD on 269, 403–5
JCS on, 272, 404
MSD and, 267–70, 402–7
of Soviet Union, 152, 225, 267–71, 402–4, 406–7
Cooperative Research Project on Arms Reductions, 273
Corneille, Pierre, 116
Cornell University, 320
"Correlation of Politics, War and a Nuclear Catastrophe, The" (Kanevsky and Shabardin), 155, 383
*Cosmos,* 457–58, 460
counterforce exchanges, 126, 197–202, 350
capabilities in, 358
targeting in, 254
TTAPS vs. latest findings on, 450
countervalue exchanges, 197–202
Covey, C., 321
Cretaceous catastrophe, late, 64–67, 175, 313, 335–36, 420–21, 458
crisis relocation, 304
Croesus, King of Lydia, 11–12, 15
Crossette, Barbara, 411
cross-targeting, 392
cruise missiles, 279, 402
arms control negotiations and, 254
MSD and, 264
Crutzen, Paul J., 41, 327, 370, 459–60
Cuban missile crisis, 69–70, 84, 87–88, 123, 224, 397
Cyrus, King of Persia, 11–12
Cyvertson, Clarence, 462

Dafni, Reuven, 71
Dante Alighieri, 62
"Darkness" (Byron), 94–95
Darling, John, 363–64
Darmstadt, firestorms in, 124
da Silva, L., 175
Daugherty, W. H., 360
"Day After, The," 374–75
Dean, Jonathan, 268
death and dying:
in Bhopal disaster, 86, 336
as cost of nuclear war, 86–88
from Doomsday Machines, 215–17
in extreme nuclear winter, 197
in famines, 161–62
from greenhouse effect, 89
in marginal nuclear winter, 194
from Nagasaki bomb, 329

death and dying *cont'd*
in nominal nuclear winter, 195, 197
in noncombatant nations, 173
in nuclear war, 52–54, 73, 86–88, 149, 159–63, 194–95, 197, 303, 331–32, 360–61
in nuclear winter, 7, 149, 184, 198–99, 288–89
in nuclear winter compared with representative human catastrophes, 198–99
from ozone layer depletion, 89
from pyrotoxins, 47–49
from radioactive fallout, 52–54, 73, 303
SCOPE on, 74
TTAPS v. latest findings on, 450
*see also* extinction, extinctions
debate, 15
*Decline and Fall of the Roman Empire, The* (Gibbon), 170
deep-underground basing systems, 243, 394–95
Defense Nuclear Agency (DNA), 458–59
on biological effects of nuclear war, 328–29
in funding nuclear winter research, 314, 338, 458, 466
deforestation, 338
deLauer, Richard, 316–17
delayed gamma rays, 52–53
Delhi Declaration, 179, 366
Dellums, Ronald V., 392
Delphic Oracle:
ambiguous prophecies of, 13–14
on Croesus's invasion of Persia, 11–12
Demack, G., 305
denial, 14
Department of Defense (DoD), U.S., 316–17, 465–66
on arms control, 278
on fighting nuclear war in nuclear winter environment, 148
and global policy implications of nuclear winter, 184, 378–79, 381
in recognizing consequences of smoke and dust, 317
on risk of conventional war in Europe, 269, 403–5
on Soviet nuclear arsenal, 280
on Soviet shelter system, 164
on survival after nuclear war, 361
TTAPS and, 314
worries about nuclear winter in, 33–34
de Seversky, Alexander, 361–62
Desired (or Designated) Ground Zeros (DGZs), 118
deterrence, 82, 145–55, 226, 357–60
and abolition of nuclear weapons, 298, 415–17
arms control and, 151, 223–24, 304
arms race and, 386
in case of marginal or nominal nuclear winter, 209–10
in case of severe nuclear winter, 210–11
in case of substantial nuclear winter, 210
against conventional armaments, 152–53, 186, 207
in Cuban missile crisis, 224
of Doomsday Machines, 215–17, 219
existential, 233, 357, 394
extended, 269–70, 285, 404
out of fear of consequences of retaliation, 146–47
and global policy implications of nuclear winter, 181–82
INF for, 305
minimum sufficiency, *see* minimum sufficiency deterrence
New Zealand's criticisms of, 367–68
nuclear weapons accumulated in name of, 239–40
nuclear winter in, 135, 145–49, 181–82, 207–11, 224, 306, 313–14, 357, 359–60, 383, 385–88
as psychological phenomena, 387
reevaluation of, 306, 313–14
SND, *see* strategic nuclear deterrent
success of, 83–84

without threat of large-scale environmental consequences, 209
TND, 207–8
and window of vulnerability, 148
de Toledano, Ralph, 141
Di Hua, 413
Dillon, C. Douglas, 88
Dinerstein, H. S., 334
dinosaurs, extinction of, 64–67, 175, 313, 420, 458
dioxins, 48
Dire Straits, 14
*Discriminate Deterrence*, 233
disease, 160–63, 175
famines and, 162
in noncombatant nations, 172, 174
from radioactive fallout, 53
after San Francisco earthquake, 29
in severe nuclear winter, 196
after Tambora volcanic explosion, 100–101
*Divinations* (Cicero), 4
*Divine Comedy, The* (Dante), 62
*Dr. Strangelove*, 214, 217–18
Dolan, P. J., 121
Donahue, Thomas M., 309
Doomsday Machines, 211, 387–88
acceptability of, 216–17
controllability of, 216
criteria of, 215–16
dismantling of, 218–19
Kahn on, 215–19, 416
in light of nuclear winter, 217–19
Drell, Sidney, 395
Dresden, firestorms in, 124
Drew, D. M., 305
*Dropshot*, 119
droughts, 194
Duffield, J., 361
Du Fu, 158
Dunn, Frederick S., 297, 415
dust, 26, 42–43
Alexandrov on, 138
climatic influence of, 23, 42, 317–18, 436–41
greenhouse effect and, 436–37, 440–41, 466–67
from groundbursts, 202
high-altitude, 108*n*
after Hiroshima attack, 106
jet engine performance in, 148, 359
in late Cretaceous catastrophe, 65, 458
on Mars, 39, 313, 456
in nuclear autumn vs. nuclear winter model, 39, 322
radioactive fallout and, 49–50
temperature declines from, 39, 318, 437, 440
TTAPS on, 460–64
TTAPS vs. latest findings on, 447–48, 453
Turco's research on, 458–59
on Venus, 457
*see also* soot
Dworkin, G., 319
Dye, Lee, 247
Dyson, Freeman J., 310, 379–80, 417–18

Earth:
average surface temperature of, 22–24, 104
extraterrestrial impacts with, 64–66, 420, 442, 458, 466
earthquakes, 27–29
Eastern Europe:
opening up of, 300
revolution in, 268–69
Eban, Abba, 71
Echikson, William, 356
Economic/Industrial (E/I) targeting, 122–23, 131–32, 185–86, 348–49
and severity of nuclear winter, 201–3
economics, xvii
of greenhouse effect, 89–90, 337–38
Egypt, 286
Ehrlich, Paul R., 43, 362, 464–65
Einstein, Albert, 70–71, 299–300, 418
Einstein/Russell Manifesto, 70, 336
Eisenhower, Dwight D., 69, 357
on deterrence value of nuclear weapons, 151–52
on MSD, 224–25, 231, 389
on strategic bombers, 252–53
El Chichón eruption, 109–11
electromagnetic pulses, 20

Eliot, T. S., 64
Ellis, Richard H., 118
Ellsaesser, Hugh W., 141
Emery, David F., 310–11
Energy Department, U.S., 338
*Enola Gay*, 106
Eocene catastrophe, late, 64, 67
*EOS*, 42, 461
equilibrium average land surface temperature, 104
Ethiopia, famines in, 14
Etna, Mt., eruption of, 112
Euripides, ix
Europe:
  balancing conventional forces in, 267–71, 402–4, 407
  Central, 268
  DoD on conventional war in, 269, 403–5
  Doomsday Machines and, 216
  Eastern, *see* Eastern Europe
  estimated casualties in, 360–61
  INF to deter Soviet attack in, 305
  JCS on conventional war in, 272
  nuclear weapons in, 208, 234, 266–68, 270, 403–4
  proposed removal of U.S. nuclear weapons from, 270
  smoke absorption optical depths in, 343–44
  Soviet withdrawal of troops and armor from, 225
  after Tambora eruption, 100–101
existential deterrence, 233, 357, 394
*Exocet*, 254, 397
extended deterrence, 269–70, 285, 404
extinction, extinctions:
  arguments against, 333–34
  decried as loose talk, 67, 329
  destruction of global civilization vs., 73–74
  of dinosaurs, 64–67, 175, 313, 420, 458
  through evolutionary complacency, 421
  exposure of public to possibility of, 71–73
  from extraterrestrial bodies striking Earth, 64–66, 420, 442, 458, 466
  history of, 63–67
  of humans, 63–78, 329–36
  in late Cretaceous catastrophe, 64–67, 175, 313, 335–36, 420–21, 458
  from near collisions of comets with Earth, 66
  possibility of, 70–76, 333–35
  seriousness of, 67–72
  and severity of nuclear winter, 441
  of species in Southern Hemisphere, 73–74
  in substantial nuclear winter, 195–96
  *see also* death and dying
extreme nuclear winter, 196–97, 201, 203

Falklands-Malvinas war, 254
Fallows, James, 358
famines:
  after Mt. Etna eruption, 112
  after Mt. Rabaul eruption, 111
  natural disasters and wars as causes of, 161
  in nominal nuclear winter, 194–95
  after nuclear war, 161–62
  predictions on, 14
  in Soviet Union, 161–62, 166
  after Tambora eruption, 100
Fascell, Dante, 408
*Fate of the Earth, The* (Schell), 72, 334–35
Federal Bureau of Investigation (FBI), 140
Federal Emergency Management Agency (FEMA):
  nuclear winter ignored by, 163, 361
  nuclear winter research commissioned by, 363
  on survival after nuclear war, 163, 361
Federation of American Scientists, 273
Fehsenfeld, Fred, 459
Feiveson, H. A., 246, 395
Feldbaum, C. B., 305–6
Fermi, Enrico, 69, 87
Fieldhouse, Richard W., 350
fighter-bombers, 398
Finland, 54

fireballs:
in ozone layer depletion, 57
radioactive fallout and, 52
firebombing, climatic anomalies due to, 124, 126
fires, 159
ashes from, 28
in Chicago in 1871, 29
in cities, 27–29, 34–35, 48, 343, 445–46
deterrence provided by, 146
in forests, 26, 38–39, 41, 126, 320
in late Cretaceous catastrophe, 65
from nuclear explosions, 23
in nuclear war simulations, 51
in petroleum facilities, 41, 49
poisons from, *see* pyrotoxins
post-nuclear, 324
as result of rainfall reductions, 175
rural, 445–46
after San Francisco earthquake of 1906, 27–29
smoke from, 23–24, 28, 38–39, 324, 343–45
TTAPS vs. latest findings on, 445–46
*see also* pyrotoxins; wildfires
firestorms, 106, 124
fissionable materials, 261–62
continuous on-site monitoring of, 262
shutting down environmentally dangerous production reactors for, 280–81
U.S. and Soviet inventories of, 281, 408–9
Five Continent Peace Initiative, 179, 367
flaming combustion, 23
food chains, threat of ozone layer depletion to, 58, 328, 337
force projection, strategic bombers for, 398
*Foreign Affairs*, 69, 303–4, 312, 321, 333, 357, 465
*Foreign Policy*, 151
forest fires, 26, 38–39, 41, 126, 320
forests, damage to, 175, 338
Forest Services, 51
Forsberg, Randall, 386
fossil fuels, massive conversion from, 89–90
*Four Quartets* (Eliot), 64
France, 126
on arms control, 134, 355–56
in causing nuclear winter, 201, 203
in city targeting, 117
medium- and/or long-range ballistic missiles of, 286
on nuclear weapons in Germany, 412
nuclear weapons inventory of, 134, 201, 203, 224, 285, 355–56
on nuclear winter, 356
and proliferation of nuclear weapons, 290–91
prospects of causing marginal or nominal nuclear winter, 203
Franck, James, 178
*Frankenstein* (Shelley), 94–95, 100
Franklin, Benjamin, 99
Franklin, H. Bruce, 43–44
Frederick, J. E., 328
Freeze movement, 278–79

Gallagher, Paul, 315–16
Galunggung, Mt., eruption of, 359
gamma rays, 49–53
Gandhi, Indira, 411
Garn, Jake, 313
Garrett, B. N., 305–6
Garwin, Richard, 246, 395–96, 408
Gayler, Noel, 118–19, 207–8, 347
Geddes, Donald Porter, 106
Gelb, Leslie H. 368
General Accounting Office (GAO), 380
Geneva Convention, Protocol I of, 410
Geological Society of America, 458
Germany:
NATO nuclear weapons in, 412
unification of, 269, 407
Gerry, Elbridge, 260
Gibbon, Edward, 170
Ginsburg, A. S., 38
Glaser, B. S., 305–6
glasnost, 155
Glasstone, S., 121
Glatzmaier, G., 328

"Global Consequences of Nuclear 'Warfare' " (Turco, Toon, Pollack, and Sagan, 1982), 42
Godwin, P. H. B., 305
Goldansky, Vitaly, 155, 370–71
Golitsyn, Georgi S., 38–39, 370, 382
Goodin, R., 319
Gorbachev, Mikhail S., 226, 261, 383
  on abolition of nuclear weapons, 298, 416–17
  arms control and, 268, 353, 356, 379, 393
  on conventional reductions, 268
  on global policy implications of nuclear winter, 183, 373
  on glut of nuclear weapons, 388
  MSD and, 247
Gordon, Michael R., 396–97
Gouré, Leon, 310, 353, 381
Graham, Kennedy, 367–68
Graham, William R., 78
gravity, Newton's law of, 423
Gray, Colin S., 149, 151–52, 319, 389–90
Great Britain, 126
  and global policy implications of nuclear winter, 181–82
  Limited Test Ban Treaty and, 409–10
  medium- and/or long-range ballistic missiles of, 286
  on MIRVing, 394
  nuclear forces of, 134, 201, 203, 224, 285, 355–56
  Nuclear Non-Proliferation Treaty and, 409
  and proliferation of nuclear weapons, 290–91
  prospects of causing marginal or nominal nuclear winter, 203
  prospects of causing nuclear winter, 201, 203
  deterrence policy of, 233
  after Tambora explosion, 100–101
Grechko, A. A., 154
Greece, 182
greenhouse effect, 15–16, 292, 301, 325, 455–57
  in calculating Earth's average surface temperature, 23
  in climate system, 427, 429, 433–36
  cost and risk of increases in, 89–90, 337–38
  dust and, 436–37, 440–41, 466–67
  impact of smoke from nuclear explosions on, 24, 436–37, 440–41
  impacts of asteroids and comets related to, 466
  invalid criticism of, 316
  in late Cretaceous catastrophe, 65
  on Mars, 39
  nuclear winter insights related to, 34, 320
  role of clouds in, 435
  and severity of nuclear winter, 440–41
  temperature increases due to, 40–41, 320, 325, 337–38, 434–36
  on Venus, 455, 457
Greenpeace, 181
Gribkov, A. I., 227
Grier, Peter, 330
Grinspoon, David, 466
Gromyko, Andrei, 231, 391
gross national product, 160
Gulf crisis, 282
*Gulliver's Travels* (Swift), 44
Gustafero, Angelo, 462

Hamburg, incendiary bombing of, 124
Happy Land Social Club fire, 326
Hardin, R., 319
Harriman, Averell, 5
Harris, D. L., 42, 125
Harrison, Roy M., 324
Hart, Jeffrey, 315
Hart, Tony, 383
Harwell, Mark, 320
Hatfield, Mark, 138, 376
Hazarika, Sanjoy, 336
Healey, Dennis, 182
heat reservoirs, 427, 432–33
Hekla, Mount, eruption of, 112
Herodotus, 11–12
high-altitude dust, 108*n*
high-altitude nuclear explosions, 20

high explosives winter, 414–15
high-level scenarios, 305
high-temperature fusion pwoer, cost of, 89–90
Hillenbrand, Martin J., 309
Hirohito, Emperor of Japan, 12
Hiroshima, atomic bombing of, xviii, 19, 23, 70
  black rains after, 447
  description of, 106–7
  de Seversky on, 362
  firestorms after, 124
  yield of weapon used in, 254
Hirschboeck, K. K., 341
*History* (Herodotus), 11–12
Hitler, Adolf, 210, 389
Hobbes, Thomas, 415
Hoeber, F. P., 306
Hoess, Rudolf, 327
Holloway, David, 350
*Homo sapiens*, 63*n*
Hoover, Herbert, 161–62
Horowitz, Dan, 306
Hoyle, Fred, 66
Huang Hua, 354
Hudson Institute, 215
Hugo, Victor, 296
human error, 304
Hunt, Lee, 458
hydrogen chloride (HCl), 48
hydrogen cyanide ((HCN), 48

Ice Ages:
  Little, 341
  nuclear winter compared with, 26
  temperatures of, 24–26, 102, 341
  Wisconsin, 102
Iklé, Fred, 335
impact winter, 64–66, 420–21
impermeable defensive shields, 131, 133
India:
  development of nuclear weapons by, 285–86, 411
  medium- and/or long-range ballistic missiles of, 286
  after Tambora eruption, 99, 101
  targeting and, 331–32
infrared radiation:
  in atmospheric models, 426–27
  in climate system, 426–29
  greenhouse effect and, 434–35
  influence of smoke and dust on, 436–37
  in normal atmosphere vs. atmosphere after nuclear war, 431
insects, 160, 175
Institute for Cosmic Research, 139
Institute of Atmospheric Physics, 38, 382
intercontinental ballistic missiles (ICBMs):
  single-warhead, 242
  strategic bombers vs., 252–53
intermediate-range nuclear delivery systems, phasing out of, 228, 267
intermediate-range nuclear forces (INF), 305
Intermediate-Range Nuclear Forces (INF) Treaty, 133, 208, 228, 230, 233–34, 260–61, 267, 305, 311, 383–84, 400, 402
*International Affairs*, 155
International Astronomical Union (IAU), 75
International Atomic Energy Agency, 411
International Control Commission, 263–64
International Council of Scientific Unions (ICSU), 40, 75, 323
International Forum for a Nuclear-Free World and the Survival of Humanity, 373
International Geophysical Year, 75
International Union of Geodesy and Geophysics, 75
Intrilligator, M.D., 387
Iran-Contra scandal, 412
Iran-Iraq war, 123, 413
Iraq, 286
  Kuwait invaded by, 282
  nuclear weapon proliferation and, 126
  Osiraq nuclear reactor of, 411
Ishikawa, Eisei, 107
Israel:
  development of nuclear weapons by, 126, 285–86, 291–92, 413–14
  medium- and/or long-range ballistic missiles of, 286, 291–92

Israel *cont'd*
raid against Iraqi nuclear reactor by, 411
*Izvestia,* 155, 370–71, 383, 388

Japan, 173
Japanese-American, reparations paid to, 86
jet engines, 148, 359
John Paul II, Pope, 139
Johns Manville Corporation, 336
Johnson, Lyndon, 232
Joint Chiefs of Staff (JCS):
on conventional forces, 272, 404
on conventional war in Europe, 272
and global policy implications of nuclear winter, 185, 381
on MSD, 233
on Soviet shelter system, 164
on Soviet vs. U.S. national security objectives, 153–54
"Joint Soviet-Finnish Declaration: New Political Thinking in Action," 418
Joint Strategic Target Planning Staff (JSTPS), 118
Jones, David C., 330–31
Jones, Eric, 459
*Journal of the British Interplanetary Society,* 329

Kahn, Herman, 149, 215–19, 336, 387, 416
Kavensky, Boris, 155, 383
Kant, Immanuel, 222, 284
Kapitsa, Sergei, 155, 319, 370–71
Kaplan, M. M., 417
Kasting, James, 442*n*
Katz, A., 361
Kaufmann, William W., 407
Kavka, G. S., 388
Keegan, George J., Jr., 164–65
Kegley, Charles W., Jr., 312
Kennan, George F., 133, 231, 390–91
Kennedy, Donald, 362, 464–65
Kennedy, Edward, 138, 376
Kennedy, John F., 69–70, 232, 289
deterrence and, 146
on targeting, 122
Kennedy, Robert F., 70, 309–10
Keyes, L., 362
KGB, Alexandrov and, 136, 140–41
Khadaffi, Muammar, 412
Khan, Imamullah, 373
Khrushchev, Nikita, 70, 165, 247, 289
Kiernan, Vincent, 399
*King Henry VIII* (Shakespeare), 206
Kissam, Edward, 18
Kissinger, Henry, 242, 375
Kistiakowsky, George, 313
Knapp, Harold A., 303
Knox, Joseph B., 319
Kokoshin, Andrei, 370, 394
Kondratyev, K. Ya., 382
Koppel, Ted., 315
*Koran, The,* 238
Kotz, Nick, 253, 392
Krakatoa eruption, 111
Krueger, A. J., 328
Kubrick, Stanley, 214, 217–18
"Kunlun" (Mao Zedong), 287
Kuwait, Iraqi invasion of, 282

Labaton, Stephen, 336
Lackey, Douglas, 358
LaMarche, V. C., 341
Lambeth, B.S., 348–49
Lange, David, 180, 367
Larouche, Lyndon, 315–16
laser winter, 401
Lawrence Livermore National Laboratory, 141, 318–19, 322
leadership targets, 350–51
Leaning, Jennifer, 362
LeMay, Curtis, 231, 348
Lenin, V. I., 166
Leonardo da Vinci, 419
"Letters from a Dead Man," 372
*Leviathan* (Hobbes), 415
Lewis, John Wilson, 347
Leis, K. N., 348–49
Liang Yufan, 354
Libya, attempts to acquire nuclear weapons by, 411–12
Lieber, Robert J., 306
*Life with King Henry the Fifth, The* (Shakespeare), 284
Limited Test Ban Treaty, 5, 52, 152
preamble to, 409–10
Lin, P. T. K., 385–86
*Literaturnaya Gazeta,* 141
"Little Gidding" (Eliot), 64

Little Ice Age, 341
Liu Yuhe, 174
logistic thresholds, 96–98, 107
London, Jack, 27–29
Long, Susan G., 313
Los Alamos National Laboratory, 57
low-level scenarios, 305
Lubin, D., 328
Lydia, Persia invaded by, 11–12, 15
Lynch, Allen, 307, 382
Lynch, H., 401–2
Lyons, J. A., Jr., 380–81

McFarlane, Robert C., 396
machine error, 304
Machta, L., 42, 125
McKay, C. P., 64, 67
McLoughlin, David, 361
McNamara, Robert S., 88–89, 135, 224, 228, 239, 289, 405
McTague, John P., 380
Maddox, J., 320–21
maladaptive behaviors, 332
Malone, R., 328
Manhattan Project, 230, 232*n*
Mao Zedong, 287–89
Marcus, Steven J., 395
marginal nuclear winter, 194, 197, 201, 203, 385
  deterrence in case of, 209–10
*Mariner 9*, 39, 456
Mark, J. Carson, 408
Mars:
  dust storms on, 39, 313, 456
  greenhouse effect on, 39
  temperatures of, 456
Maryland, University of, 38
Mearsheimer, J., 319
*Medea* (Euripides), ix
medical systems, 163
Melville, Herman, 46, 296
Mendlovitz, S., 385–86
*Merchant of Venice, The* (Shakespeare), 419
methyl isocyanate, 48
middle-level scenarios, 305
Migule, Albrecht, 411
Mike Pacific test, 125
mild nuclear winter, 304, 416
military facilities, targeting of, 199–202
Mill, John Stuart, 219
Milton, John, 192, 445
minimum sufficiency deterrence (MSD), 224–35, 304, 388–93
  and abolition of nuclear weapons, 297–99, 415–18
  arms control and, 227–29, 232–33, 242, 259
  communications, command, and control of forces of, 244
  conventional forces and, 267–70, 402–7
  cost of, 246–49, 251, 255
  definition of, 230
  early recognition of, 230–31, 390–91
  early U.S. debates on, 260
  flexible targeting in, 224
  how to get to, 259–73, 398–408
  in light of nuclear winter, 230, 233, 250
  limited terminal defenses for, 265–66
  MIRVs and, 241–44, 250–51, 255, 259, 264, 279, 396
  near-term strategic steps for, 277–82, 408–10
  nuclear weapon yields and, 254–55
  origin of term, 226
  and proliferation of nuclear weapons, 286, 290–93
  size of forces for, 230, 232–35, 391–93
  strategic bombers and, 251–55, 279
  strategic defense and, 244–46, 259
  strategic force structures and, 239–55, 267, 394–98
  tactical nuclear weapons and, 267–68, 270
  targeting and, 224, 241, 243, 270–71
  verification and, 259–65, 281, 399
*Les Misérables* (Hugo), 296
Mishkin, Ellen, 316
missiles, 19
  proliferation of, 286–92, 413–14
  testing of, 262
  *see also specific kinds of missiles*
Mitchell, Parren J., 377

Mitterrand, François, 355–56
mobile land-based missiles, 242–44, 250, 255
*Moby Dick* (Melville), 46, 296
Molina, Mario, 56
Montreal Protocol, 56
Moore, Molly, 396
Moran, William, 458
Moscow Summit, 254
immediate aftermath of, 106–7
yield of weapon used in, 254

Natan, Otto, 71
National Academy of Sciences (NAS), 323
on ozone layer depletion caused by nuclear war, 317
on stratospheric dust injection from nuclear war, 317–18
TTAPs and, 313–14, 458–60, 462, 465
National Aeronautics and Space Administration (NASA), 111
Ames Research Center of, 456–57, 460–62
Jet Propulsion Laboratory of, 456
TTAPs and, 313–14, 456–57, 460–63, 465
National Center for Atmospheric Research (NCAR), 35, 37, 57, 136–38
National Command Authority, 159
National Commission on Integrated Long-Term Strategy, 268
National Intelligence Estimates, 269
National Research Council (NRC), 458–59
on possibility of climatic catastrophe after nucler war, 323
urban smoke injection profile of, 37, 323
*National Review, 314–15*
National Science Foundation, 313–14, 338
National Security Agency, 5
National Security Council, 5
National Security Decision Memorandum-242 (NSDM-242), 117–18
National Strategic Target Data Base (NSTDB), 118
National Strategic Target List (NSTL), 118, 351
National Strategic Target List (NSTL) Directorate, 118
Natural Resources Defense Council, 229–30
*Nature,* 320–21
Navier-Stokes equations, 422
Navy, U.S., 85–86
Nazi death camps, poison gas used in, 48, 327
Neak Luong, accidental bombing of, 337
near-infrared radiation, greenhouse effect and, 433
Nehru Memorial Symposium: Towards a Nuclear Weapon-Free and Non-Violent World, 383
neutron bomb, 348, 352, 356
*New England Journal of Medicine,* 162
Newman, James R., 215*n*
Newton, Isaac, 423
*New York Times,* 70, 106, 185, 311–12, 336–37, 368
*New York Times Book Review,* 317
New Zealand:
nuclear deterrence criticized by, 367–68
nuclear weapons policies criticized by, 180, 368
nuclear winter study funded by, 181
public opinion on nuclear winter in, 187
U.S. Nuclear weapons banned by, 368
Nigeria, 368
acquiring of nuclear weapons by, 285, 412
*Nightline,* 315, 376
nitrogen oxides, 51, 57
Nix, H. A., 324–25
Nixon, Richard, 117–18, 122
nominal nuclear winter, 194–97, 201–4, 240, 385
and abolition of nuclear weapons, 415
deterrence in case of, 209–10

noncombatant nations:
- arms control and,172, 179–80, 369
- consequences of nuclear winter in, 27, 171–76, 210, 364–66, 387
- and global policy implications of nuclear winter, 179–86, 366–67
- in marginal nuclear winter, 194
- SCOPE on, 172–73
- starvation in, 172
- in substantial nuclear winter, 195
- targeting of, 171–72

Norden, Heinz, 71

North Atlantic Treaty Organization (NATO), 360, 406
- in city targeting, 117
- conventional force dilemma of, 292
- conventional formations of, 186, 267–69, 271–72, 277, 282, 403–7
- in conventional war in Europe, 272
- Doomsday Machines and, 216
- German-bsed nuclear weapons of, 412
- INF of, 305
- MSD and, 233
- perceived conventional superiority of WTO over, 186
- reduction, balancing, and substantial pull-back and demobilization of, 229, 267
- and seriousness of extinction of human species, 67
- tactical nuclear weapons and, 266
- on TND and SND, 208

Northern Hemisphere, 173–76, 340
- agriculture in, 77
- ozone layer depletion above, 57–58
- severe nuclear winter in, 385
- smoke absorption optical depths in, 108–9, 343
- in substantial nuclear winter, 195
- as vulnerable to nuclear winter, 225

*Notebooks of Leonardo da Vinci, The*, 419

nuclear autumn, 35–40, 77, 321–23

nuclear first strikes, deterrence of, *see* deterrence

nuclear fuel facilities, targeting of, 73

Nuclear Non-Proliferation Treaty, 152, 290
- Article VI of, 281, 409

nuclear power reactors, xviii, 53–54, 85–85, 166

nuclear war, 329
- abstractions in planning of, 6
- anticipated casualties of, 52–54, 73, 86–88, 149, 159–63, 194–95, 197, 303, 331–32, 360–61
- climate impacted by, 5–6, 20–21, 23, 26, 98, 131–33, 304–5, 323, 339–40
- command and control of nuclear forces in, 148
- connection between number of warheads, target categories, and severity of, 199–201
- consequences of, 6–7, 33–34, 78, 90, 159–67, 172, 316–19, 352, 360–64
- as containable, 225–27, 245
- counterforce, *see* counterforce exchanges
- difficulties due to nuclear winter in fighting of, 148
- dramatizations of aftermath of, 375
- dust generated in, 42–43, 313, 317–18
- energy budget of normal atmosphere vs. atmosphere after, 431
- extinction of human species in, 67–72
- famines after, 161–62
- with few weapons, slow pace, or under favorable meteorological conditions, 131–35, 353
- forest damage in, 175
- late Cretaceous catastrophe compared to, 64–67
- major strategic attacks in, 304–5
- Mao on, 288
- minimizing consequences of, xvii-xviii, 78, 352

nuclear war *cont'd*
without nuclear winter, 193–94, 201
nuclear winter as probable outcome of, 41
ozone layer depletion from, 50–51, 55–59, 174, 317, 328
psychological effects of, 332–33
radioactive fallout in, 20, 27, 52, 146, 172–73, 194–96, 450–51
risk of, *see* risk of nuclear war
smoke from, 43, 50–51, 317
in spite of nuclear winter, 307
sub-threshold, 304
in summer, 193, 384
survivability of, 160–63, 188–89, 209, 361–62
TTAPS on, 20–21, 316–17, 445–46, 448, 450, 460–64
underscored by nuclear winter, 74–76
U.S. and Soviet Union eluded regarding consequences of, 33
as winnable, 5, 183–84, 189, 382
in winter, 193, 384
nuclear weapons, 19, 268
abolition of, 261, 297–301, 414–18
accumulated in name of deterrence, 239–40
accuracy of, 131–32, 202, 244, 246, 304
assembly plants for, 53
burrowing, 304
calculating radioactive yields of, 52–53
as catastrophe waiting to happen, 85, 367–68
of China, 134, 201, 285, 354–55, 411
command and control of, 148
cost of, 20
delivery systems for, *see specific delivery systems*
in destroying petroleum facilities, 119–26, 197–203, 240, 347
destructive power of, 19, 68–69
as deterrent to conventional armaments, 152–53, 186, 207
as essential for national security, 117
in extreme nuclear winter, 203
of France, 134, 201, 203, 224, 285, 355–56
freezing arsenals of, 278–79, 313, 315
of Great Britain, 134, 201, 203, 224, 285, 355–56
international respectability associated with, 19
of Israel, 126, 285–86, 291–92, 413–14
moral critiques of, 308–10
for MSD, 230, 232–35, 240, 243–51, 254–55, 286, 290–93, 391–93
nations without, 287
in nominal nuclear winter, 202–4
number of, xviii, 6–7, 19–20, 87, 131–35, 199–203, 224, 353, 388
policy on, *see* politics, policymakers
in preventing nuclear winter, 131–32
proliferation of, xviii, 126, 282–94, 298, 304–5, 410–14, 416–17
proposed U.S. removal from Europe of, 270
protocols to inspect dismantling and destruction of, 261
public opinion on deterrent use of, 152–53
relationship between severity of nuclear winter and number of, 199–203
safety and reliability of, 85, 361–62
in severe nuclear winter, 203
in Soviet arsenal, 201, 239–40, 250–51, 280, 285, 287–88, 395
Soviet vs. U.S. policy on, 151, 153–55
space-based surveillance of, 261–62, 265
storage facilities for, 53
strategic, *see* strategic nuclear weapons
in substantial nuclear winter, 202
tactical, *see* tactical nuclear weapons
targeting of, *see* targeting
TTAPS vs. latest findings on, 450

in U.S. arsenal, 239–40, 250–51, 395
on U.S. soil, 386
uses of, 147, 207, 385–86
verifiable limits on number of, 87
yields of, 57, 131–32, 165, 184–85, 240, 244–46, 254–55, 304, 328, 380
*see also* arms control

Nuclear Weapons Employment Policy-1987 (NUWEP-87), 185–86

nuclear weapons testing, 20
climatic anomalies due to, 124–26
radioactive fallout from, 337
reduction and phasing out of, 229, 279, 408

nuclear winter:
acute phase of, 441
acute vs. chronic effects of, 193, 362
apocalyptic speculations associated with, 308–9
asymmetry of effects about, 149–50
asymmetry of perception about, 149, 359–60
basic scientific and quantitative aspects of, xix
chronic phase of, 441
coining of term, 465
confirmation of, 34–40, 309–13
controversy over, 33, 312–19
criticisms of, 34–40, 313–16, 320–21, 364
discovery of, 6, 39, 188
duration of effects of, 26, 362–63
early forerunners of, 43
early research on, 5–6
experimental reproduction of, 50
extreme danger of nuclear war underscored by, 74–76
global biological and ecological effects associated with, 74–77, 449–50
latest findings on, 21, 40–41, 324, 445–53
as liability in all conceivable cases, 209
with no significant environmental effects, 193–94, 201
possibility of, 119, 121, 123–26, 339, 349, 352
premonitions of, 95
propaganda asymmetry of, 188
proposed expanded international research program on, 410
as sobering reality, 307
strengthening scientific basis of, 34–42, 324, 445–53
survivors in combatant and noncombatant nations at risk from, 27
validity of, 21–22

"Nuclear Winter: Global Consequences of Multiple Nuclear Explosions" (Turco, Toon, Ackerman, Pollack, and Sagan), 303
nuclear winter, prevention of, 131–42, 352–57
and ability to deter nuclear war, 135, 357

*Nucleus*, xxii

Nye, Joseph S., Jr., 307, 330

oceans:
decreases in surface temperatures of, 363
as heat reservoirs, 427, 432–33
influence of smoke and dust on surface temperatures of, 440–41
and severity of nuclear winter, 441
temperatures in predicting nuclear winter, 366
thermal inertia of, 322

"Odes" (Santayana), 62

Office of Science and Technology Policy, 77

Ogarkov, Nikolai V., 80, 353

Oppenheimer, J. Robert, 68–69

optical depths;
absorption, *see* absorption optical depths
calculation of, 102, 341–44
influence of smoke and dust on, 437–39
as logarithmic, 439
scattering, *see* scattering optical depths
volcanic eruptions and, 102

ozone, difference between ordinary oxygen and, 55
ozone layer depletion, 16, 41–42, 301
CFCs in, 51, 55–57, 88–89, 303
chlorine in, 55–57
cost and risk of, 88–89, 337
decried as loose talk, 330
experiments simulating nuclear war's impact on, 50–51
food chains threatened by, 58, 328, 337
from high-altitude nuclear explosions, 20
in late Cretaceous catastrophe, 65
from nuclear war, 50–51, 55–59, 174, 317, 328
PAN-EARTH project on, 320
predictions on, 14
in severe nuclear winter, 196
ultraviolet radiation allowed by, 56, 58–59, 328, 337

Pakistan, development of nuclear weapons by, 285–86, 411
Palme Olof, 181, 366–67
Papandreou, Andreas, 182, 186
Papp, D. S., 305
*Paradise Lost* (Milton), 445
Pascal, Blaise, 339
Payne, Keith, 149, 151–52
Perez de Cuellar, Javier, 180
Perkovich, G., 406
Perle, Richard N., 184, 349, 378–79
Permian catastrophe, 67
permissive action links (PALs), 85
*Perpetual Peace* (Kant), 222, 284
Pershing II missiles, 313
Persian Empire, Croesus's invasion of, 11–12, 15
petroleum facilities:
climatic sensitivity of, 201, 203
effects of blast overpressure on, 121
fires in, 41, 49
targeting of, 119–26, 197–203, 240, 347
threatened release of pyrotoxins in nuclear ignition of, 49
world refining capacity of, 121
Petrovsky, Vladimir, 182
phytoplankton, ultraviolet radiation sensitivity of, 58, 328, 337
*Pilgrim's Progress* (Bunyan), 130
*Pioneer Venus* probes, 455
Pipes, Richard, 382–83
Pittock, A. Barrie, 150, 174–75
Plutarch, 112
plutonium refineries, 53, 281
politics, policymakers, xvii, 305, 389
and abstractions in planning nuclear war, 6
Alexandrov on, 139
arms control and, 223, 227, 229, 278–79
in assessing validity of prophecies, 15
confusion regarding threshold concept among, 339
consequences of nuclear war elusive to, 33–34, 316–19
deterrence and, 146–51, 153–55, 359–60
implicatons of nuclear winter for, xix, 26, 179–89, 303–12, 316, 321, 323, 366–84
Kahn, and, 215
necessity of understanding in, 12–13
and nuclear winter classifications, 197
on possibility of human extinction, 72–73, 76, 330
predictability of, xviii
relationship between scientists and, 299–300
in response to prophecies, 12–13, 15
and risk of nuclear war, 82–88
sanity of, 210–11, 218, 387
shelters and, 165–66
significance of nuclear winter minimized by, 34, 319
land survivability of nuclear war, 209
Pollack, James B., 20, 64, 67, 110, 303, 311
*see also* TTAPS
polychlorinated biphenyls (PCBs), 48
Pontifical Academy of Sciences, 139

population density, increases in, 16
*Portland Oregonian*, 363–64
post-traumatic stress, 74, 334
Powers, Thomas, 33–34, 306
*Pravda*, 353, 383, 418
Predictive Assessment Network for Ecological and Agricultural Responses to Human Activities (PAN-EARTH) project, 320
Presidential Review Memorandum, 10, 73
Press, Frank, 459–60
Priam, King of Troy, 13
probability, laws of, 84–85
prompt gamma rays, 52
prophecies, 11–16
  assessing validity of, 15
  in Croesus's invasion of Persia, 11–12, 15
  current resistance to, 14
  on fall of Troy, 13–15
  of planetary catastrophies, 16
  policy and, 12–13, 15
Proxmire, William, 147, 377, 379
public opinion:
  on abolition of nuclear weapons, 297
  arms control and, 262–64
  on arms race, 20
  on nuclear winter, 184, 187, 311, 373–76
  on possible extinction of human species, 71–73
  on use of nuclear weapons as deterrent, 152–53
pyrotoxins, 47–51, 59, 326–27, 329
  in nominal nuclear winter, 194
  in noncombatant nations, 172
  in nuclear war, 51, 159
  in severe nuclear winter, 196
  in substantial nuclear winter, 195

Qian Qichen, 355
quality of life, 160
quantitative prediction, 15
Quester, George, 330

Rabaul, Mt., eruption of, 111
Rabi, I. I., 69, 87
radioactive fallout, 20, 27, 41–42, 49–54, 59, 159
  from aboveground nuclear testing, 337
  agriculture and, 59, 160, 328–29
  calculating intensity of, 52–53
  from Chernobyl accident, 53–54
  death and dying from, 52–54, 73, 303
  decried as loose talk, 329–30
  deterence provided by, 146
  distribution of, 52
  forest damage due to, 175
  gamma rays and, 49–53
  after Hiroshima and Nagasaki explosions, 107
  in nominal nuclear winter, 194
  in noncombatant nations, 172–73
  secondary illnesses due to, 53
  in severe nuclear winter, 196
  in substantial nuclear winter, 195
  TTAPS vs. latest findings on, 450–51
*Rainbow Warrior*, 181, 369
rainfall, 174–75, 195
rainout scavenging, TTAPS vs. lastest findings on, 446–47
Ramaswamy, V., 321
Rand Corporation, 215
R & D Associates (RDA), 457–58, 463
*Reader's Digest*, 362
Reagan, Ronald, 152, 188, 261, 353, 368, 382, 395, 398, 462
  on arms control, 278
  awareness of nuclear winter of, 185
  window of vulnerablity, doctrine of, 148
  on winning nuclear war, 5, 183–84
Redoubt, Mt., eruption of, 359
refrigeration, nuclear war's disruption of, 90
religion, xvii
Reule, F. J., 305
Reykjavik Summit, 133, 355
Richardson, Elliot, 349
Ride, Sally, 254, 397
risk of nuclear war, 81–91, 336–39
  and continued improvement in reliability and safety of nuclear weapons, 85
  cost of human life in calculation of, 86–88

risk of nulear war *cont'd*
probable cost of nuclear war considered in, 82–86, 90–91
replacement cost of everything on Earth made by humans in calculation of, 86–88
severity of nuclear war considered in, 82, 336
Roberts, Walter Orr, 362, 464–65
Robock, Alan, 38–39, 363, 383–84
Romania, 368
Roosevelt, Franklin, 230, 299
Rotblat, Joseph, 460
Rowland, Sherwood, 56
Rowling, Sir Wallace, 368
Royal Society of Canada, 321
rural fires, TTAPS vs. latest findings on, 445–46
Rusk, Dean, 289
Russell, Bertrand, 44, 70–71, 170, 214, 336
Ryzhkov, Nikolai I., 407

Sagdeev, Roald Z., 247–48, 394
Sahel, famines in, 14
Sakharov, Andrei, xviii, 69, 74, 123, 165, 299, 332
*Samson Agonistes* (Milton), 192
San Francisco, Calif., 1906 earthquake in, 27–29
Santayana, George, 62
Sasser, James R., 377
Saudi Arabia, medium- and/or long-range ballistic missiles of, 286, 414
scattering optical depths, 104, 108, 342
absorption optical depths compared with, 437–39
from El Chichón eruption, 110
from Krakatoa eruption, 111
from Mt. Rabaul eruption, 111
for sulfuric acid droplets, 102, 109–11
from Tambora eruption, 110
temperature declines related to, 110–13
Scheer, Robert, 183–84, 334
Schell, Jonathan, 72, 312, 334–35
Schlesinger, James, 358
Schmidt, Michael, 18
Schneider, Stephen H., 35–40, 138, 321–23, 333, 339
*Science*, 42, 125, 139, 303, 308, 310, 316–17, 359, 457, 465
science fiction, nuclear winter in, 43–44
*Scientific American*, 215*n*, 371–72
Scientific Committee on Problems of the Environment (SCOPE), 40, 385, 449, 465
and asymmetry of perception about nuclear winter, 360
on biological implications of nuclear winter, 74–77
on effects of worldwide disruption in agriculture, 76–78, 172–73
on noncombatant nations, 172–73
urban smoke injection profile of, 37, 321, 323
on vulnerability of human life, 421
scientists, relationship between political leaders and, 299–300
Snowcroft, Brent, 217–18, 233
Scrimshaw, Nevin, C., 162
Sederberg, Peter C., 309–10, 339
Seitz, Russell, 313–14
self-lofting smoke, 362–63, 440, 449
severe nuclear winter, 196–97, 201, 203, 385
deterrence in case of, 210–11
Shabardin, Pyotr, 155, 383
Shakespeare, William, 206, 284, 419
Shallow Undersea Mobiles (SUMs), 395
Shamir, Yitzhak, 414
Shaw, George Bernard, 80
Shawcross, William, 337
*Sheffield*, sinking of, 254, 397
Shelley, Mary Wollstonecraft, 94–95, 100
Shelley, Percy Bysshe, 95
shelters, 304, 363–64
political role of, 165–66
of Soviet Union, 163–66, 185
Shelyug, V., 334
Shen, Dingli, 394
Shenfield, Stephen, 227, 381–82, 413
Shukurov, A. K., 39
Shute, Nevil, 54

Siberia, forest fires in, 38
sigmoidal thresholds, 96–98, 107
Simon, Herbert A., 310
Single Integrated Operational Plan (SIOP), 73, 122
  elimination of E/I targets from, 186
  nuclear winter ignored in, 364
Single Integrated Operational Plan (SIOP) Directorate, 118
single-warhead delivery systems, 241–43
skin cancer, 58, 89
Slavsky, Efim, 123
Slingo, T., 320
Small, Richard D., 37, 343, 350
Smil, Vaclav, 303
Smith, Adam, 276
Smith, R. Jeffrey, 139, 316–17, 396
Smith, Samantha, 187
smoke, 43, 317, 459–60
  absorption optical depths of, 96–99, 102–9, 117, 125–26, 341–44, 352, 439
  antigreenhouse effect and, 435–36
  attempts to experimentally simulate nuclear war's release of, 50–51
  from burning vegetation, 202
  climatic influence of, 23–24, 98–99, 317–18, 324, 339–40, 436–41
  from fires, 23–24, 28, 38–39, 324, 343–45
  greenhouse effect and, 24, 436–37, 440–41
  after Hiroshima attack, 106
  jet engine performance in, 148
  in marginal nuclear winter, 194
  after Nagasaki attack, 106–7
  over noncombatant nations, 173
  nonvolatile particles in, 345
  in nuclear attacks producing nuclear winter, 96–99, 102–3, 108–9, 117, 125–216, 309. 341–45, 352
  nuclear autumn vs. nuclear winter simulations on, 35–40, 322
  in ozone layer depletion, 50, 57
  from post-nuclear fires, 324
  and preventing nuclear winter, 131
  self-lofting, 362–63, 449
  in severe nuclear winter, 196
  in substantial nuclear winter, 195
  targeting and, 148
  temperature declines due to, 24, 41–43, 437, 440
  TTAPS on, 35–40, 460–64
  TTAPS vs. latest findings on, 446–48, 453
smoke injection profiles, 37, 321–23
smoke plumes, physics of, 35
smoldering combustion, 23
Snow, D. M., 305
society, societies, xvii
  long-term consequences of nuclear war for, 6–7
  SCOPE on losses in, 76
  SND and, 209
Sokolovskii, V. D., 164
soot, 44, 104
  absorption optical depths for, 102–3, 109–10, 341, 447
  Alexandrov on, 138
  climatic influence of, 437, 440
  in late Cretaceous catastrophe, 65
  in marginal nuclear winter, 194
  reaction of stratospheric ozone with, 328
  self-lofting, 440
  TTAPS vs. latest finding on, 446–49
  TTAPS vs. nuclear autumn simulations on, 35
South Africa, Republic of, 285–86, 414
Southeast Asia, targeting in, 332
Southern Hemisphere:
  agriculture in, 77
  consequences of nuclear winter in, 73–74, 77, 173–76, 195, 365–66, 448
  in nominal nuclear winter, 195
  ozone layer depletion above, 57–58
  in substantial nuclear winter, 195
  survival in midlatitudes of, 73–74
  targeting in, 174–76, 365–66
  TTAPS vs. latest findings on, 448

Soviet Academy of Sciences, 38, 136, 139, 188, 273, 369–70, 382
Soviet Union, 6, 245, 299, 399
and abolition of nuclear weapons, 298, 415–18
in acknowledging nuclear winter hypothesis, 309
agriculture of, 149–50
arms control and, 133–35, 204, 223–25, 229–30, 232–35, 260–64, 277–81, 311, 354–56, 383, 390, 393, 400, 409–10
ASAT testing and, 399
after Chernobyl accident, 54
consequences of nuclear war elusive to, 33
conventional forces of, 152, 225, 267–71, 402–4, 406–7
criticisms by noncombatant nations of, 180
in Cuban missile crisis, 70, 87–88, 123, 224
deterrence and, 147, 153, 386
Doomsday Machines and, 216
in ending nuclear weapons testing, 279
estimated nuclear war casualties in, 361
famines in, 161–62, 166
feasibility of winning nuclear war against, 5
fissionable materials inventory of, 281, 408–9
Geneva Convention and, 410
and global policy implications of nuclear winter, 181–89, 369–73, 380–84
INF in retaliation against, 305
INF Treaty signed by, 133
internal ethnic conflicts of, 233, 393
and invalid criticisms of nuclear winter, 315–16
JCS on, 153–55, 164
Limited Test Ban Treaty and, 409–10
medium- and/or long-range ballistic missiles of, 286
MIRVs and, 241–42, 394
MSD and, 231–35, 240, 246–49, 259, 262, 277–81, 391–95
Nuclear Non-Proliferation Treaty and, 409
nuclear weapons in arsenal of, 201, 239–40, 250–51, 280, 285, 287–88, 395
on nuclear weapons in Germany, 412
and nuclear winter classifications, 197
nuclear winter studies in, 135–42, 149, 339, 357, 359–60
opening up of, 300
political predictability of, xviii
on possibility of human species extinction, 70–71, 334
projected nuclear winter casualties in, 288
and proliferation of nuclear weapons, 290–93, 413
public opinion on nuclear winter in, 187
public opinion on use of nuclear weapons against, 152–53
in reversing arms race, xviii-xix
safety precautions taken by, 85
SDI and, 401
shelter system of, 163–66, 185
SIOP for, 73
smoke absorption optical depths in, 343–44
tactical nuclear weapons and, 266
targeting and, 73, 117, 119–26, 203, 347–52
vs. U.S. on nuclear weapons policy, 151, 153–55
U.S relations with, xvii-xix
*Venera* spacecraft of, 455
Sparks, Brad, 314–15
species, extinction of, *see* extinctions, extinctions
Spector, Leonard S., 411
Spohr, Carl W., 43
"Spring Prospect" (Du Fu), 158
Squire, R. K., 306
Stalin, Joseph, 210, 287–88
starvation, 160
in nominal nuclear winter, 195
in noncombatant nations after nuclear war, 172
in substantial nuclear winter, 195
Star Wars, *see* Strategic Defense Initiative

State Department Policy Planning Staff, 231
Stealth bomber program, 279
Steinbruner, John D., 391
step function thresholds, 96–98, 101–2
Stockholm Accords, 230, 261, 390
Stoessinger, John, 397
Stommel, E., 340
Stommel, H., 340
Stonier, Tom, 125
Stothers, Richard B., 111
Strategic Arms Limitation Talks I (SALT I):
  Interim Agreement, 261
  Treaty, 261
Strategic Arms Limitation Talks II (SALT II), 261
  Treaty, 385
Strategic Arms Reduction Talks (START), 204, 234–35, 247, 254, 259, 280–81, 311, 390, 393
  Treaty, 133, 242, 355
Strategic Arms Reduction Talks II (START II), 407
strategic attacks, climatic consequences, of, 304–5
strategic bombers:
  in context of nuclear winter, 252, 397
  for force projection, 398
  justifications for, 252–54
  MSD and, 251–55, 279
strategic defense:
  MSD and, 244–46, 259
  uncertainty generated by, 265–66
  varieties of nuclear winter caused by, 265, 401–2
Strategic Defense Initiative (SDI), 133, 165, 278–79, 304, 398
  abandonment of, 229
  connections between nuclear winter and, 389–90
  cost of, 249
  as destabilizing, 265, 401
  and global policy implications of, nuclear winter, 184, 379
  MSD and, 261
  permeability of, 389–90
  technical, strategic, and political deficiencies of, 389
  terminal defenses compared with, 265–66
strategic nuclear deterrent (SND), 207–9
  Doomsday Machine as, 217
  ways nuclear winter can enhance, 208–9, 387
strategic nuclear weapons, 304–5
  delaying production and deployment of, 277–79
  MSD and, 239–55, 267, 394–98
  negotiations on reduction of, *see specific treaty negotiations*
*Strategic Review*, 306
submarines, 279
  MIRVs in, 243–44, 250–51, 395–96
  MSD and, 243–44, 250–52, 264–65, 395
substantial nuclear winter, 195–97, 201–3, 210, 385
sub-threshold wars, 304
sulfuric acid droplets, 457
  from El Chichón eruption, 109–11
  from Mt. Rabaul eruption, 111
  scattering optical depths for, 102, 109–11
Sun Tzu, 144
Surikov, Boris Trofimovich, 369–70
Swain, David L., 107
Sweden, 54
Swift, Jonathan, 44
Szilard, Leo, 68, 87, 230, 264, 299, 390–91

Taal, Mt., eruption of, 109
tactical nuclear deterrent (TND), 207–8
tactical nuclear weapons, xviii, 292
  in Europe, 266–68, 270, 403–4
  gradual elimination of, 228–29
  justifications for, 266–68, 402
Tadzhikistan, dust storms in, 39
Taiwan, 286
Tallent, William H., 77–78
Tallent Report, 77–78
Tambora eruption, 99–101, 103, 107, 110, 340–41, 416
tanks, low temperature performance of, 148

targeting, xviii, 117–27, 308, 346–52
  and abolition of nuclear weapons, 416
  China and, 73, 117, 332, 346–47
  of cities, 73, 117–26, 197–203, 210, 240, 270–71, 330–31, 346–47, 349–52, 364, 397
  counterforce, 254
  cross-, 392
  E/I, *see* Economic/Industrial targeting
  in extreme nuclear winter, 203
  flexible, 224
  and incidence of opaque smoke, 148
  India and, 331–32
  leadership, 350–51
  in light of nuclear winter, 185–86, 364, 381
  likelihood of nuclear winter due to, 119, 121, 123–26, 349, 352
  of military facilities, 199–202
  MSD and, 224, 241, 243, 270–71
  in nominal nuclear winter, 202–4
  of noncombatant nations, 171–72
  NSTL on, 118, 351
  of nuclear fuel facilities, 73
  of petroleum facilities, 119–26, 197–203, 240, 347
  In preventing nuclear winter, 131–32
  relationship between severity of nuclear winter and, 197–204
  in severe nuclear winter, 203
  SND and, 209
  in Southeast Asia, 332
  in Southern Hemisphere, 174–76, 365–66
  Soviet Union and, 73, 117, 119–26, 203, 347–52
  in substantial nuclear winter, 202–3
  treaties on, 304
  of war-sustaining industries, 123–26, 349
  and window of vulnerability, 148
Taylor, Maxwell, 146
Taylor, Theodore B., 399
*Technology Review*, 384
Teller, Edward, 68, 318–19
temperatures:
  average, 22–24, 104
  climate defined in terms of, 419–21
  in climate system, 428–29
  with and without greenhouse effect, 24
  of Ice Ages, 24–26, 102, 341
  during Little Ice Age, 341
  of Mars, 456
  natural extremes in, 98, 340
  nuclear autumn vs. nuclear winter simulations on, 35–40, 321–22
  in predicting nuclear winter, 366
  seasonal changes in, 429–32
  short-term changes in, 429
temperatures, declines in:
  in absorption of light by smoke, 97–99, 102–4, 107
  agriculture impacted by, 22, 43, 99–102, 341
  Alexandrov on, 137–38
  and cancellations of greenhouse effect, 440–41
  from dust, 39, 318, 437, 440
  from dust injections from volcanic vs. nuclear explosions, 318
  from El Chichón eruption, 110
  from forest fires, 38–39, 126
  from Krakatoa eruption, 111
  in marginal nuclear winter, 194
  from Mt. Etna eruption, 112
  in nominal nuclear winter, 194–95
  in nuclear winter, 25–26, 40, 43–44, 194–96, 325, 448, 453
  of oceans, 363
  in relation to number of nuclear explosions at lower Northern latitudes, 174
  scattering optical depths related to, 110–13
  as seasonally related, 98
  in severe nuclear winter, 196
  from smoke, 24, 41–43, 437, 440
  from Tambora eruption, 99, 101, 340–41
  tank performance and, 148
  TTAPS vs. latest findings on, 448, 453

from volcanic eruptions, 24, 99–102, 110–12, 318, 340–41, 457
temperatures, increases in:
Alexandrov on, 138
greenhouse effect and, 40–41, 320, 325, 337–38, 434–36
terminal defenses, 265–66
Third World:
nuclear war viewed as political blackmail by, 385–86
*see also* noncombatant nations
Thomas, Lewis, xvii, 312, 375–76
Thompson, Starley, L., 35, 321–22, 333, 339
"Threads," 71, 375
thresholds:
definition of, 96
logistic, 96–98, 107
step function, 96–98, 101–2
thresholds, climatic effects, 339
for absorption of light, 96–99, 102–3, 341–44
absorption optical depths related to, 103–9, 113
for disruption of agriculture, 98–99, 101, 103, 340–41
MSD and, 240
scattering optical depths related to, 112–13
for smoke, 117, 125–26, 352
TTAPS on, 108–9
Tibbet, Paul W., 106
*Times* (London), 375
Tipton, Karla, 364
"To Cap the Volcano" (Bundy), 357
Tokyo, firestorms in, 124
Tolstoy, Leo, 455
Toon, Owen Brian, 20, 64, 67, 110, 303, 311
*see also* TTAPS
toxic pollution, 41–42
transportation systems, nuclear war's disruption of, 90
tree-ring dating, 112
Trident missile submarine program, 279
tritium, present inventories of, 408–9
Troy, prophecied fall of, 13–15
TTAPS (Turco, Toon, Ackerman, Pollack, and Sagan), 20–21, 197, 308, 371–72
Alexandrov and, 137–38
on antigreenhouse effect, 435
atmospheric models of, 425
Blue Book of, 464–65
DNA and, 314, 459, 466
on dust, 460–64
early topics considered by, 460–61
errors in calculation sought by, 21
independent reviews of nuclear winter study of, 462, 466–67
initial motivation for, 313
invalid criticisms of, 313–15
re latest findings on counterforce nuclear strikes, 450
re latest findings on declines in temperatures, 448, 453
re latest findings on dust, 447–48, 453
re latest findings on fire ignition by nuclear detonations, 445–46
re latest findings on human survival and agriculture, 449–50
re latest findings on nuclear weapons, 450
latest findings on nuclear winter compared with original conclusions of, 21, 445–53
re latest findings on radioactive fallout, 450–51
re latest findings on rainout scavenging, 446–47
re latest findings on smoke, 446–48, 453
re latest findings on soot, 446–49
re latest findings on spread of nuclear war perturbations to Southern Hemisphere, 448
re latest findings on urban damage, 445–46
NASA and, 313–14, 456–57, 460–63, 465
NAS and, 313–14, 458–60, 462, 465
nuclear autumn model compared with nuclear winter model of, 35–40, 321–22
nuclear winter introduced by, 41–42
origin of name, 20, 303

TTAPS *cont'd*
in pointing out elusive consequences of nuclear war, 316–17
on possibility of human species extinction, 333
publication of nuclear winter study of, 465
refinement, extension, and strengthening original work of, 34–37, 42, 324, 445–53
on self-lofting smoke and climate feedbacks, 362
short history of nuclear winter study of, 455–67
on smoke, 35–40, 460–64
on thresholds for smoke optical depths, 108–9
urban smoke injection profile of, 37, 321–22
validity of discovery of, 21–22
TTAPS (Turco, Toon, Ackerman, Pollack, and Sagan) II, 324
on smoke absorption optical depths, 343–44
Turner Broadcasting System, 375
"Twilight at Noon: The Atmosphere After a Nuclear War" (Crutzen and Birks), 41

Ullman, R. H., 246, 395
ultraviolet radiation, 159, 329, 427
in noncombatant nations, 172–73
ozone layer depletion and, 56, 58–59, 328, 337
in severe nuclear winter, 196
in substantial nuclear winter, 195
Union Carbide pesticide plant accident, Bhopal, India, 48–49, 86, 336
United Nations, 40, 268, 300, 324–25
and asymmetry of perception about nuclear winter, 360
disarmament conferences at, 354
and global policy implications of nuclear winter, 180–81
uranium miners, radiation damage suffered by, 337
uranium refineries, radioactive fallout from, 53
Urey, Harold, 66

Vanunu, Mordechai, 291
Velikhov, Yevgeny P., 136, 166, 183, 383
Veltishchev, N. N., 38
Venus, 39
greenhouse effect on, 455, 457
sulfuric acid clouds of, 457
*Viewpoint*, 375
*Viking* lander, 39
vinyl chloride ($C_2H_3Cl$), 48
visible light:
in atmospheric models, 426–27
in climate system, 427–32
greenhouse effect and, 433–35
influence of smoke and dust on, 436–41
in normal atmosphere vs. atmosphere after nuclear war, 431
volcanic eruptions, 26, 43, 160, 313, 320, 442
agriculture impacted by, 99–101, 111–12, 442
climate models employed in nuclear winter in predicting cold and dark produced by, 39
climatic anomalies due to, 42, 125–26
of El Chichón, 109–11
Hiroshima and Nagasaki explosions compared with, 107
of Krakatoa, 111
of Mt. Etna, 112
of Mt. Galunggung, 359
of Mt. Hekla, 112
of Mt. Rabaul, 111
of Mt. Redoubt, 359
of Mt. Taal, 109
optical depths and, 102
stratospheric dust injection from nuclear war compared to, 42–43, 317–18
on Tambora, 99–101, 103, 107, 110, 340–41, 416
temperature declines after, 24, 99–102, 110–12, 318, 340–41, 457
volcanic winters, 99–102, 109–13, 160
scattering optical depths in, 102, 110
von Hippel, F., 246, 361, 395
von Neumann, John, 42–43, 317

Voznesensky, Andrei, 372
Vyshinsky, Andrei, 231

Wagner, Richard, 308
Wallace, Henry, 231, 390
Waller, Douglas C., 279
Warner, Frederick, 75
Warnke, Paul C., 339
Warsaw Treaty Organization (WTO), 225, 402
  conventional formations of, 186, 229, 267–69, 271–72, 277, 282, 403–7
  in conventional war in Europe, 272
  on European nuclear weapons-free corridor, 268
  perceived conventional superiority of, 186
  reduction, balancing, and substantial pull-back and demobilization of, 229, 267
war-sustaining industries, targeting of, 123–26, 349
Washington Summit, 254
*Wealth of Nations, The* (Smith), 276
weather:
  climate vs., 419
  models for prediction of, 423
Weather Bureau, 125
Webster, William H., 404
Weinberger, Caspar N., 184, 349, 377–79, 381
Welch, David A., 87–88, 289
Welch, Larry, 186, 392
Wendt, Gerald, 106
Wexler, Harry, 38
White, Carol, 315–16
White, Thomas D., 253
Wickramasinghe, Chandra, 66
wildfires:
  in China, 38–39
  TTAPS vs. latest findings on, 446
Wilson, Ben Hur, 43
Wilson, Edwin, 411–12
Wilson, Robert R., 390
window of vulnerability, 148
Winter, Sidney G., 309
Wirth, Timothy, 378
Wisconsin Ice Age, 102
Wittkopf, Eugene R., 312
Wood, Lowell, 318, 379
Woodwell, George M., 463
Worcester, Dean, 109
World After Nuclear War: The Conference on the Long-Term Biological Consequences of Nuclear War, The, 465
World Court, 300
World Health Organization (WHO), 331–32
World Meteorological Organization, 40, 324

Xue Litai, 347

Zhang Aiping, 346–47
Zhukov, Georgi K., 289
Zuckerman, Lord Solly, 228, 266, 402
Zyklon B, 48, 327

# ABOUT THE AUTHORS

Carl Sagan is the David Duncan Professor of Astronomy and Space Sciences and Director of the Laboratory for Planetary Studies at Cornell University. Richard Turco is Professor of Atmospheric Sciences at the University of California at Los Angeles (UCLA), and has been awarded the MacArthur Prize Fellowship.

With their colleagues, Drs. Turco and Sagan are co-recipients of the American Physical Society's Leo Szilard Award for Physics in the Public Interest for the discovery of nuclear winter. They have been active in advising the Congress, the Department of Defense, other government agencies, and the leaders of other nations on the long-term consequences of nuclear war.

For his work in planetary exploration, Dr. Sagan has been awarded the NASA Medals for Exceptional Scientific Achievement and (twice) for Distinguished Public Service, and the Tsiolkovsky Medal of the Soviet Cosmonautics Federation. He is a former President of the Planetology Section of the American Geophysical Union, former Chairman of the Division for Planetary Sciences of the American Astronomical Society, and for twelve years Editor-in-Chief of *Icarus*, the leading professional journal of planetary studies. Dr. Sagan is a Pulitzer Prize–winning author, and his *Cosmos* television series has been seen in sixty countries by over 400 million people. He is a recent recipient of the Oersted Medal of the American Association of Physics Teachers. Dr. Sagan has served as Feinstone Lecturer, U.S. Military Academy, West Point; Forrestal Lecturer, U.S. Naval Academy, Annapolis; Distinguished Lecturer, U.S. Air Force Academy, Colorado Springs; and, from 1984 to 1986, Keystone Lecturer, National War College, National Defense University, Washington, D.C.

For almost a decade and a half Dr. Turco worked on the effects of nuclear war for R & D Associates, a leading defense contractor. He is also an expert on the atmospheres of other planets and on the Earth's ozone layer and its depletion, and leads UCLA's atmospheric chemistry program, which includes studies of air pollution in the Los Angeles basin. Dr. Turco serves as an Associate Editor of the American Geophysical

Union's *Journal of Geophysical Research* and is President-elect of its Atmospheric Sciences Section. Both Drs. Turco and Sagan are studying the atmosphere of Titan, Saturn's largest moon, where some of the building blocks of life seem to be forming.